Urban Recovery

This book calls for re-conceptualising urban recovery by exploring the intersection of reconstruction and displacement in volatile contexts in the Global South. It explores the spatial, social, artistic, and political conditions that promote urban recovery.

Reconstruction and displacement have often been studied independently as two different processes of physical recovery and human migration towards safety and shelter. It is hoped that by intersecting or even bridging reconstruction with displacement, we can cross-fertilize and exploit both discourses to reach a greater understanding of the notion of urban recovery as a holistic and multi-layered process. This book brings multidisciplinary perspectives into conversation with each other to look beyond the conflict-related displacement and reconstruction and into the greater processes of crises and recovery. It uses empirical research to examine how trauma, crisis, and recovery overlap, coexist, collide, and redefine each other. The core exploration of this edited collection is to understand how the oppositional framing of destruction versus reconstruction and place-making versus displacement can be disrupted; how displacement is spatialized; and how reconstruction is extended to the displaced people rebuilding their lives, environments, and memories in new locations. In the process, displacement is framed as agency, the displaced as social capital, post-conflict urban environments as archives, and reconstructions as socio-spatial practices.

With local and international insights from scholars across disciplines, this book will appeal to academics and students of urban studies, architecture, and social sciences, as well as those involved in the process of urban recovery.

Howayda Al-Harithy is a Professor of architecture at the department of architecture and design at the American University of Beirut and a Research Director at the Beirut Urban Lab. She is the editor of *Lessons in Post-War Reconstruction: Case Studies from Lebanon in the Aftermath of the 2006 War* published by Routledge in 2010.

Routledge Critical Studies in Urbanism and the City

This series offers a forum for cutting-edge and original research that explores different aspects of the city. Titles within this series critically engage with, question and challenge contemporary theory and concepts to extend current debates and pave the way for new critical perspectives on the city. This series explores a range of social, political, economic, cultural and spatial concepts, offering innovative and vibrant contributions, international perspectives and interdisciplinary engagements with the city from across the social sciences and humanities.

Socially Engaged Art and the Neoliberal City
Cecilie Sachs Olsen

Peri-Urban China
Land Use, Growth, and Integrated Urban–Rural Development
Li Tian and Yan Guo

Spatial Complexity in Urban Design Research
Graph Visualization Tools for Communities and their Contexts
Jamie O'Brien

Housing Displacement
Conceptual and Methodological Issues
Edited by Guy Baeten, Carina Listerborn, Maria Persdotter and Emil Pull

Urban Recovery
Intersecting Displacement with Post War Reconstruction
Edited by Howayda Al-Harithy

For more information about this series, please visit www.routledge.com/Routledge-Critical-Studies-in-Urbanism-and-the-City/book-series/RSCUC

Urban Recovery

Intersecting Displacement with Post War Reconstruction

Edited by
Howayda Al-Harithy

Routledge
Taylor & Francis Group
LONDON AND NEW YORK

First published 2022
by Routledge
2 Park Square, Milton Park, Abingdon, Oxon OX14 4RN

and by Routledge
605 Third Avenue, New York, NY 10158

Routledge is an imprint of the Taylor & Francis Group, an informa business

British Library Cataloguing-in-Publication Data
A catalogue record for this book is available from the British Library

Library of Congress Cataloging-in-Publication Data
A catalog record has been requested for this book

ISBN: 978-0-367-55042-4 (hbk)
ISBN: 978-0-367-55043-1 (pbk)
ISBN: 978-1-003-09170-7 (ebk)

Contents

Figures

Tables

Contributors

Howayda Al-Harithy is a Professor of Architecture at the department of architecture and design at the American University of Beirut and a Research Director at the Beirut Urban Lab. She is the editor of *Lessons in Post-War Reconstruction: Case Studies from Lebanon in the Aftermath of the 2006 War* (Routledge, 2010).

Jennifer Hyndman is a Professor of Geography at York University in Toronto. Her research focuses on the geopolitics of forced migration, the biopolitics of humanitarian responses to displacement, and refugee resettlement. Her most recent book (with Giles) is *Refugees in Extended Exile: Living on the Edge* (Routledge, 2017).

Diane Davis is the Charles Dyer Norton Professor of regional development and urbanism at Harvard's Graduate School of Design (GSD), and a co-director of the risk and resilience track of the Master in design studies program. Research interests include relations between urbanization and national development, socio-spatial practice in conflict cities, and new territorial manifestations of sovereignty.

Mona Fawaz is a Professor of Urban studies and planning at the department of architecture and design and a research director at the Beirut Urban Lab at the American University of Beirut. She is the author of over 50 scholarly publications addressing issues of urban history and historiography, social and spatial justice, informality, law, housing, land, property, space, and planning theory.

Mona Harb is a Professor of urban studies and politics and a Research Director at the Beirut Urban Lab at the American University of Beirut. Her research examines the roles of multiple actors in the spatial production and governance of the city and investigates how oppositional politics and urban activism may lead to political change.

Carla Al-Hage is an architect and urbanist, and a researcher at the Beirut Urban Lab at the American University of Beirut. She holds a Bachelor in Architecture and an MPhil in Architecture and Urban Studies from the University of Cambridge, UK. Her research at the Lab focuses on the urbanization of forced population displacement in secondary Lebanese cities.

Loren Landau is a Professor at Oxford and Witwatersrand Universities. A political scientist by training, his interdisciplinary scholarship explores mobility, multi-scale governance, and the transformation of socio-political community across the global south. He has published widely in the academic and popular press and works closely with civil society, local government, and international organisations to promote inclusion amidst widespread mobility and dislocation.

Ghiath Al Jebawi is a development systems researcher. Fascinated about digital systems and their architecture, Al Jebawi is educated as an architectural engineer, and urbanist from Damascus University (2010) and Politecnico di Milano (2016). He investigates in communities in motion, socioeconomics, and local governance. Al Jebawi is dedicated to the spatial, cultural, social, environmental, and digital narration of development.

Cathrine Brun is a Professor and the Director of the centre for development and emergency practice at Oxford Brookes University. She is a geographer researching protracted displacement and the ethics and politics of humanitarianism, particularly focusing on housing and home, place and temporality from the perspective of what people do when living with displacement.

Joana Dabaj is a co-founding architect, urban designer, and researcher at CatalyticAction, a UK charity that empowers communities through innovative community-led interventions. Building on her extensive experience in design in development, migration, and human rights in the Middle East, she has recently been working closely with displaced and host communities in Lebanon to co-design and implement interventions of strategic value.

Camillo Boano is a Professor of urban design and critical theory at The Bartlett Development Planning Unit at UCL, UK, and a Professor of Architectural and urban design at the dipartimento interateneo di scienze, progetto e politiche del territorio, politecnico di Torino, Italy. He is a Co-Director of the UCL urban laboratory and the author of *The Ethics of a Potential Urbanism: Critical Encounters Between Giorgio Agamben and Architecture* (Routledge, 2017).

Namariq Al-Rawi is an architect and independent urban researcher who obtained her master's degree from Stuttgart University in international urban studies with a focus on integrated urbanism and sustainable design. She investigates in her researches the effects of physical and social destruction and the possibilities of reconstructing stricken cities in her homeland Iraq.

Rohit Jigyasu is a conservation architect and risk management professional from India, currently working at ICCROM as Project Manager on urban heritage, climate change and disaster risk management. He is also the

Vice President of ICOMOS and its International Scientific Committee on Risk preparedness.

Dina Mneimneh is a practicing architect and researcher based in Beirut, and a graduate of the American University of Beirut (AUB). Her design and research contributions span across the fields of public architecture, sustainable urban development in historic environments, and post-war reconstruction of Beirut's city centre.

Suzanne Harris-Brandts is an Assistant Professor of architecture and urbanism at Carleton University and co-founder of the design-research practice collective domain. Her research brings together design and the social sciences to explore issues of power, equity, and collective identity in the built environment.

David Sichinava is an Assistant Professor of human geography at Tbilisi State University and co-founder of the design-research practice collective domain. His research focuses on the social, spatial, and temporal aspects of inequality, the politics of urban development, and the role of civil society in urban policy. David is also a Research Director at the Caucasus Research Resource Centre (CRRC-Georgia).

Chiara De Cesari is an Associate Professor in European studies and cultural studies at the university of Amsterdam. Her wide-ranging re-search explores how forms of memory, heritage, art, and cultural politics are shifting under conditions of contemporary globalization and state transformation.

Eva Ziedan is an archaeologist, heritage specialist, and community development expert with more than ten years of field experience in the Middle East and Europe. Ziedan's work examines how living heritage is about re-adapting to develop concrete socio-economic benefits and ensuring the continuity of human interaction.

Elena Isayev is a historian and archaeologist focusing on migration, hospitality, and displacement. She has written on Migration Mobility and Place in Ancient Italy, and on agency for the Red Cross. She currently leads Imagining Futures through Un/Archived Pasts (AHRC, GCRF Network+), and is a Professor of ancient history, university of Exeter, UK.

Judith Naeff is an Assistant Professor at the Leiden University Institute for Area Studies (LIAS) working in the field of cultural studies with a focus on the contemporary Middle East. She has extensively studied and published a monograph on urban imaginaries of post-civil-war Beirut in 2018.

Omar Mismar is an interdisciplinary artist based in Beirut, Lebanon. He has participated in exhibitions in San José Museum of Art, San José (2018), MoMA, New York (2018), Homeworks 8, Beirut (2019), and Oakland Museum, California (2020) among others. Mismar is Assistant Professor of Visual Arts at the American University of Beirut.

Acknowledgement

This book would not have been possible without the inspiration, help, and support of so many students, researchers and colleagues. I would like first to thank the Maroun Semaan Faculty of Engineering and Architecture at the American University of Beirut (AUB), the Ford Foundation, the RELIEF Centre at the Institute for Global Prosperity, University College London (UCL), and L'Institut Français du Proche-Orient for funding the City Debates conference that I organised in April 2019. I take this opportunity to thank all of those who helped in conceptualising and organizing the conference that led to the publication of this edited volume, including all my graduate students who took the City Debates seminar of 2019 and my colleagues Mona Harb, Mona Fawaz and Camillo Boano.

This book brings together some of the impressive scholars who participated in City Debates 2019 to contribute unique insights into the timely topic of urban recovery across different dimensions, expertise, and disciplines. I would also like to extend my thanks to my colleagues Robert Saliba and Jala Makhzoumi for their support with the book proposal.

I am very grateful for my research team that helped me with the organisation of City Debates 2019 and the preparation of the manuscript of this book: Najmeh H. Viki, Batoul Yassine, and Ali Khodr. Special thanks go to Abir Eltayeb who played a key role in bringing the conference and the edited volume to light. Her dedication through organizing, communicating and editing ensured the completion of both.

Howayda Al-Harithy
Beirut
March 2021

1 Re-conceptualizing urban recovery in the age of protracted displacement

Howayda Al-Harithy

Introduction

In today's world an unprecedented rate of one person every two seconds is forcibly displaced due to disasters and violent conflicts. Calling the period of 2010–2019 a decade of displacement, the United Nations High Commissioner for Refugees (UNHCR) reports, 'At least 100 million people were forced to flee their homes during the last 10 years, seeking refuge either within or outside the borders of their country'.[1] Displacement and reconstruction are more critical than ever with the total number of displaced people worldwide reaching 79.5 million by the end of 2019 due to conflicts, violence, and persecution (UNHCR, 2019). This edited book therefore frames the following question: What does urban recovery look like in the age of mass and protracted displacement? It thus aims to re-conceptualize urban recovery at the intersection of *displacement* and *reconstruction*.

Reconstruction and displacement form two independent scholarly trajectories that do not intersect often enough to enrich or challenge one another. While reconstruction is mostly focused on the physical aspects of the built environment, displacement emerges as a human-centred discourse that encompasses the social and temporal dimensions of human migration towards safety and shelter. And, while displacement has occupied a central focus in research across historical, urban, anthropological, geographical, cultural, and refugee studies, emerging threads call for more interdisciplinary and spatial reflections. This book therefore brings together different disciplinary perspectives into conversation with each other and with empirical case studies. These perspectives look beyond conflict-related displacement and reconstruction to explore the greater processes of crises, trauma, and recovery, and examine how they overlap, co-exist and collide and hence, redefine each other. The goal is to understand how the oppositional framing of destruction versus reconstruction and place-making versus displacement can be disrupted; how displacement is spatialized; and how reconstruction is extended to displaced people rebuilding their lives, environments, and memories in new locations. In the process, displacement is framed as an agency, the displaced as social capital, post-conflict urban environments as archives, and reconstructions as socio-spatial practices.

The idea for this book was initiated in the wake of City Debates 2019, an international conference on contemporary urban issues that I organized in April 2019 at the American University of Beirut, titled *Urban Recovery at the Intersection of Displacement and Reconstruction*. The conference examined the spatial modalities that inhabit and represent these intersections. It was premised on the following: 'While reconstruction has long been debated, its intersections with protracted and mass displacement call for more critical conversations'.[2] The book is designed to bring together the variety of approaches and perspectives that were debated during the conference and share them with the wider academic and professional community throughout the region of the Middle East and beyond.

This book is therefore an invitation to re-conceptualize urban recovery by exploring how reconstruction and displacement intersect across volatile contexts from a Global South perspective. It is hoped that by intersecting and even colliding reconstruction with displacement, we can cross-fertilize and contaminate both discourses positively towards a greater understanding of the notion of urban recovery as a holistic and multi-layered process; one that goes beyond the physical and the humanitarian and proposes a more inclusive and multidisciplinary approach. This approach transcends the formal and the immediate to include informal practices and tune to the complexity of the accumulation and archiving of history. The book is, thus, an in-depth exploration of spatial, social, artistic, virtual, and political modalities that promote the process of urban recovery. It is at the same time an interrogation of the strategies and discourses surrounding recovery, reconstruction, and displacement. These explorations and interrogations are directed towards understanding how violence and protracted displacement impact urban practices; posing urban recovery as a complex process that challenges existing conventions, tackles displacement as urban agency, and engages reconstruction as being equally occupied with the intangible.

The content of this book and its thematic structure explore narratives of displacement and modalities of reconstruction through multiple lenses such as time, space, locality, gender, sectarianism, and memory. The thematic intersections between reconstruction and displacement are organized around understanding systems and scales of governance; housing the displaced; conceiving of cultural heritage in the recovery process; and spaces and imaginaries of post-crisis recovery(s).

Conceptualizing recovery

Existing literature presents wide-ranging differences in the definitions and conceptualizations of recovery that have provoked the entry point of this book. While diverse definitions co-exist, recovery as a construct evolved over the past three decades from a term that is synonymous with reconstruction to an iterative process that is intertwined with development, reconstruction, or resilience, to then emerge as an open-ended participatory process that is locally informed and socially anchored. When used in relation to post-conflict or post-disaster, some scholars reduce recovery to the physical aspects of the built and natural environments, often equating it with reconstruction while calling for the implementation of comprehensive urban planning schemes. Other scholars think of recovery as an ongoing act of resilience that starts before a crisis ends, while others link it to bottom-up strategies that involve contextual and participatory approaches. These differences in conceptualizations lie largely in the disciplinary or sub-disciplinary framing; of particular interest here are the differences that emerge in the conception of recovery among the three strands of post-disaster recovery, post-war reconstruction, and humanitarian aid and relief.

This section of the chapter offers an overview of the different definitions and conceptualizations of recovery and identifies the main theoretical shifts that have taken place in the literature. It then positions the book and its thematic tracks in relation to such trajectories. It is important, however, to note that the different definitions of recovery have developed and continue to co-exist in parallel. Recovery remains open to further conceptualizations upon which the invitation of this book is extended.

Early research equated recovery with reconstruction of the physical environment but paved the way for more critical conceptions of recovery in relation to reconstruction (Haas, Kates, and Bowden, 1977; Bolin, 1976, 1982; Bolin and Trainer, 1978; Drabek and Key, 1984; Oliver-Smith, 1986; Bates and Peacock, 1993). The distinction between recovery and reconstruction was sharpened when the literature shifted from a focus on pure tangible reconstruction schemes to the recovery of intangible dynamics and networks. This significant shift became apparent after the end of the post-Yugoslav collapse and the siege of Sarajevo in 1996. The sole focus was no longer on the reconstruction of the built environment as was the case following the World War II. Recovering memories, practices, and a sense of place became crucial in the aftermath of violent conflict, as well as state rebuilding and community reintegration. For example, when the physical impact of urban trauma is considered, authors such as Shaw (1996) and Zetter (2005) focus on the need to tie recovery to reconstruction, and hence development. They suggest a push for more sustainable economic structures as part of the reconstruction agenda. Stanley-Price (2007) ties reconstruction to the recovery of cultural heritage. He views it as a priority, along with the provision of shelter, basic needs, and services for people. It encompasses elements of both tangible and intangible heritage, including the restoration of symbolic monuments, cultural practices, and markers of national identity by looking at the significance of home and land in the minds of people. In addition to material losses, he further argues that cultural heritage is grounded in people's personal and collective histories and their visions for their neighbourhoods. He, thus, focuses on reviving the cultural fabric of a place, including the identities of individuals, their communities, and sense of belonging. Similarly, Gotham and Greenberg (2008) consider the activation of cultural institutions as paramount to recovery, alongside the provision of humanitarian aid and immediate housing.

Addison (2003a, 2003b) ties reconstruction to recovery through his conception of 'broad-based recovery'. To Addison, achieving recovery in Africa from an empirical point of view requires a set of parallel actions that go beyond the act of rebuilding damaged physical infrastructure. These actions include peace building, securing political stability, strengthening the functions of the state, resettling refugees, and rebuilding the social and economic sectors. Addison (2003b) argues that moving from conflict to recovery in Africa can only be secured when peace is sought and broad-based recovery is designed to improve the livelihood of all social groups, especially the disadvantaged. In this case, achieving recovery includes two requirements. The first requirement is the balanced and active role of all the stakeholders involved in the recovery process. Involvement in recovery is illustrated by a 'reconstruction and growth' pyramid where national and international actors are key players. At the base of this pyramid are local communities. To Addison (2003b), success lies in the ability of communities to rebuild and strengthen their livelihoods. The second requirement proposes legislative and political reform to accompany the reconstruction of physical, natural, and human capital. Here, Addison (2003b) draws a parallel between reconstruction initiatives and the legislative and political reform. While both are imperative to achieving recovery, they need to be addressed in tandem: prioritizing one over the other can lead to a narrow rather than broad-based recovery. Recovery for Addison (2003a, 2003b) is therefore a two-fold condition that is reached through reconstruction and reform.

Over the last three decades, the term recovery further evolved as an alternative to reconstruction that does not assume a return to the previous state that was linked to the roots of conflict. To Barakat and Zyck (2009, p.1072):

> Although the term 'reconstruction' has gained considerable momentum, it has also been criticised for suggesting a return to the *status quo ante* which had been implicated in the cause of the conflict. The constructive urge, it was implied, had failed to adequately deconstruct the weaknesses and vulnerability factors evident in the pre-conflict environment. In the light of such debates, 'recovery' gained favour for its pre-existing use in economic, social, medical and psychological realms and because of its apparent distance from the infrastructure-oriented connotations and origins of reconstruction.

As such, the conceptualization of recovery shifted from framing it as a return to the *status quo ante* to a 'new normal'. According to Chang (2010), recovery produces an environment that becomes the 'new normal'. In this case, building back as before either becomes impossible due to irreversible social, political, and economic changes or reproduces the conditions that created the disaster. It is, however, important to note that adapting to the 'new normal' might not always lead to a 'new better'. Chang (2010) therefore conceptualizes recovery as an adaptive process that negotiates the reconstruction of pre-disaster systems with major alterations of those systems.

In parallel, scholars also challenge the conception of recovery as linear. They advance a more critical take that acknowledges recovery as iterative with overlapping phases at multiple scales from local, regional, to national (Berke, Kartez, and Wenger, 1993; Neal, 1996; Chang, 2009). Post-disaster scholars begin to conceptualize recovery as a process that is versatile, complex, and without an endpoint. They no longer conceive of disaster as an interruption in development and recovery as a reversal process seeking to restore the pre-disaster normality. Rather, recovery begins to be conceived as complex and nonlinear and results in a 'new normality'; an environment that was altered and reconfigured by the disaster. In the process, recovery prioritizes re-establishing people's lives over physical rebuilding (Olshansky, 2005). Scholars even add that a successful recovery improves a community's disaster resilience rather than re-establish their vulnerabilities (Mileti, 1999; Wisner et al., 2004). As a result, recovery emerges as a process of rebuilding both the physical and the social. It involves more than the reconstruction of the built environment and is more appropriately conceptualized as a social process that is shaped by both pre- and post-disaster conditions. It acknowledges the challenges faced by people as well as the recovery of the built and natural systems. As such, recovery in post-disaster literature is defined as the process of restoring, rebuilding, and reshaping the spatial, socio-cultural, economic, and natural environments through pre-event planning and post-event actions (Smith and Wenger, 2007).

This invited the conceptualization of recovery in relation to the urban and to resilience. Davis (2005) focuses on resilience while examining the impacts of Mexico City's disastrous earthquake on the spatial, physical, social, and political character of the capital city. She questions how a city can recover from a disaster of this magnitude by being more resilient. Davis (2005) defines recovery through resilience as more than mere physical rebuilding. It is understood in terms of the enduring qualities of places, people, and their social practices and values. It allows cities to respond to the stresses of disaster by transforming and adapting to a new state. Furthermore, Davis's (2005) work proposes a link between recovery as a process and the practice of urban citizenship. The earthquake in Mexico City exposed the hidden corruption embedded in the city and acted as a catalyst that mobilized citizens to search for social justice. It expanded the understanding of recovery to become an opportunity for people to seek answers to what they need for a better future, with lessons learned from the past. Davis (2005, p. 260) makes the following propositions while distinguishing between the resilience of the people and the resilience of corrupt systems that produce a recovery shaped by the 'logic of money and power': 'a city is more than its buildings and, thus, resilience must be understood as more than physical reconstruction'. As such, 'reconstruction is not necessarily recovery (or vice versa), and resilience is not always a good thing'. Similarly, Vale and Campanella (2005) define resilience as the capacity of a city and its people to rebound from a disaster. Campanella (2006) explores recovery in New Orleans after hurricane Katrina and its ability to rebound from a disaster. As Davis (2005), Campanella (2006) links resilience to recovery, not only through the physical recovery of the city itself, but also to the recovery of the social fabric. For Campanella (2006), recovery is defined in terms of the resilience of the residents to reclaim their neighbourhoods after a disaster. Cities, to Campanella, are not just buildings; they are sets of communities with people, culture, and heritage that define resilience and possible recovery. Recovery does not necessarily mean bringing back the city to its initial state; it is also not about physical improvements to neighbourhoods. Recovery is about ensuring that people maintain their bonds and community relationships. As such, Campanella (2006) argues that recovery is difficult when the post-disaster reconstruction of a city is imposed on the residents, leading to relocation and further damage. He stresses the importance of local bonds between the residents to ensuring recovery; recovery is then linked to the state of mind of the residents living in the city where social bonds outlast disaster. Campanella (2006) further argues that a city is as resilient as its residents who are willing to come back and claim it as their own.

Another shift occurred when recovery is conceptualized as an open-ended process within a participatory framework, as opposed to a top-heavy one. It has been argued that urban planners can play a vital role in instigating change by adopting resilience as a strategy and augmenting the collaborative role of communities in the recovery process. Authors such as Davoudi (2012) and Owen (2015) stress the importance of resilience in pre- and post-urban planning. Davoudi (2012) emphasizes that a positive change is possible through dialogue between urban planners, NGOs, humanitarian agencies, and residents. Tag-Eldeen (2017) highlights the importance of the local application of recovery using participatory urban planning tools and involving affected communities in the process. In this case, urban planners employ their professional skills to learn from communities and understand their priorities, roles, and expectations in reaching successful recovery and development goals. Therefore, participatory urban planning reinforces communication and provides a platform for negotiation among the different involved actors for effective decision-making as an end goal. Long-term community outreach in decision-making is also fundamental to sustaining recovery and resilience. Brown (2013) and Tag-Eldeen (2017) therefore advocate for an interdisciplinary recovery discourse, combining urban planning with political reform, disaster risk and mitigation, and humanitarian relief to formulate strategic plans in pre- and post-conflict and disaster conditions.

Once recovery is conceptualized as an open-ended process that focuses on the intangible within the participatory framework, some scholars begin to define it as social recovery. Tierney and Oliver-Smith (2012) call for an integrated approach to recovery, conceptualized as an all-encompassing social process that is conceived, implemented, and experienced within physical, natural, cultural, economic, and institutional environments. The authors, thus, focus on the social recovery of households, neighbourhoods, and communities, involving both civil societies and government actors. As a result, physical reconstruction and social recovery become intertwined in the larger recovery process. In fact, Tierney and Oliver-Smith (2012, p. 124) define social recovery as the following:

> Disasters exist as complex material events and, at the same time, as a multiplicity of interwoven and often conflicting social and cultural constructions. Social recovery is therefore inextricably linked to the recovery of structures and infrastructural elements, ecosystems, organizations and institutions, economic activity and culture, making recovery a truly holistic process. Thus, some of the most crucial and interesting research challenges concern the interfaces among these different environments.

March and Kornakova (2017) advocate for people-oriented and holistic development schemes within the social recovery agenda. They argue that physical recovery is not independent of the social, economic, and environmental concerns that constitute human settlements. According to the authors, neglecting community concerns can increase socio-economic inequalities among vulnerable groups affected by conflicts and disasters. Andrews and Caron (2016) highlight the need to mitigate community vulnerabilities through more socially just recovery schemes. They consider the challenges faced by all people as well as the built and natural systems. Therefore, the literature reached a conceptualization of urban recovery as an inclusive social process that is shaped by both pre- and post-crisis conditions.

Structure of the book: Re-conceptualizing urban recovery in the age of protracted displacement

This edited book recognizes the multi-dimensional facets of recovery in the literature, including the physical, economic, social, political, and cultural. These dimensions are interrelated in complex ways to provide a more holistic and participatory conceptualization of recovery. There is, however, a need to advance the discourse to explore the role of gender and minority groups in post-rupture recovery. Furthermore, although emerging trends in the literature are moving towards social recovery in the Global South, they tend to overlook the agency of displaced communities as an economic capital in the recovery efforts and decision-making processes.[3] This book hopes to advance the conversation and re-conceptualize urban recovery in relation to displacement. This is part of a greater effort to address urban ruptures that started in the City Debates 2019 conference and will continue in the research and work pursued in the newly established Beirut Urban Lab at the American University of Beirut. The Urban Recovery Platform at the Beirut Urban Lab aims to advance recovery through research and capacity-building workshops, as a process that is triggered by different types of ruptures and violence and is impacted by displacement and local and national politics and power relations. The Platform conceives of urban recovery as going beyond restoration or regeneration. It is neither a 'post' condition nor a physically bounded process. It is a process that is, at times, intertwined with conflict and displacement across temporal and geographic moments. In its extreme form, it is a process of reconfiguration that addresses all urban vulnerabilities and injustices to the displaced communities, and is an opportunity to investigate how the spatial, socio-cultural, and the imaginative dimensions of urban existence are included and translated. In practice, urban recovery to the Beirut Urban Lab operates within a framework that is bottom-up, participatory, socially just, and inclusive. It is also people-centred, heritage led, and place specific.

The authors in the book offer new interpretations of urban recovery. Jennifer Hyndman, for example, extends the notion of urban recovery in scale and nature. She frames recovery within the international (geo)political context and discusses the recovery of refugee-migrant subjectivity and refugees as city makers and challenges the status quo of 'occidentalist humanitarianism'. Namariq Al-Rawi focuses on the urban recovery of Baghdad's neighbourhoods in the aftermath of the US invasion of Iraq in 2003 and the rise of local sectarianism. She views urban recovery as an act of resistance and reversal to sectarianism through the reactivation of the physical and social networks of the city. Urban recovery in the work of Ghiath Al Jebawi emerges as an experimental initiative by a small community of activists in the Syrian countryside of Azaz. Chiara de Cesari proposes creative institutionalism in promoting local culture as a process of social recovery even before a conflict is over. Isayev explores recovery through imagined constructions of communities in Palestine and Rome, and their relationship to land beyond nation-states and spatially contained citizenship.

This book is structured around four thematic intersections that facilitate for critical and cross-dimensional conversations on overlapping narratives of displacement and spatial modalities of reconstruction: understanding systems and scales of governance, housing the displaced, conceiving of cultural heritage in the recovery process, as well as space and imaginaries in framing post-crisis recovery(s). The thematic intersections incorporate the work of local and international scholars and experts on displacement and reconstruction, as well as those who can locate these themes in a longer historical trajectory. It engages them in an interdisciplinary reflection on the challenges, approaches, and politics involved in the complex operations and networks of local and international actors, as well as communities involved in urban recovery.

Part I. Understanding systems and scales of governance

As cities receive and accommodate waves of displaced people over time in large numbers, global agreements and compacts across nations aim to strengthen the international response to large movements of refugees while alternative models of local governance emerge to meet their daily needs for shelter and public services. The chapters in this section of the book explore: (1) the role of Global Compacts in normalizing extended exile; (2) the scales at which institutional risks and resilience strategies associated with refugee migration unfold; (3) the informal management of displacement in the face of deficient and outdated urban planning systems; (4) the impact of refugee mobility and multilocality in generating new forms of political engagement on the municipal level; and (5) experimental activism that balances the work of government units in generating socio-economic benefits.

Part II. Housing the displaced

Displaced people fleeing conflict are faced with the immediate challenge of adapting to sudden change while recreating a house for themselves in a new environment. The chapters in this section of the book tackle the question of housing for the displaced as agents of urban transformation across multiple scales, ranging from the building, the neighbourhood, to the city. They explore (1) the theoretical debates around home-making under protracted displacement, primarily the practice of carrying one's memories and lived experiences of a previous home to a new one; (2) the active agency of displacement in the making of urban territories across peri-urban residential areas and agricultural fields; and (3) the role of foreign invasion and local sectarian violence in reinforcing socio-spatial divisions between different neighbourhoods and the resiliency of the residents in reactivating their city.

Part III. Conceiving of cultural heritage in the recovery process

Cultural heritage is defined as a social construct that incorporates both the tangible and intangible elements of the built and natural environments. It is recognized as an open-ended process of production and transformation of symbols and meanings throughout history, sustained by its rooted links to communities, their socio-spatial practices, and local identities. In post-conflict and post-disaster environments, cultural heritage becomes closely intertwined with processes of historical editing and rewriting and the need to safeguard people's projected memories of the past for future generations. The chapters in this section of the book explore (1) the value of mainstreaming cultural heritage in the greater recovery process; (2) the commodification of heritage to project a sterile image of post-war national unity; (3) the redevelopment of a legacy of heritage buildings through a selective reading of history; (4) the role of creative institutionalism in reimagining state functions and negotiating opposing claims by different groups to the histories of old cities; and (5) bottom-up approaches by local activists aiming to revive damaged natural heritage and social practices to generate socio-economic benefits.

Part IV. Space and the imaginaries of post-crisis recovery(s)

Urban recovery is conceived as an ongoing process that equally addresses all levels of human vulnerabilities and injustices while navigating omnipresent politics and power relations. To counter these barriers to urban recovery, the chapters in this section of the book delve into the imagined, sometimes whimsical, narratives of space from an artistic point of view. They explore (1) the imagined community and its ties to land under spatio-temporal conditions of post-exile; (2) the design of shopping malls that inspire a sense of amnesic disorientation and echo conditions of protracted displacement; and (3) the visual representation of legal documents to discourse the politics of post-war reconstruction.

Notes

1. UNHCR, Global Trends Forced Displacement in 2019. https://www.unhcr.org/globaltrends2019/.
2. City Debates 2019 Concept Note by Howayda Al-Harithy.
3. The 2006 Reconstruction Unit at The American University of Beirut, co-founded by the editor of this book, was a key moment that allowed, for the first time, the participation of the Lebanese public in the physical, social, and psychological reconstruction and recovery efforts following the 2006 war, which resulted in the death of 1,183 civilians, the indiscriminate bombardment and destruction of hundreds of villages, the levelling of 612 schools, and the displacement of 1,000,000 people. Public exposure and engagement with the reconstruction and post-war planning policies meant that a much more holistic approach to recovery was attempted and reflected on in a scholarly publication. See *Lessons in Post-War Reconstruction: Case Studies from Lebanon in the aftermath of the 2006 War*, edited by Howayda Al-Harithy (Planning, History and Environment Series. Routledge, 2010).

References

Addison, T. (ed.). (2003a). *From Conflict to Recovery in Africa*. Oxford: Oxford University Press.

Addison, T. (ed.). (2003b). From Conflict to Recovery? In *From Conflict to Recovery in Africa*, edited by T. Addison. Oxford: Oxford University Press.

Andrews, C., and Caron, C. (2016). Long-term Disaster Recovery Planning in Urban Centers: The Role of Land Tenure and Housing in Reducing Vulnerability. In *Scaling Up Responsible Land Governance*. Washington, DC: World Bank Land and Poverty Conference.

Barakat, S., and Zyck, S. A., (2009) The Evolution of Post-conflict Recovery, *Third World Quarterly*, *30*(6), pp. 1069–1086.

Bates, F. L., and W. G. Peacock. (1993). *Living Conditions, Disasters and Development: An Approach to Cross-Cultural Comparisons*. Athens, GA: University of Georgia Press.

Berke, P. R., Kartez, J., and Wenger, D. (1993). Recovery after Disaster: Achieving Sustainable Development, Mitigation, and Equity. *Disasters, 17*, pp. 93–109.

Bolin, R. C. (1976). Family Recovery from Natural Disaster: A Preliminary Model. *Mass Emergencies, 1*, pp. 267–277.

Bolin, R. C. (1982). *Long-term Family Recovery from Disaster*. Boulder, CO: Natural Hazards Center, Institute of Behavioral Science, University of Colorado.

Bolin, R. C., and Trainer, P. (1978). Modes of Family Recovery Following Disaster: A Cross-National Study. In *Disasters: Theory and Research*, edited by E. L. Quarantelli, pp. 234–247. Thousand Oaks, CA: Sage Publications.

Brown, B. J. (2013). *Disaster Preparedness and the United Nations: Advance Planning for Disaster Relief*. Amsterdam: Elsevier.

Campanella, T. J. (2006). Urban Resilience and the Recovery of New Orleans. *Journal of the American Planning Association*, *72*(2), pp. 141–146.

Chang, S. E. (2009). Urban Disaster Recovery: A Measurement Framework and Its Application to the 1995 Kobe Earthquake. *Disasters 34*, pp. 303–327.

Chang, S. (2010). Urban Disaster Recovery: A Measurement Framework and its Application to the 1995 Kobe Earthquake. *Disasters*, pp. 303–327.

Davis, D. (2005). Reverberations: Mexico City's 1985 Earthquake and the Transformation of the Capital. In *The Resilient City, How Modern Cities Recover from Disaster*, edited by Vale, L. and Campanella, T., pp. 255–280. Cambridge, MA: MIT Press.

Davoudi, S. (2012) Resilience: A Bridging Concept or a Dead End? 'Reframing' Resilience: Challenges for Planning Theory and Practice Interacting Traps: Resilience Assessment of a Pasture Management System in Northern Afghanistan Urban Resilience: What Does It Mean in Planning Practice? Resilience as a Useful Concept for Climate Change Adaptation? The Politics of Resilience for Planning: A Cautionary Note. *Planning Theory & Practice*, *13*(2), pp. 299–333.

Drabek, T. E., and Key, W. H. (1984). *Conquering Disaster: Family Recovery and Long-term Consequences*. New York, NY: Irvington.

Gotham, K., and Greenberg, M. (2008). From 9/11 to 8/29: Post-Disaster Recovery and Rebuilding in New York and New Orleans. *Social Forces*, *87*(2), pp. 1029–1062.

Haas, J. E., Kates, R., and Bowden, M. (1977). *Reconstruction Following Disaster*. Cambridge, MA: MIT Press.

March, A., and Kornakova, M. (2017). Planning for Recovery: Ideas and Problematics. In *Urban Planning for Disaster Recovery*, edited by March A. and Kornakova M., pp. 231–248. Oxford and Cambridge: Butterworth-Heinemann.

Mileti, D. S. (1999). *Disasters by Design: A Reassessment of Natural Hazards in the United States*. Washington, DC: Joseph Henry Press.

Neal, D. (1996). Reconsidering the Phases of Disaster. *International Journal of Mass Emergencies and Disasters, 15*, pp. 239–264.

Oliver-Smith, A. (1986). *The Martyred City: Death and Rebirth in the Andes*. Albuquerque, NM: The University of New Mexico Press.

Olshansky, R. B. (2005). How Do Communities Recover from Disaster: A Review of Current Knowledge and an Agenda for Future Research. Paper presented at the 46th Annual Conference of the Association of Collegiate Schools of Planning, Kansas City.

Owen, G. (2015). City of Risk: Organization and Individualization in the Urban Recovery of New Orleans. *Minnesota Review*, (84), pp. 83–91.

Shaw, T. M. (1996). Beyond Post-conflict Peacebuilding: What Links to Sustainable Development and Human Security? *International Peacekeeping*, *3*(2), pp. 36–48.

Stanley-Price, N. (2007). The Thread of Continuity: Cultural Heritage in Postwar Recovery. In *Cultural Heritage in Postwar Recovery*, edited by Stanley-Price N., pp. 1–16. Rome: ICCROM Forum.

Smith, G. P., and Wenger, D. (2007). Sustainable Disaster Recovery: Operationalizing an Existing agenda. In *Handbook of Disaster Research*, pp. 234–257. New York, NY: Springer.

Tag-Eldeen, Z. N. (2017). Bridging Urban Planning Knowledge into Post-Disaster Response: Early Recovery Road Map within the International Humanitarian Cluster System. *International Journal of Disaster Risk Reduction*, *24*, pp. 399–410. DOI:10.1016/j.ijdrr.2017.05.023.

Tierney, K., and Oliver-Smith, A. (2012). Social Dimensions of Disaster Recovery. *International Journal of Mass Emergencies & Disasters*, *30*(2), pp. 123–146.

UNHCR. (2019). *Global Trends: Forced Displacement in 2019*. Copenhagen: UNHCR. https://www.unhcr.org/globaltrends2019/.

Vale, L., and Campanella, T. (eds.). (2005). *The Resilient City, How Modern Cities Recover from Disaster*. Cambridge, MA: MIT Press.

Wisner, B., Blaikie, P., Cannon, T., and Davis, I. (2004). *At Risk: Natural Hazards, Peoples' Vulnerabilities and Disasters*. Routledge: London.

Zetter, R. (2005). Land, Housing and the Reconstruction of the Built Environment. In *After the Conflict: Reconstruction and Development in the Aftermath of War*, edited by Barakat S., pp. 155–172. London and New York: I.B. Tauris & Co Ltd.

Part I

Understanding systems and scales of governance

2 Global Compacts or containment?

Geopolitics by design

Jennifer Hyndman

While academics, billionaires, politicians and United Nations agencies staff across the world promote a suite of *solutions* to the *crises* of displacement, this chapter traces some of these 'big ideas' and shows how they are not so new. Ultimately, the proposed fixes do little to redesign humanitarian responses for refugees-migrants or reorganize responsibility for protecting displaced people compelled to leave their home countries.[1] To be more precise, the Global Compacts that were launched with such fanfare in 2016 serve to instantiate the status quo: money from the wealthiest states fund major host countries in the global South and Middle East to support refugees, reproducing the containment of displaced people in their *regions of origin*. Complementary regional measures to keep people at or near home figure alongside these international agreements. For example, the European Union's (EU) migration 'partnerships' include individual *compacts* that have 'priority partners' with countries such as Lebanon and Jordan; such compacts aim to provide support for refugees and host communities in refugees' *regions of origin* and thwart onward journeys to the EU (European Commission, 2016).

The language of 'solutions' is not new, and yet this latest round of proclamations and prescriptions might be considered 'recipes for recovery', but not for those who find themselves in conditions of protracted displacement. Rather, many of the solutions are aimed at helping major refugee host states to continue to cope with these responsibilities. At the end of the day, the solutions preferred are in the interests of the wealthy global North states that do not want any more asylum seekers on their territories. How can 'recovery' be recast to address the status quo of containing displacement 'over there', offshore, not in our backyard? While I have no easy answer to this question, this chapter also aims to address the intransigence of states – all of which signed on to the New York Declaration (NYD) for Refugees and Migrants in 2016 – but have agreed to take little action since the release of the final 2018 texts of Global Compacts for Migrants and on Refugees. Urban recovery in this book is defined as a process that is multi-layered and bottom up. Recovery that includes the urban, but invokes multiple scales, including the global, will have to address the messier politics of colonial legacies, the shift away from Cold War priorities, and the Orientalist portrayals of migrants and refugees who are represented as 'not us' (Kyriakides et al., 2018).

Since the end of the Cold War, when the ideological and political value of the refugee diminished dramatically, there has been little appetite for receiving displaced people across borders (Reynolds and Hyndman, forthcoming). The majority of the world's governments may have signed onto the main international refugee treaties and they may have legal obligations to receive asylum seekers *on paper*, through the 1951 Convention Relating to the Status of Refugees and its 1967 Protocol. In practice, however, increased border enforcement and *preclusion* – the prevention of the arrival of asylum seekers by 'pushing the border out' (Arbel and Brenner, 2013; Reynolds and Hyndman, forthcoming) – has been the rule among global North governments for some time. As early as the 1990s, the Canadian Government posted airline liaison officers (ALOs) in locations as far away as Jomo Kenyatta International Airport in Nairobi to examine documents before passengers boarded planes *en route* to Canada. Borders moved outward transnationally, so that monitoring and preventative measures could block potential asylum seekers from reaching Canada or other territories.

In the spring of 2015, however, public opinion began to change when a ship full of 800 migrants capsized off the Italian coast (*The Guardian*, 2015). In early September, the tragic image of Alan Kurdi, a well-dressed Syrian toddler drowned and lying face down on a Turkish beach affected people deeply across the globe. This tragic image of this child was notable for its ability to transcend the us-them binary, to get beyond Orientalist representation (Gökarıksel and Secor, 2015). By the end of the year, one million people had arrived in Europe by sea, with some 4000 suspected to have died (Clayton and Holland, 2015).

The immense loss of life at sea among asylum seekers to Europe that began in 2015 was also seen by many EU governments as an immense threat. Germany and Sweden opened their ports of entry to asylum seekers, especially from Syria, while other member-states built walls and detention centres to deter migrant-asylum seekers from landing on their territory. The Dublin Regulation III provides rules for which country should register and process asylum seekers, normally the first country in which an asylum seekers lands. Yet this regulation largely fell apart in 2015, and the internal borders of the EU abolished by the Schengen Agreement emerged once more, as Germany opened its doors, and Hungary, Macedonia, Slovenia, Croatia and others successively closed their borders.

This short overview provides only the basic context from which both the EU agreements on migrants (also confusingly called compacts) and Global Compacts stem, but is essential to understand their genesis. Many scholars have published more detailed, geographically specific, portraits of these crossings (Crawley et al., 2018; Vaughan-Williams and Pisani, 2018). Almost immediately, in early 2016, the EU had created compacts with several of the transit countries through which Syrians were moving; I highlight examples of this 'migration partnership' between the EU and Lebanon as well as Jordan. Turkey receives the most funding from the EU in its opaque deal that keeps asylum seekers in Turkey and out of the EU, but that case study is beyond the scope of this chapter.

In the fall of 2016, the UN Summit on Refugees and Migrants was launched and the NYD was rolled out; then President Obama convened a donor meeting. Some 50 states committed $4.5 billion, $1b from the United States alone (UNHCR, 2016). The first substantive paragraph of the NYD, signed by 193 states in September 2016, reads:

> Since earliest times, humanity has been on the move. Some people move in search of new economic opportunities and horizons. Others move to escape armed conflict, poverty, food insecurity, persecution, terrorism, or human rights violations and abuses. Still others do so in response to the adverse effects of climate change, natural disasters (some of which may be linked to climate change), or other environmental factors. Many move, indeed, for a combination of these reasons.
>
> (UNGA, 2016)

This excerpt acknowledges the multiple motivations to move and the dimensions of power, violence and privation that shape different combinations and permutations of human movement. In a subtle way, it unsettles the categories and any easy distinction between refugees and migrants, just as the term 'survival migrants' highlights the partiality of the international legal refugee definition (Betts, 2013). While the narrowness of the 1951 Convention Refugee definition is not explicitly addressed, the need for a 'big tent' under which many vulnerable people on the move could find support is clear.

Understanding the relationship between the large-scale arrival of mostly Syrian asylum seekers, vivid deaths so close to home in the Mediterranean and the European 'crisis' it generated is vital to tracing the international solidarity that emerged, namely the NYD. From this declaration in 2016 to the final texts of the Global Compacts in 2018, solidarity has waned, I contend, but global North 'cash for containment' flows generously.[2]

This chapter proceeds in four parts. First, I analyze the NYD and in an effort to historicize refugee protection, I note how closely it echoes, in parts, the Universal Declaration of Human Rights of 1948. I trace how the NYD lays out a plan for two international compacts, and show how the Compacts stray from the ideals of the NYD. Moreover, I emphasize how the EU agreements with Lebanon and Jordan (also confusingly called 'compacts') were concluded quickly to stem the flow of asylum seekers from these countries, even before the UN summit in fall 2016 began.

Second, I provide a critical overview of the Global Compacts and their genesis. As a foundation for the analysis, I reiterate the case that Wenona Giles and I have made that *solutions* as technical fixes to clearly diagnosed problems are not the answer to the geopolitics and disparate power relations of widespread or protracted displacement (Hyndman and Giles, 2017). Drawing on our analysis and findings, I argue that the conversation needs to change from 'fixes' and 'solutions' to a reframing of language and power that can genuinely address 'responsibility sharing' for asylum seekers-refugee-migrants among states, unsettling the disparate North–South divide. At the risk of oversimplifying the global refugee regime, the wealthy global North states currently *pay* support for displaced persons, while global South countries *host* the vast majority of people forced to flee.

Third, I critically review and assess some salient 'big ideas' (Crawley, 2018), or solutions, that have been presented by academics and others over the past few years to address the refugee '*crisis*' and forced migration more generally. From the isolationism of 'Refugee Nation' to the utopic if more grounded idealism of *Refugia*, or a reinvented twenty-first century version of the Nansen passport (which helped create new homes for Europe's displaced after World War II [WWII]), I outline some of the concrete options for change on offer at the current juncture. I do so with trepidation, as the entire project of finding 'solutions' for displacement is state-centric and geopolitically fraught.

Fourth, I acknowledge scales of analysis and action that are not international and state-centric, as the compacts are. In Beirut, refugees are 'city makers' and self-authorize and actualize their own protection (Fawaz et al., 2018). In Southern African cities, asylum seekers shed their monikers and find work, housing and education for their children (Landau and Duponchel, 2011; Kihato and Landau, 2017). Moreover, other scholars in countries located outside the parameters of international refugee law have pushed back against the Global Compacts, noting their exclusions and flaws. The chapter concludes by highlighting the Kolkata Declaration as a subaltern response to and pushback against the Global Compact on Refugees (GCR), not a solution. India, like much of South and Southeast Asian as well as the MENA region or 'Middle East', lies outside the legal frame of the Eurocentric GCR. The Kolkata Declaration is a manifesto for self-authorized notions of security and protection for people who are displaced but are also workers who must forge livelihoods under conditions not of their own making. It also recognizes that sovereign states act to protect displaced persons in need, even without the (weak) obligations codified in international refugee law.

One can make the case that the rise of the NYD and the Global Compacts that followed emerged from the anxiety of the 'migrant crisis' in Europe, stemming largely but by no means solely from the Syrian civil war. Acute civil war in Syria since 2011 displaced millions of people both internally and as refugees well before 2015. As others have argued, the declaration of a 'crisis' performs political work, creating a space for extraordinary measures, both in the name of security and humanitarian response (Crawley and Skleparis, 2018). Accordingly, 'crisis' has given rise to international solidarity, as outlined in the NYD, and new ideas for making the international refugee regime work better. 'Crisis' has also done a great deal of political work in the name of national security: further securing borders in wealthier countries and creating grounds for returning migrants to their places of departure through what the EU used to call bilateral readmission agreements; these are now called 'compacts', somewhat confusing the regional EU compacts and the global ones on refugees and for migrants. The less visible and more protracted displacement of refugees is not as spectacular, or always as fatal as crossing the Mediterranean has been; it is often a 'slow violence' of deprivation and precarious impermanent status in a host state (Hyndman, 2019). Any new protection strategies must address not only those asylum seekers-refugee-migrants *en route* to Europe or North America, but also those precariously perched in conditions of displacement for years and decades (Mixed Migration Platform, 2017).

The New York Declaration and promise of change

The NYD of 2016 proclaims its commitment to migrants and refugees as bearers of the same human rights in Section 1.6:

> Though their treatment is governed by separate legal frameworks, refugees and migrants have the same universal human rights and fundamental freedoms. They also face many common challenges and have similar vulnerabilities, including in the context of large movements.
>
> (UNGA, 2016)

This excerpt underscores the nuance required in understanding people on the move. Categorical distinctions are blurred. Commonly held rights for all – refugees, migrants and people who do not move at all – are affirmed.

The document is inspiring; it genuinely appears to want to make a difference and change the status quo:

> We declare our profound solidarity with, and support for, the millions of people in different parts of the world who, for reasons beyond their control, are forced to uproot themselves and their families from their homes.
>
> (from Section 1.8)

> Refugees and migrants in large movements often face a desperate ordeal. Many take great risks, embarking on perilous journeys, which many may not survive. Some feel compelled to employ the services of criminal groups, including smugglers, and others may fall prey to such groups or become victims of trafficking. Even if they reach their destination, they face an uncertain reception and a precarious future.
>
> (from Section 1.9)

These passages are worth rehearsing at length because they resonate with, and echo in spirit the pledges and politics of the post-WWII period, when displacement was rampant, and many people were stateless. The pledges continue:

> 10. We are determined to save lives. Our challenge is above all moral and humanitarian. Equally, we are determined to find long-term and sustainable solutions. We will combat with all the means at our disposal the abuses and exploitation suffered by countless refugees and migrants in vulnerable situations.
>
> 11. *We acknowledge a shared responsibility to manage large movements of refugees and migrants in a humane, sensitive, compassionate and people-centred manner. We will do so through international cooperation*, while recognizing that there are varying capacities and resources to respond to these movements. International cooperation and, in particular, cooperation among countries of origin or nationality, transit and destination, has never been more important; 'win-win' cooperation in this area has profound benefits for humanity. *Large movements of refugees and migrants must have comprehensive policy support, assistance and protection, consistent with States' obligations under international law.* We also recall our obligations to fully respect their human rights and fundamental freedoms, and we stress their need to live their lives in safety and dignity. *We pledge our support to those affected today as well as to those who will be part of future large movements* (emphasis added in italics).

Upon rereading the 1948 Universal Declaration of Human Rights (UDHR), the NYD echoes similar language. 'All human beings are born free and equality in dignity and rights', reads article one of the UDHR (UNGA, 1948, Article 1). 'All human beings are born free and equal in dignity and rights', reads Section 1.13 of the NYD, which goes on to add an important contemporary rejoinder in the context of displacement:

> ...We recall that our obligations under international law prohibit discrimination of any kind on the basis of race, colour, sex, language, religion, political or other opinion, national or social origin, property, birth or other status. Yet in many parts of the world we are witnessing, with great concern, increasingly xenophobic and racist responses to refugees and migrants.
>
> (from Section 1.13)

The excerpt above from the NYD is more contemporary, and addresses race, gender, and other axes of difference head on in comparison to the European WWII Universal Declaration of Human Rights (UDHR). The UDHR referred to displacement in Europe while the NYD is more global South and

West Asia/Middle East in its scope. Yet the spirit of both documents is very similar. While the NYD represents an impressive international consensus about the plight of the displaced, it is not without precedent.

The NYD also provides a framework for the Global Compacts, one *for* migrants and the other *on* refugees (Clark-Kazak, 2018; UNGA, 2016). The choice of prepositions is interesting. Perhaps the biggest promise made on the refugee front albeit in rather vague language is this:

> 68. We underline the centrality of international cooperation to the refugee protection regime. We recognize the burdens that large movements of refugees place on national resources, especially in the case of developing countries. *To address the needs of refugees and receiving States, we commit to a more equitable sharing of the burden and responsibility for hosting and supporting the world's refugees*, while taking account of existing contributions and the differing capacities and resources among States.
>
> (from Section 4; emphasis added.)

How exactly responsibility can be shared more equitably is not elaborated in the Declaration. Such detail was seemingly left to be worked out in the GCR. At the Summit, all 193 Member States of the United Nations unanimously adopted the NYD for Refugees and Migrants (Resolution 71/1), a striking consensus at a time when Europe was managing a million asylum seekers. Even as countries signed on to this non-binding and sorely needed expressions of concern, many details still needed to be worked out.

Containing human displacement through regional compacts

The scale, scope and protracted nature of human displacement are more acute today than at any time since WWII. After WWII, an international refugee regime was established to manage human displacement in Europe. A major international treaty was signed, the *1951 Convention Relating to Refugees*, and the Office of the United Nations High Commissioner for Refugees (UNHCR) was formed (with a temporary mandate) to oversee the protection of European refugees and find new and permanent homes for them. By the 1960s, it became clear that human displacement was not going away, nor did it affect only Europe, and the 1967 Protocol to the Convention was signed. The Protocol extended the Convention beyond Europe and lifted the temporal restriction of post-WWII displacement, but it kept the same persecution-focused definition of refugee derived largely from the violence and conflict in Europe during WWII.

When the Cold War ended two decades ago, it realigned the geopolitics of asylum altogether, devaluing the 'refugee' dramatically. During the ideological battle between capitalist and communist superpowers and the bitter proxy wars that it spawned, admitting refugees was evidence of one side's superiority over the other. Whereas refugees were once proof of ideological superiority by a largely capitalist superpower, they became seen as a 'burden' on host countries and the international community. And human displacement became more and more protracted (UNHCR, 2012). Alternative, innovative, unique and new approaches to refugee protection are sorely in demand and are being developed by practitioners, policy makers and scholars across world regions.

More than 65 years after its inception, the international refugee regime still exists, but its efficacy and influence have been compromised since the end of the Cold War. Its architecture and membership, generated through a 'First World'/Western Cold War lens that resulted in the United Nations system, excluded many parts of the globe, rendering it a partial constellation of protection in two senses: the legal part of the regime relates to about half of the world's displaced people; and its Western conception and legal frameworks focus on civil and political rights at the expense of social, economic and cultural ones (Hathaway, 1995).

The UNHCR, the UN's refugee agency, provides material humanitarian assistance and protection against forced return (i.e. *non-refoulement*) where possible. UNHCR's three 'durable solutions', however, fail to deliver much protection for refugees in the current context: 1) voluntary repatriation is at a 30-year low; 2) local integration into a nearby host country is not an option for most refugees, with 85% of the world's total living in global South/Middle Eastern and North Africa (MENA) regions; and 3) refugee resettlement to countries like Canada is modest, affecting about 1% of refugees worldwide (UNHCR, 2012, 2018). Hence, in 2016, the UN released the NYD and launched Global Compacts on which all UN member countries were unanimous: change is needed. Refugee protection needs are largely unmet and are unfairly shouldered by poorer countries in the global South and Middle East.

Protracted refugee displacement is the norm, not the exception, among the world's 25 million refugees (UNHCR Executive Committee, 2009). At least 65% of refugees are in what the UNHCR calls 'protracted refugee situations' (PRS).

Well before the NYD was signed, however, the EU was organizing a different kind of response to the asylum seekers arriving mostly from Syria, but also Afghanistan, Iraq, the Horn of Africa and other regions. The NYD even signals some of this prior response in relation to *readmission agreements*:

> We note that cooperation on return and readmission forms an important element of international cooperation on migration. Such cooperation would include ensuring proper identification and the provision of relevant travel documents. Any type of return, whether voluntary or otherwise, must be consistent with our obligations under international human rights law and in compliance with the principle of non-refoulement. It should also respect the rules of international law

> and must in addition be conducted in keeping with the best interests of children and with due process. While recognizing that they apply only to States that have entered into them, we acknowledge that existing readmission agreements should be fully implemented. We support enhanced reception and reintegration assistance for those who are returned.
>
> (from Section 3; Commitment for Migrants, number 58, p. 728)

Readmission agreements aim to ensure safe and orderly return of asylum seekers deemed ineligible to remain in a given state. Yet such agreements are also insisted upon by global North states, normally in relation to less powerful and poorer states, to prevent and/or ensure return of uninvited migrants from global South countries. Such agreements may seem like 'deals' if one accounts for the 'aid' wealthier states employ to incentivize them. For example, Australia has given the Indonesian Government billions of 'development' dollars in exchange for bilateral 'cooperation' in preventing boat arrivals on Australian shores and providing temporary housing and essential support for asylum seekers on various Indonesian islands, such as Java and Lombok (McPhail, Ochoki Nyamori, and Taylor, 2016).

The rise of the Global Compact on Refugees

From the 'zero draft' to the final third draft in June 2018, the GCR saw many iterations and went through a number of consultations at various regional, national and international scales. Perhaps the strongest language in the GCR comes through in its emphasis on combatting all forms of discrimination, promoting gender equality, empowering women and girls, and ending all forms of sexual and gender-based violence. Refugees are to be treated as participants in their own protection. Delivering on these laudable goals through a framework of loans and aid to refugee-migrant host countries will be more difficult.

The GCR is convention-centric, by which I mean it performs a *legal orientalism* (Ruskola, 2013) that renders states that do the hard work of protecting displaced people beyond the view and treaties of the international refugee regime. The protection of millions of people in this region is sequestered off to individual states through the compacts (or deals) with the EU. Lebanon, Jordan and Turkey, along with vast world regions located in South and Southeast Asia, are simply left off the GCR plans for Comprehensive Refugee Response Frameworks (CRRF), a major tool used to show the application and relevance in 13 case studies (none of which are in the Middle East and North Africa – MENA – region or South or Southeast Asia, but instead were all in Sub-Saharan Africa and Central and South America). Countries like Lebanon have specific agreements, also called compacts, with the EU, which supports the refugees Lebanon hosts but also provides development funds for Lebanese governments and vulnerable segments of its civil society (European Commission, 2017).

A brief note about the EU-Lebanon compact illustrates how this Migration Partnership Framework trickles down. Since 2011, the European Commission (EC) has provided more than €1.6 billion in assistance to Lebanon for both bilateral assistance (€292 million) and particularly in response to the Syrian crisis (over €1.3 billion). The EC lists the important projects which these funds contribute to: humanitarian protection for the most vulnerable, including but not limited to refugees; ensuring decent living conditions for refugees and access to basic services for them *in Lebanon*; and stimulating as well as supporting economic growth for all Lebanese and refugee groups in the country: 'The aim is to foster the inclusive socio-economic development of Lebanon and increase the refugees' contribution to its economic growth' (EC, 2019). While never explicit, onward movement to Europe is to be thwarted.

Giles and I have argued that the problem of protracted displacement is 'crisis' by design, not an anomaly to world politics. While international refugee law and human rights law may provide basic protection for refugees and other migrants, a *geopolitical* consensus among the wealthiest countries to support and finance management of human displacement in 'haven countries' prevails (Hyndman and Giles, 2017). Betts and Collier (2017) galvanize and build upon this consensus, arguing that more support from the global North should support refugees in host countries located in their *regions of origin*, called 'havens'. They aim to improve conditions for refugees and host states in this offshoring model, but do little to reimagine protection and security for displaced people, especially those facing precarious protracted conditions.

The culmination of the Global Compact for Migration

The Global Compact for Migration, specifically called the Compact for Safe, Orderly and Regular Migration, is both aspirational and inspirational. It is, however, largely derivative of existing commitments, especially the UN's Agenda for Sustainable Development 2030. The lead agency for this compact is the International Organization for Migration (IOM), a recent addition to the United Nations associated organizations. Until it was incorporated as such, IOM was a contractor that states could hire to manage various dimensions of migration, from detention to accompaniment (Ashutosh and Mountz, 2012). It still lacks a specified mandate and mission. What is problematic about this new arrangement and IOM's lead on the GCM is that IOM can no longer be paid by UN organizations for specific 'jobs', yet it has no guaranteed funding stream from the UN coffers. Like UNHCR and UNICEF, IOM needs to raise funds to operate, but no longer from 'member states'. This creates, in my view, a conflict of interest in generating the recommendations in the GCM. Almost half of the GCM recommendations represent functions and tasks that IOM has done in the past and has competence to perhaps do in future. In short, the GCM looks like, in part, a funding proposal for IOM.

The GCM is the default document for states that have not signed on to international refugee law, which amounts to much of the MENA region, as well as most of South and Southeast Asia. In other words, the GCR is highly exclusionary because of its legal focus on the 1951 Convention and 1967 Protocol, as well as the states that have signed these treaties.

Big ideas, muted hope

This section briefly outlines some 'big ideas' to solve the protracted human displacement conundrum, namely 1) a GCR-friendly approach; 2) a transnational citizenship choice for those who prefer to live in the diaspora; and 3) based in the historical aftermath of WWII, an idea that worked for many European refugees, more mobility for displaced people to choose a safe destination employing a Nansen-like passport.

I did consider including a fourth idea, the Refugee Nation approach first pitched well before the others in 2015 by Jason Buzi, an Israeli-born Silicon Valley billionaire, but it does not belong in the same comparison with the others. Refugee Nation (Lam, 2015) envisions an island, financed by Buzi, where all the world's refugees could find a common home. Robin Cohen (2015) engages with Buzi's proposal, but ultimately departs from it and founds the more transnational idea elaborated below of Refugia. Buzi's 'island solution', one defined by refugee apartheid and isolation was then improved upon by the Egyptian billionaire, Naguib Sawiris, who suggested that any islands to be purchased should be part of the EU, likely Greek islands since that country was bankrupt at the time. In this way, the islands would come with built-in rights and a solid system of governance. Isolating displaced people from dozens of different nations, speaking many languages, with no obvious basis for 'nation-ness' (Anderson, 1991, p. 3), autonomy or livelihood is not an option that can be taken seriously.

The three ideas outlined below are important to consider, given the scale and protractedness of displacement. I heard the first one proposed at the City Debates forum at the American University of Beirut (AUB) in 2016; it touted the promise of refugee employment in 'special economic zones' in Jordan. It was an anomalous presentation in a sea of carefully researched, critically informed papers. Refugees could be put to work in garment factories, what some might call 'sweatshops'. The idea that refugee host countries would like (more) World Bank concessionary loans in return for their good deeds is rather insulting, if perhaps practical where foreign exchange is in short supply (Zamore, 2018). The idea that international financial institutions (IFIs) should incentive refugee-hosting in global South countries, called 'havens' by some scholars, is one of the most unusual applications of the export-oriented industrialization strategy since its debut in the early 1980s (Dicken, 2015). Such practices have been the 'recipe' of IFIs since the 1980s, but using the international loans to finance factories to hire refugees was a new twist. It is also rather twisted to consider putting refugees to work by day in factories, without any clear provision for residency status or a pathway to such status.

I call this option 'the GCR-friendly model', one outlined in the GCR, but before that by Oxford academics, Betts and Collier (2017). Proponents of such zones argue that World Bank loans can incentivize the hosting of refugees. Such zones have been operating for more than three years; but have they worked? The jury is out, but no conclusive evidence that it has worked has been presented. Leah Zamore, co-author of the third option in this section, has argued that fierce competition in lending terms has put the IFIs out of business when many global South countries need access to loans. Chinese concessionary loans are the norm in much of Sub-Saharan Africa and South Asia (excluding India, a country concerned about this trend).

Betts and Collier (2017) separate the world into 'fragile states' that produce refugees, 'havens' for their safety and 'honeypots' – the wealthiest countries – seemingly reserved largely for those already in them. The two scholars say that the main distinction between a migrant and refugee is the 'psychological impetus': 'Migrants are lured by hope. Refugees are fleeing fear. Migrants hope for honeypots; refugees need havens' (Betts and Collier, 2017, p. 30). This apparent categorical clarity so oversimplifies the conditions that precipitate people's mobility, it is hard to take their typology seriously Crawley, 2017. Betts (2013) himself argues that the Convention refugee definition in law is too narrow to capture the wide gambit of political conditions and violence that define 'survival migration'. The very idea that migrants are greedy and refugees are worthy pits these two groups against one another when Crawley (2017) and Skleparis (2018) among others, show how the membership in these two categories is overlapping, changing over time and anything but clear-cut. The book also demonizes the displaced as though they were a threat and/or a contagion. In one example, '[t]he displaced swarmed across Syria's borders to Turkey, Jordan and Lebanon' (Betts and Collier, 2017, p. 104).

Haven countries do the hard work of hosting the vast majority of the world's refugees, or so the theory goes. They should be supported by the so-called honeypots for all their good work, but there is no challenge to the status quo, whereby more than 85% of refugees reside in global South countries. A majority of the world's refugees have found asylum in just a few 'haven states', located in the global South (Betts and Collier, 2017). Instead of low-interest loans to help these overburdened states kick-start their economies and get further into debt, how can protection be reimagined? How can responsibility be renegotiated to redistribute support for those displaced? The disparate and sharing unequal responsibility of refugees is acute. Zamore (2018) has criticized this position, stating that at the very least, existing host countries should be receiving debt relief, not more loans. Cash transfers or debt write-offs to assist these indebted states are more reasonable prospects, creating more equity between so-called global North 'honeypots' and 'havens' in the global South, including in the MENA region.

The second big idea is one still in its infancy to the extent that it has not been fully fleshed out in published form. I saw it presented in 2018 by another Oxford scholar, Nicholas Van Hear, at the International Association for Studies in Forced Migration conference (see also Van Hear et al., 2018). The objective is to facilitate *Refugia*, a voluntary global network of refugium or diaspora groups linked together formally by shared transnational citizenship: 'a confederal, transnational polity emerging from the connections built up by refugees, with the help of sympathizers' (Van Hear, 2017). The displaced would have permission to reside in the host country where their refugium is located, but would hold citizenship transnationally with other members of their specific diaspora worldwide. All 'refugees' would not be lumped together on an island as with Refugee Nation described above, but would connect with their own self-defined communities.

Based on Van Hear et al. (2018), the idea of Refugia could build upon transnational governance structures, such as those developed in the Tamil diaspora of people displaced from Sri Lanka. After the end of military conflict in Sri Lanka, with the defeat of the Tamil Tigers and the massacre of tens of thousands of Tamil civilians, some diasporic Tamils who envisioned a separate state of Tamil Eelam launched a Transnational Government of Tamil Eelam (TGTE). While many canvassed in the Tamil diaspora see the idea as one that has not worked in the ensuing decade since, it models a form of transnational civic engagement, but not citizenship.

This second option has been criticized for letting host states off the hook in terms of providing rights, social inclusion and potential citizenship, essentializing all refugees as somehow sharing histories, traits and conditions of displacement, and reinforces containment rather than cosmopolitanism (Crawley, 2018; Baldwin-Edwards, Blitz, and Crawley, 2019).

The third idea is most compelling to me, though it is an approach that worked in the context of post-WWII Europe; it will be harder to promote in a contemporary context. Aleinikoff and Zamore's (2019) proposal is, in one sense, the more legal, applied analysis of the three. It is an antidote to protracted displacement and refugee rightlessness in comparison to my own more theoretically driven and social science-oriented treatment of the same issues in a book written with Wenona Giles and published in 2017. Both books identify the protection gap for refugees whose safety comes second to state interests. Each aims to scale down the state-centric 'solutions' for refugees and foreground protection strategies identified by those forcibly displaced, by including them. Both books endorse humanitarian action as a vital intervention in the short-term to save lives, but criticize it as a long-term prospect.

The authors argue that the concept of refugee has expanded, but the protection afforded to them has diminished. The international refugee regime 'constructs a bargain: hosting states will keep their borders open and house refugees in exchange for cash and camps and the international community will turn a blind eye to protection of rights and granting of membership'

(Aleinikoff and Zamore, 2019). This trade-off distorts humanitarian assistance by making it into a 'care and maintenance' situation of dependency for refugees. And, it raises the vital question: how can the humanitarian footprint be decreased without eroding further protection for the displaced? The authors offer an innovative and expansive definition of international protection and possible ways forward by arguing that *mobility* can offer increased protection to refugees.

The core principle of mobility is the critical centrepiece of Aleinikoff and Zamore (2019):

> Indeed, mobility sounds like a radical suggestion only because our thinking is locked into the idea of a system where a refugee is granted protection in a country of first asylum and then must be 'resettled' elsewhere if he or she seeks to leave (and refugees who depart a country of first asylum otherwise become 'illegal migrants'). But in reality, a solution of mobility asks less of other states — a right for a refugee to enter and take up employment or go to school or join a family member. It is not a request for membership.

Enabling mobility and decision-making by refugees is indeed radical in the current context where most refugees are faced with conditions of protracted displacement. This idea also removes the Eurocentric definition of refugee, based on persecution as one's reason for leaving. Still, the consensus of states needs to be secured; the unanimity of support by UN member countries for the non-binding NYD would seem elusive. How many states would sign on, and which ones? Nonetheless, this idea has worked before for Europeans who are displaced. Can it be made workable again?

Concluding thoughts in a minor key: A subaltern declaration on humane mobility

Despite the remarkable consensus of the NYD in 2016, the GCR in 2018 does little more than improve the status quo. Likewise, the Global Compact for Migration consolidates important but existing commitments to the sustainable development goals (SDGs) and other milestones of development. It represents the default compact of states that host large numbers of forced migrants, but which have not signed on to international refugee law. The narrowness of the GCR, then, is an important flaw: both in its use of the Convention definition of refugee and in its geographical-geopolitical exclusion of major countries from the Non-aligned Movement during the Cold War in the global South. Patrolling the narrow legal boundaries around who is a 'refugee' does little to expand protection for the wider, more nuanced, people who must flee their homes and countries. Aleinikoff and Zamore's (2019) proposal becomes the most viable: it argues for a broader conception of protection for (all of) those forcibly displaced; 'fleers of necessity' are not adjudicated on their reasons for leaving home but by the action that they have done so (see also Reynolds and Hyndman, forthcoming).

The Centre for Refugee Studies convened a workshop in June 2018 to highlight alternative solutions to the Global Compacts and the status quo for the protection of displaced people. While that work is elaborated elsewhere (Hyndman and Reynolds, 2020), a major outcome of the research meeting was the 'Humane Mobility Manifesto' at humanemobility.org. This declaration interrupts the business-as-solution and business-as-usual sensibilities of the GCR, in particular. It created a platform for more discussion in Greece a month later at the International Association of Studies in Forced Migration and in Kolkata, India at the Mahanirban Calcutta Research Group's international conference in November 2018. One important subaltern response and corrective to the GCR must be noted.

I was invited to participate in the final discussions and wordsmithing of the Kolkata Declaration in India in November 2018. A critique of, and response to, the NYD of 2016, the Kolkata Declaration demands for the displaced to be included in wide-ranging dialogues about their futures (Mahanirban Calcutta Research Group, 2018). It recognizes the multiple scales at which protection and security are enacted. The document captures salient, if minor, key issues that prevail in South Asia, and in Southeast Asia: statelessness, the importance of migrant rights in the absence of the UNHCR-approved 'refugee' definition and so many others (Mahanirban Calcutta Research Group, 2018).

The Kolkata Declaration, like the GCR, is not legally binding and is authored by and for civil society. Nonetheless, it accomplishes at least three important objectives: 1) it exposes the legal orientalism (Ruskola, 2013) of the GCR, demonstrating that only the 'other' Global Compact (for migration) can apply to a country not 'signed up' to international refugee law, a product of the West, for the West and by the West in a bipolar world after WWII; 2) it situates the issue of labour rights and forced migration in context, elucidating the regional tensions and struggles of people who have been displaced through political *and* economic projects; and 3) it reminds all readers and authors of who gets to speak, when and how. Those whom we cite and those left out are also vital to any analysis of the complementary academic work around the Global Compacts. So much more remains to be done.

The Humane Mobility Manifesto and the Kolkata Declaration in 2018, along with engaged and activist research, such as *Refugees as Citymakers* (Fawaz et al., 2018), that demonstrates the protagonism of displaced people acting under conditions not of their own making all speak to a very different form of recovery than the solutions discourse analyzed here.

Whereas solutions look for 'fixes' to maintain the status quo where containment of displaced groups, these interventions aim to interrupt, undo and remake processes of recovery at multiple scales. The Humane Mobility Manifesto pushes for the recovery of the subjectivity and self-authorized protection of the migrant-refugee-asylum seeker (Hyndman and Reynolds, 2020). It shows how the global refugee regime is increasingly irrelevant and how recovery will be incumbent on fishing communities in Banda Aceh who encounter Rohingya asylum seekers on the high seas; civil societies that can and already do much to provide material and political safety for those displaced from their homes; and individual states that must determine their own humanitarian policies and practices towards refugees in the absence of international law and obligation.

The Kolkata Declaration makes a much more subtle claim: from its subaltern location, it reminds us all that its very subjectivity needs to be 'recovered'. In this context, the subaltern cannot speak because it has not signed on the grid of legibility, namely the Occidentalist 1951 Convention Relating to Refugees and its 1967 Protocol. Nonetheless, it proclaims that the GCR left out South and Southeast Asia, that its companion compact, the GCM, is hopeful but without resources and mechanisms to realize the goals outlined. It outlines concrete concerns about statelessness, labour migrants and other displaced persons whom are being cared for *outside* the confines of the international refugee regime. This multi-scalar rendering of recovery, as the subaltern, the self-authorized notion of protection, the state that did not sign the Refugee Convention, and against the paradigm of geopolitical containment is critical to shifting the ground and changing the status quo.

Acknowledgments

Many thanks to Amy Drury, Johanna Reynolds and Myriam Ouellet who offered invaluable feedback on this chapter, and to Dr. Howayda Al-Harithy for inspiring me to write this chapter.

Notes

1. I use the term 'refugees-migrants' as an imperfect moniker that aims to capture the range of categories and monikers that states apply to people on the move. This is a huge debate and discussion among many actors long before this chapter was written. Many states are not signatories to the 1951 Convention Relating to the Status of Refugees or its 1967 Protocol, and are thus not bound by its provisions. Nonetheless, many of these states provide protection for people forced outside the borders of their countries of origin. For analysis and discussion of these terms, see Kyriakides (2017), Crawley and Skleparis (2018), Zetter (2007), and humanemobility.net.
2. I thank Jesse Thomson, Vice-President for Partnerships at CARE International in Ottawa, for sharing this term with me, and for her candour in calling out states that engage in these practices.

References

Aleinikoff, T.A., and Zamore, L. (2019). *The Arc of Protection: Reforming the International Refugee Regime*. Stanford University Press: Stanford.

Anderson, B. (1991). *Imagined Communities: Reflections on the Origin and Spread of Nationalism*. Verso: London.

Arbel, E., and Brenner, A. (2013). *Bordering on Failure: Canada- U.S. Border Policy and the Politics of Refugee Exclusion*. Harvard Immigration and Refugee Law Clinical Program: Boston, MA.

Ashutosh, I., and Mountz, A. (2012). The Geopolitics of Migrant Mobility: Tracing State Relations Through Refugee Claims, Boats, and Discourses. *Geopolitics*, *17*(2), 335–354. DOI: 10.1080/14650045.2011.567315.

Baldwin-Edwards, M. Blitz, B.K., and Crawley, H. (2019). The Politics of Evidence-based Policy in Europe's 'Migration Crisis'. *Journal of Ethnic and Migration Studies*, *45*(12), 2139–2155.

Barnard, L. (2017). Sawiris Raw as Deal for Greek Islands for Refugees Falls Flat. Feb 2, *The National*, https://www.thenational.ae/business/property/sawiris-raw-as-deal-for-greek-islands-for-refugees-falls-flat-1.50242.

Betts, A. (2013). *Survival Migration: Failed Governance and the Crisis of Displacement*. Cornell University Press: Ithaca and London.

Betts, A., and Collier, P. (2017). *Refuge: Transforming a Broken Refugee System*. Penguin Press: London. DOI: 10.1093/ijrl/eey015.

Calcutta Research Group (2018a). Discussion on Re-imagining 'Fortress North America' and 'Fortress Europe' in Context of Recent Migrations. Nov 25–30, *The State of the Global Protection System for Refugees and Migrants Conference*, held in Kolkata, India, https://www.youtube.com/watch?v=WKJOAccP97k&feature =youtu.be.

Clark-Kazak, C. (2018). 'A Refugee is Someone who Refuses to be Oppressed': Re-centering Refugees' Agency and Intersectionality. Presentation to Workshop on Alternative Solutions Workshop, York University, June, Toronto.

Clayton, J., and Holland, H. (2015). Over One Million Sea Arrivals Reach Europe in 2015. Dec. 30, *UNHCR*, https://www.unhcr.org/news/latest/2015/12/ 5683d0b56/million-sea-arrivals-reach-europe-2015.html.

Cohen, R. (2015). Refugia: The Limits and Possibilities of Buzi's Refugee Nation. July 30. A blog on Nando Sigona's website. https://nandosigona.info/2015/07/30/refugia-the-limits-and-possibilities-of-buzis-refugee-nation.

Crawley, H. (2017). Migration: Refugee Economics. *Nature*, *544*, 26–27. DOI: 10.1038/544026a.

Crawley, H., F. Duvell, K. Jones, S. McMahon and N. Sigona (2018). *Unravelling Europe's 'Migration Crisis': Journeys Over Land and Sea*. Bristol UK: Policy Press.

Crawley, H. (2018). Why We Need to Protect Refugees From the 'Big Ideas' Designed to Save them. *The Independent*, July 28.

Crawley, H., and Skleparis, D. (2018). Refugees, Migrants, Neither, Both: Categorical Fetishism and the Politics of Bounding Europe's 'Migration Crisis'. *Journal of Ethnic and Migration Studies*, *44*(1), 48–64. DOI: 10.1080/1369183X.2017.1348224.

Crisp, J. (2003). No Solutions in Sight: the Problem of Protracted Refugee Situations in Africa. Working Paper 75. *UNHCR Evaluation and Policy Analysis*, Jan, https://www.unhcr.org/3e2d66c34.html.

Dicken, P. (2015). *Global Shifting: Mapping the Changing Contours of the World Economy*. Sage: London.

European Commission (2016). Commission Announces New Migration Partnership Framework. EC Press Database, https://europa.eu/rapid/press-release_IP-16-2072_en.htm.

European Commission (2017). EU-Lebanon Partnership: The Compact. https://ec.europa.eu/neighbourhood-enlargement/sites/near/files/lebanon-compact.pdf.

European Commission (2019). Responding to the Syrian Crisis: EU Support to Resilience in Lebanon. See EC fact sheet at https://ec.europa.eu/neighbourhood-enlargement/sites/near/files/eu-support-to-lebanon-factsheet.pdf.

Fawaz, M., Gharbieh, A., Harb, M., and Salamé, D. (eds.) (2018). *Refugees as City-Makers*. The Social Justice and the City Program at the Issam Fares Institute for Public Policy and International Affairs, American University of Beirut: Beirut, Lebanon.

Gökarıksel, B., and Secor, A. (2015). Post-secular Geographies and the Problem of Pluralism: Religion and Everyday Life in Istanbul, Turkey. *Political Geography*, *46*, 21–30.

Hathaway, J. C. (1995). New Directions to Avoid Hard Problems: The Distortion of the Palliative Role of Refugee Protection. *Journal of Refugee Studies*, *8*(3), 288–294.

Hyndman, J. (2019). Unsettling Feminist Geopolitics: Forging Feminist Political Geographies of Violence. *Gender, Place and Culture*. DOI:10.1080/0966369X.2018.1561427.

Hyndman, J., and Giles, W. (2011). Waiting for What? The Feminization of Asylum in Protracted Refugee Situations. *Gender, Place & Culture*, *18*(3), 361–379. DOI: 10.1080/0966369X.2011.566347.

Hyndman, J., and Giles, W. (2017). *Living on the Edge: Refugees in Extended Exile*. Routledge: New York.

Hyndman, J., and Mountz, A. (2007). Refuge or Refusal: Geography of Exclusion. In D. Gregory and A. Pred (eds.), *Violent Geographies*, Routledge: New York, 77–92.

Hyndman, J., and Reynolds, J. (2020). Beyond the Global Compacts: Re-imagining Protection, *Refuge*, *36*(1), 66–74, https://refuge.journals.yorku.ca/index.php/refuge/article/view/40768/36573.

Jones, M. (2018). Developing a Law of Asylum in the Middle East and Asia: Engaging the Law at the Frontiers of the International Refugee Regime. Presentation at the Alternative Solutions Workshop, Centre for Refugee Studies, York University, June.

Kihato, C.W., and Landau, L.B. (2017). Stealth Humanitarianism: Negotiating Politics, Precarity and Performance Management in Protecting the Urban Displaced. *Journal of Refugee Studies*, *30*(3), 407–425.

Kyriakides, C. (2017). Words Don't Come Easy: Al Jazeera's Migrant-Refugee Distinction and the European Culture of (Mis)trust. *Current Sociology*, *65*(7), 933–952.

Kyriakides, C., Bajjali, L., McLuhan A., Anderson, K. (2018). Beyond Refuge: Contested Orientalism and Persons of Self-Rescue. Canadian Ethnic Studies. Special Issue. *50*(2), 59–78.

Lam, B. (2015). The Man Who Wants to Crowdfund a New Nation for Refugees. *The Atlantic*, Oct 15, https://www.theatlantic.com/business/archive/ 2015/10/jason-buzi-refugee-nation/410728.

Landau, L.B., and Duponchel, M. (2011). Laws, Policies, or Social Position? Capabilities and the Determinants of Effective Protection in Four African Cities. *Journal of Refugee Studies*, *24*(1), 1–22, DOI: 10.1093/jrs/feq049.

Mahanirban Calcutta Research Group. (2018). *Kolkata Declaration 2018: Protection of Refugees and Migrants.* Nov 30, http://www.mcrg.ac.in/RLS_Migration/Kolkata_ Declaration_2018.pdf.

McConnell, F., Kuus, M., Jeffrey, A., Crawley, H., and Vaughan-Williams, N. (2017). Interventions on Europe's political futures. *Political Geography*, *60*, 261–271. DOI: 10.1016/j.polgeo.2017.02.003.

McPhail, K., Ochoki Nyamori, R., and Taylor, S. (2016). Escaping Accountability: A Case of Australia's Asylum Seeker Policy. *Accounting, Auditing & Accountability Journal*, *29*(6), 947–984. DOI: 10.1108/AAAJ-03-2014-1639.

Mixed Migration Platform. (2017). https://reliefweb.int/sites/reliefweb.int/ files/resources/MMP-and-the-Global-Compacts.pdf.

Reynolds, J., and Hyndman, J. (forthcoming). Shifting Grounds of Asylum in Canadian Public Discourse and Policy. In T. Phu & V. Nguyen (eds.), *Critical Refugee Studies*, University of Toronto Press: Toronto.

Ruskola, T. (2013). *Legal Orientalism: China, the United States, and Modern Law.* Harvard University Press: Cambridge.

Ryan, D., Crawford, T., and Lee, J. (2015). B.C. Family Mourns Drowned Syrian Refugee Boy Trying to Get to Canada. *The Vancouver Sun*, Sept 4, http://www.vancouversun.com/news/metro/family+mourns+drowned+syrian+refugee+trying+canada/11337946/story.html.

The Guardian. (2015). UN Says 800 Migrants Dead in Boat Disaster as Italy Launches Rescue of Two more Vessels. April 19, https://www.theguardian.com/world/2015/apr/20/italy-pm-matteo-renzi-migrant-shipwreck-crisis-srebrenica-massacre.

UNGA. (1948). Universal Declaration of Human Rights. *UN General Assembly*, *302*(2). http://www.verklaringwarenatuur.org/Downloads_files/Universal%20Declaration%20of%20Human%20Rights.pdf.

UNGA. (2016). *New York Declaration for Refugees and Migrants: Resolution/adopted by the General Assembly, 3 October 2016.* A/RES/71/1, Annex II: Towards a Global Compact for Safe, Orderly and Regular Migration.

UNHCR. (2012). *State of the World's Refugees: The Search for Solutions.* Oxford University Press: Oxford.

UNHCR. (2016). UNHCR Welcomes New Support for Refugees at Leaders' Summit. Sept 20, https://www.unhcr.org/news/latest/2016/9/57e2481e8/unhcr-welcomes-new-support-refugees-leaders-summit.html.

UNHCR. (2018). 'Zero Draft' of The Global Compact on Refugees, January 31, http://www.unhcr.org/Zero-Draft.pdf.

UNHCR (2019). *Global Trends: Forced Displacement in 2018.* https://www.unhcr.org/statistics/unhcrstats/5d08d7ee7/unhcr-global-trends-2018.html.

UNHCR Executive Committee. (2009). *Conclusion on Protracted Refugee Situations, No. 109 (LXI).* Dec 8, https://www.unhcr.org/excom/exconc/4b332bca9/conclusion-protracted-refugee-situations.html.

Van Hear, N. (2017). Imagining Refugia: Could a New Transnational Polity Help Solve the Refugee Crisis. Oct 17, *Foreign Affairs Newsletter*, https://www.foreignaffairs.com/articles/world/2017-10-17/imagining-refugia.

Van Hear, N., Barbelet, V., Bennett, C., and Lutz, H. (2018). Imagining Refugia: Thinking Outside the Current Refugee Regime. *Migration and Society: Advances in Research*, *1*, 175–194. DOI: https://doi.org/10.3167/arms.2018.010116.
Vaughan-Williams, N., and Pisani, M. (2018). Migrating Borders, Bordering Lives: Everyday Geographies of Ontological Security and Insecurity in Malta. *Social & Cultural Geography*, 1–23. DOI: 10.1080/14649365.2018.1497193.
Zamore, L. (2018). Refugees, Development, Debt, Austerity: A Selected History. *Journal on Migration and Human Security*, *6*(1), 26–60. DOI:10.1177/233150241800600102.
Zetter, R. (2007). More Labels, Fewer Refugees: Remaking the Refugee Label in an Era of Globalization. *Journal of Refugee Studies*, *20*(2), 172–192. DOI: 10.1093/jrs/fem011.

3 Refugees, resettlement, and the territorial correlates of resilience

Diane E. Davis

Introduction

The dramatically increasing numbers of people forcibly displaced by conflict and violence has resulted in the expansion and transformation of spaces of migrant refuge across the globe. From the US border, where Central American migrants desperately seek asylum from horrifying conditions of chronic mafia violence in their home countries, to countries across Europe and the Middle East which have been receiving flows of Syrian refugees at alarming rates as the administration of Bashar Al-Assad continues his ruthless persecution of political opponents, no part of the world seems immune from the problem of refugees. The sites where migrants land in their desperate search for security may differ dramatically, depending on the conditions of their transit, which national or international authorities lend assistance in resettlement, and how permanent their emigration is intended to be. Even so, refugees in myriad settings are receiving increasing attention on the global security agenda, in large part because their existence and how they manage to survive can both enhance individual resilience and, at the same time, call into question the longer-term stability of the sites and regions in which they locate. To the extent that refugee populations have become the locus of sustained humanitarian and securitization interventions by multilateral, national, and local actors, investigating the 'spaces' and practices of their resettlement allows us to combine current security-migration discussions with emergent debates about resilience.

Camps and other spaces of refugee settlement constitute a key site of inquiry for uncovering the multi-scalar and cross-sectoral implications that massive influxes of migrants have on security and resilience. On the one hand, refugee camps are designed to provide legal and physical protection, as well as enhance community resilience. Insofar as they reduce both individual and collective vulnerabilities caused by forced migration, they may be deemed manifestation of positive resilience. On the other hand, in many contexts – particularly in protracted displacement situations – these same camps serve as amplifiers of violence, social unrest, and inequalities, as well as bearers of new forms of exclusion, deprivation, or suffering. In that sense, they also can be construed as manifesting negative resilience. Alternatively, when refugees settle in cities rather than camps, they may have more capacity to integrate into the local economy and find sources of work that sustain their capacities to move forward with a modicum of normalcy. Yet cities can also be expensive and unwelcoming environments, and stigmatization of refugees – particularly when they come from different ethno-national, class, religious, or racial backgrounds than the population in the host city – can undermine resilience. In this sense, certain territorial sites may enable or constrain resilience, or some combination of these adaptive mechanisms.

To shed light on the complex relationships between sites of resettlement and the likelihood of resilience, and to situate this discussion in a closer understanding of the spaces and territorialities of migrant refuge, this chapter adopts a 'scales of risk and resilience' framework. It assesses the multiple scales at which the various risks and attendant securitization strategies associated with refugee migration unfold, ranging from global security risks and challenges to national sovereignty to a range of (in)securities occurring in camps, and at the urban, national, and regional scales. It then asks whether strategies to minimize risk by fostering resilience at one of these scales might negatively affect another. Empirical evidence is initially drawn from the Middle East and the Levant, where the current political context has dramatically heightened migratory flows and security preoccupations in recent years.[1] Findings from Lebanon and Jordan launch our analysis because those countries have been disproportionately affected by the Syrian refugee crisis,[2] and because the Lebanese and Jordanian experiences reveal the inter-connectedness of causalities between the regional, national, and urban scales, as well as the complex layering of governance mechanisms. However, in order to widen the analytical scope of our analysis, we also focus on Colombia – a country with the world's second highest number of internally displaced peoples (IDPs), owing to its longstanding civil war and the numerous failed attempts to secure a stable peace. To the extent that Colombian government policy has been dealing with the question of forced displacement and victim resettlement for decades, and has struggled in the past 15 years to create programs and policies to foster resettlement and urban as well as rural re-integration of peoples expulsed from their homes due to long-standing violence between armed guerrilla factions sand the state, it provides a slightly different reference point for thinking about spaces of migrant refuge, the rescaling of resilience, and even reconstruction – particularly if we conceive of policies to reintegrate and re-settle as part of national (and not merely international humanitarian) efforts to reconstruct new livelihoods, new economies, and a new politics of co-existence.

Before beginning, a few notes on the concept of resilience, and its relationship to risk, are in order.[3] The notion of resilience has taken the policy, design, and urban planning worlds by storm. From the Rockefeller Foundation's new global initiative (where cities around the world compete to be branded as among the world's 100 most resilient), to the barrage of academic conferences and new study centers hosted by universities and multilaterals, to the programs sponsored by technology corporations and design firms (including the joint IBM-AECOM recent piloting of a disaster resilience scorecard), and to the US State Department's appropriation of resilience as a thematic rationale for international aid (built around the understanding that environmental vulnerabilities, broadly defined, are linked to problems of poverty, conflict, and violence), *resilience has become the watchword of our times.* It is promoted as the rationale for a new repertoire of tools – from novel technologies to reconfigured mapping and building products – that will guide us to a secure urban and global future. It also has recently entered into the humanitarian aid and security studies world (Bourbeau, 2017).

But, resilience is a tricky word, readily veering into the ideological, such that it often is deployed to celebrate the desperate acts of citizens who have no other recourse. Defined as the ability to cope and adapt so that individuals or communities survive and thrive, resilience is all about optimism. Whether conceived in terms of 'bouncing back to normal' after a disaster, or as a means of re-establishing system equilibrium after a shock, embracing resilience means having faith that with enough attention and effort, the future can be better. However, such assumptions often ignore the inter-relationalities of risk. There are trade-offs among forms and patterns of resilience, not just among different residents or locations in the same city but also in terms of immediate versus long-term gains in livability, such that coping strategies in some domains (say environment) may actually reinforce the structural problems that create risks in other domains (say inequality). To the extent that urban, social, economic, and environmental ecologies are connected, both locally and across various scales that link cities to regions and beyond, any resilience strategy must be grounded in an appreciation of the entire territoriality of connections. This is precisely why the issue of scale is so central to our understanding of resilience and how it might operate in the context of refugee resettlement.

In what follows, I start with a review of humanitarian and security studies literature on refugees and displaced peoples, identifying the ways that risk initially emerged as a framing concept. I then examine the more recent turn to resilience (in both the security field and studies of displaced peoples,) discussing the ways that this concept is seen as opening new opportunities for understanding coping and adapting strategies used by those with refugee status. In a third section, I discuss how these two literatures, despite their advances, nonetheless fail to understand that both risk and resilience unfold on multiple spatial and territorial scales spanning city, region, and nation, and in fact interact with each other. In the final section, I scale 'down' our spatial analysis from the concern with the cross-territorial dynamics of risk and resilience to the grounded neighborhood-level conditions of settlement, focusing on the ways that housing, land tenure, and property rights also mediate the dynamics of risk and resilience.

Security studies and the refugee question: Rethinking territories of risk

The last decade has seen a sharp increase in numbers of displaced individuals, many of whom are officially identified as refugees. The causes for displacement in today's world are complex and intertwined. They include not only conflict, war, and state as well as citizen-led terror, but also environmental degradation – which is increasingly producing 'climate refugees' (Black et al., 2011) as well as poverty and ever-widening socio-economic disparities that make it difficult for residents to 'stay put' (Weinstein, 2018). Often, these causes are interconnected, with environmental scarcities fueling war and conflict in the first place. In the face of these conditions, mobility can occur across international borders (Bourbeau, 2015). Yet many individuals are also forced to move within their own countries, and the numbers of IDPs – on the rise since the decolonization wars of the late 1960s and 1970s – are now estimated to be roughly double[4] the numbers of refugees (IDMC, 2015). Significantly, evidence shows that distressed populations – be they refugees or IDPs – increasingly seek safe haven outside, rather than inside, conventional humanitarian camps (Jacobsen, 2006). This has resulted in the expansion and transformation of both the scales and the forms of refugee settlements worldwide. In addition to the neatly demarcated zones of transit – ranging from border crossings and reception or detention centers to on land or offshore detention facilities – spaces of refuge are myriad, taking diffused, scattered, and often invisible physical configurations. They can come in the form of camp-cities, (Figure 3.1), informal tented or squatter settlements in dense urban neighborhoods or unoccupied lands on the urban periphery, or even rent-subsidized or free market-provided apartments in host communities. They can be conceptualized as ephemeral or fixed. And whatever the intent of their creators or residents, this wide range of refugee settlements comprises what we call the new geographies of migrant refuge.

These sites are receiving increasing attention in the global security and humanitarian worlds, in large part because their existence and operations call into question the longer-term stability of the cities and regions in which they are located, thus further creating conditions of risk (see Potvin, 2019). For precisely this reason, disciplines historically concerned with national and global political stability have started to examine the security-migration nexus. Those who have concerned themselves with securitization of migration generally frame migration and migrants in a language of threat, risk, and existential urgency, as opposed to integrating these issues into 'normal' political, social, cultural, or economic debates (Bigo, 2002; Hammerstad, 2012). Through such speech, some elements of social life are constructed as threats in order to prompt an exceptional counteracting response, one that allows urgent measures to be taken ('by any means necessary'), with the main objective to transfer issues out of

(a)

(b)

Figure 3.1 (a) Zaatari refugee camp, Jordan; (b) Syrian Refugees, Ouzai Complex, Saida, Lebanon

the sphere of normal politics and into the sphere of policing and defense (Bourbeau, 2013). This understanding of securitization stems from critical security studies, which posits that 'security threats are not objectively given realities, but inter-subjectively constituted through social processes' (Hammerstad, 2012, p. 4).

The security literature has traditionally focused on the securitizing practices of state actors. As such, it follows that the significant portion of securitization literature that deals with migration commonly does so from the perspective of the receiving states. Migration is predominantly treated as an exogenous shock or a disturbance to a given host society at the receiving end of the migration flow, and whose security and sovereignty is at risk (Brown, 2010). Even in critical security studies, the focus has generally been placed on the disruptions that migration brings about, rather than on the turmoil it stems from. Furthermore, migrants are typically dealt with as largely homogenous groups. The distinctions between economic migrants, conflict or climate refugees, and asylum seekers are blurred – which in turns 'renders [these categories] less intelligible in processes of enforcement and exclusion' (Mountz and Hiemstra, 2014, p. 384).

To be sure, the relationship between security and migration is historically specific and its conceptualization has evolved over time. In the context of Europe, at least, scholars generally begin their historical accounts with the emergence of political concerns over the large migratory flows that derived from the disintegration of the Ottoman and Habsburg empires in the wake of World War I. These massive population movements caused disturbances that heightened inter-state tensions and threatened the national security of the fragile European state system. Because these migratory events exceeded the management capacities of individual states, they incited the creation of an international refugee regime in the early 1920s (Loescher and Milner, 2005). This nascent international framework of institutions was further formalized following World War II, notably with the creation of the United Nations High Commissioner for Refugees (UNHCR) in 1950 and the United Nations Convention Relating the Status of Refugees in 1951 (Isotalo, 2009, p. 63).

Then, during the Cold War era, forced migration was tightly connected to a broader set of geopolitical struggles, and the superpowers attended to migrants as part of their respective strategic agenda, providing ideological and material support, especially in the case of people fleeing persecution in the Eastern bloc and seeking asylum in the West (Goodwin-Gill and McAdam, 2007; Hammerstad, 2000). During that period, the general assumption was that threats to a state security were primarily military in nature and existed outside its national borders (Loescher and Milner, 2005). These approaches reveal the enduring narrative about the securitization of migration in Western Liberal Democracies, which hinges on the sanctity of sovereignty and also uses the nation state as the primary unit of analysis (Isotalo, 2009; Hammerstad, 2012). In the post-Cold War era, the framing of the security-mobility nexus once again transformed, with the focus shifting to questions of refugee management (2000) and more recently on climate migration (2015).

Notably, the post-cold war was also characterized by the spread of new types of organized violence, which some have termed 'New Wars' (Kaldor, 1999), including the conflicts that erupted in the Balkans, the Middle East, East Timor, and the African Great Lakes region. It was also marked by new demographic make-ups and policy regimes in Western states, resulting in a restrictive asylum climate and a growing hostility towards migration, portrayed as massive, illegal, and threatening to an alleged status quo in the receiving countries. An important theme that emerged within the security literature produced in the late 1990s and during 2000s is how sovereignty started giving way to new practices and tools that challenged the Westphalian logic of state sovereignty, especially with the appearance of the notion of the responsibility to protect, which made it legitimate for a state (or a community of states) to intervene in another state's internal affairs (Loescher and Milner, 2005; Methmann and Oels, 2015; Chandler, 2012). Yet even beyond the political and governance discourse related to sovereignty, scholars also have construed development as yet another agent of security. Since underdevelopment can equally be a root cause of migration, the development of migrants' home country and migrant communities in immediately adjacent host countries is intended as another type of security infrastructure.

In parallel to these preoccupations with the political economy of governance and development, in recent years we have been witnessing a scalar redefinition of security in terms of human security as part of a shift towards the subnational scales, including the urban or city scale, and how daily conditions at the site of settlement impact security at the human scale (Chandler, 2012; Davis, 2012; Methmann and Oels, 2015). A concern with the security of the individual is often posed in terms of prevention, empowerment, distance from the military realm, and 'freedom from want' in addition to 'freedom from fear' (Chandler, 2012). And in a further focus on the individual, professionals in post-conflict recovery operations have appropriated the concept, recognizing that human beings share common psycho-social needs of identity, recognition, participation, and autonomy, and that ensuring human security requires giving importance to a sense of home, a link to community and a positive sense of the future (Leaning, 2009).

What I am describing here is the evolution of a territorial shift in terms of interpretative scales, with the focus slowly moving away from the national security scale towards the transnational scale, on the one hand, and towards the community and human scales, on the other. It is important to highlight these shifts in scales, as well as the new vernacular of human security, because they coincide with the emergence of the concept of resilience in the security literature.

Resilience and the refugee question: Are all adaptation and coping strategies alike?

The growing popularity of claims that resilience is a new paradigm for scholars who seek to examine the security-migration nexus owes partly to the growing preoccupation with environmental risk and not just political risk. In their genealogy of climate-induced migration, Methmann and Oels (2015, p. 58) explain how migration, when examined through the lens of resilience, no longer gets portrayed as a security issue but is instead seen as an appropriate means of adaptation – a positive form of resilience, so to speak. This is evidenced by the language employed for talking about forced migration caused by climate change, which has gone from a language of 'climate refugee' (i.e. securitization) to that or 'climate change induced migration' (i.e. more neutral or de-securitized).

More broadly, going back to internal relations and conflict research, the discourse of resilience is seen as offering a new conceptual framework for shedding new light on the contemporary security world as well as on the interventions that advance the securitization processes (Bourbeau, 2013; Chandler, 2012). In particular, security scholars concerned with migration are now pointing to new theoretical avenues that offer a useful rebuttal to conventional state-centric framings. In the context of South African mobilizations against Zimbabwean refugees in 2008, for example, scholars have shown that immigration was not initially perceived as a security threat by the South African political elite, but was experienced as such by disenfranchised citizens, mostly the urban poor. Rather than having the political elites be the primary securitizing actors, security measures were first defined and enacted at the grassroots level, through vigilantism, violence, and riots against a perceived refugee threat, and in contravention of the non-securitizing position espoused by the government. In fact, it was only when the popular response turned so violent that it threatened the stability of the political regime, that the state began recasting the refugees as a risk factor (2012). The point here is that the concept of 'securitization from below,' which grows out of recognition of individual and group adaptation to conflict, opens up new ways of thinking about resilience from the perspective of the host population and of the multiple scales of governance actors involved in dealing with refugee influx.

The discussions that most explicitly and rigorously connect questions of resilience to the security-migration nexus are at a nascent stage, and for this reason still lack robust, in-depth, theorization. Confronted with the task of examining these complex linkages, one must disentangle multiple premises of inquiry: whose security is at risk with increased refugee flows, whose resilience is at play, and at what scales do security and resilience practices get enacted when it comes to the question of refugees? Of course, use of the term refugee deserves to be scrutinized and problematized.[5] But here, I adopt an expansive and inclusive understanding of the term, to include a wider range of 'survival migrants' who face legal or ethical threats not only to their lives but also to their human dignity, fleeing violence, and the deprivation of basic rights (Betts, 2013, p. 23). And once one widens the conceptual framing in this way, we are forced to acknowledge that refugee movements are inherently dynamic processes, and they involve multiple actors and audiences, including not only the refugees themselves and the state or non-state bodies that regulate them, but also the communities of origin and those of destination. Likewise, studying refugee flows requires the capacity to look at multiple sites simultaneously: from the points of departure, to the spaces of transit and the localities of arrival or resettlement.

Yet, it is in the interaction of these various temporal and spatial scales that resilience must be understood. Indeed, recent scholarship (Gemenne and Blocher, 2016, p. 1) shows that migration itself is often conceptualized as adaptation – a key foundational idea buttressing the concept of resilience – and that migrant adaptations 'can range from short-term coping to longer-term, deeper transformations (… which) may or may not succeed in moderating harm or exploiting beneficial opportunities.' In other words, through the 'resilience' lens, the hefty security challenges posed by climate change, environmental degradation, and the like are presented as being insurmountable, so much so that migration becomes a necessary, if not the sole, form of resilience. This framing is problematic because it allows states or governments to eschew their political responsibility, and it places the onus on the refugee, the 'resilient subject,' to become an entrepreneur and master of her own fate (Cavelty et al., 2015; Kaufmann, 2016). Yet just as important, we must be prepared to recognize that all migrant adaptations are not positive.

Lischer (2005), for example, describes how the political organization – that is, the responsive political capacities – of refugee communities can affect their propensity to engage in violence, thereby drawing new links between securitization and negative and positive forms of resilience. In this framing, the 'securitization of refugees' through settlement or encampment policies can even motivate negative resilience. At the same time, of course, another strand of this literature sees the resettlement of refugees as laying the potential for more positive types of refugee resilience against the 'humanitarian sovereigns' (Agier, 2011; see also Sanyal, 2014; Herz, 2011). Through this lens, resilience occurs through the new 'politics of the refugees, their claims, their rights and their potential action, their wishes, their exercise and their evasion of power' (Weizman, 2011, p. 62).

Resilience, risk, and refugees: The territorial scaling of positive and negative outcomes

The point here is that the linkages between refugee migration, security, and resilience are multivariate and complex. By their very nature, refugee and forced displacement 'crises' lay bare the intricate connections that exist between simultaneous yet at times contradicting scales of security and resilience. Stated more clearly, refugee flows – and the security challenges they either derive from or prompt, including in particular the widespread construction of refugee camps – *can be both positive and negative* depending on whose resilience is at stake.[6] On the one hand, refugee camps are designed to provide legal and physical protection of individuals, often for the purposes of human security, and to relocate entire communities so as to keep bonds of solidarity. Insofar as they reduce both individual and collective vulnerabilities caused by forced migration, they may be deemed manifestations of positive resilience. On the other hand, in many contexts these same camps serve as amplifiers of violence, social unrest and inequalities, as well bearers of new forms of exclusion, deprivation, or suffering. In that sense, they also can be construed as manifesting negative resilience.

All this suggests that if one wants to employ the notion of resilience in the study of displaced peoples, it is important to move beyond meanings and normative values given to resilience in the fields of psychology and ecology – where the idea of resilience refers to the capacity of 'bouncing back' after a shock, and to the ability of individuals and ecological systems, respectively, to recover from or adjust to adversity, conflict, failure, or profound change in a successful way (see Walker and Cooper, 2011). Rather, one must question the assumption that disturbance or shocks are always bad or negative, as well as the idea that adaptation to shocks – resilience – is therefore always a positive thing. What follows from this position, then, is that resilience itself ought to be interpreted as being socially constructed, historically specific, and contentious. In fact, rather than debating the one true definition of 'resilience,' it is more useful to speak of the myriad practices, forms, and patterns of resilience.

In my own prior work on chronic urban violence, I have suggested a framing of resilience that displays positive, negative, and equilibrium qualities. Positive resilience occurs when adaptation to change is realized in a constructive way, and when the collective social, political, and economic capacities of institutions or groups are invigorated. In contrast, negative resilience happens when adaptive responses strengthen or perpetuate the forces, conditions or systems responsible for insecurity in the first place. Equilibrium resilience is a more tenuous form of resilience, where only some groups are able to cope, or only some institutions merely strengthened, without any effect on the root causes of violence or insecurity (Davis, 2012). Moreover, I have found that these different degrees of resilience can occur simultaneously, at various scales, thus implying a high level of trade-offs between each other.

Evidence from the Syrian refugee crisis: City vs. camp

To shed light on these complex relationships in the context of migration and security, the Syrian Refugee crisis is instructive. Even though the war in Syria had already caused alarming levels of population displacement by early 2013, in the last several years Turkey, Lebanon, Jordan, and Iraq have been dealing with a massive Syrian refugee influx into their national territories. In fact, 2016 data suggests that this region hosts a total of over 4 million refugees, approximately 90 percent of whom live in non-camps settings[7] (UNHCR, 2016), although a good number are also in camps. So, the question is whether the resettlement of refugees in camps – as opposed to the city, or other sites we consider to be non-camp settings – is associated with a certain form of resilience. What do the non-encampment vs. camp policies tell us about the resilience of refugees or their host societies? Is there evidence that the crisis has generated greater insecurity? For whom, where and at what scale? Alternatively, is there evidence that the less than desirable conditions have produced new social or spatial relationships or conditions that could be considered to lay the foundation for a better future? I try to address these questions by comparing Lebanon to Jordan.

First, the case of Jordan. Since the outbreak of the Syrian refugee crisis, Jordan has favored a strong encampment policy. From the point of view of the Jordanian authorities, the encampment of Syrian refugees has allowed (in theory at least) the maintenance of the status quo. In addition to fulfilling security aims, camps have enabled the spatial segregation of refugees of certain economic classes whom the authorities deemed surplus to labor market requirements. Political scientist Lewis Turner persuasively argues that what is most feared in the Jordanian context is not the radicalization or militarization of refugees, but the impact that refugee labor could have on the national economy and the national labor market. The contention being that the choice between security and segregation (camps) is the results of economic policy rather than the suppression of dissent or expansion of humanitarian control (Turner, 2015). It has been in the authorities' interest to impose restrictions on freedom of movement and freedom to work (Newhouse, 2015). Moreover, the existence (and visibility) of camps has allowed the government to benefit from international aid (Figure 3.2) (Turner, 2015, p. 387). In contrast, from the perspective of refugee communities themselves, the lack of access to the labor market, and the opposition to the creation of micro-markets, inside or outside camps, had been harmful. Such policies go against the idea that camps can be engines of social and economic dynamism, and entrepreneurship, allowing localize economies to flourish and strengthening the resilience of encamped refugees (Beehner, 2015). Here, one observes a direct contradiction between the Jordanian state and the refugees' own resilience practices.

In Lebanon, the situation is somewhat different. Given its violent history with Palestinian Refugee camps, the government decided early on a non-encampment policy (Fawaz et al., 2014; Turner, 2015). As a result, refugees have sought shelter wherever they could: with host families, in makeshifts settlements on agricultural lands, or in the low-income neighborhoods of major cities (Figure 3.3). The country is now host to over 1.2 million

Figure 3.2 Streetscape refugee housing, Jordan

Figure 3.3 Palestinian camp, Lebanon

registered refugees[8] (UNHCR, 2016). Of the Syrian refugees that chose to settle in rural areas, many initially did so because of existing ties with local Lebanese communities. In fact, prior to the war, many Syrians were employed as low-wage seasonal labor for Lebanese business and landowners. When violence erupted, Syrian workers originally found shelter among the host communities that they knew. At first glance, this adaptation can be seen as a sign of positive resilience at the individual and community level. However, as the war dragged on and refugees' financial resources began to dwindle, new labor-for-shelter arrangements soon appeared whereby refugees began having to work for landowners in exchange for rent (Nayel, 2014; Fawaz et al., 2014; Turner, 2015). This is an example of resilience devolving from positive to negative, as the new arrangements developed into 'means for the exploitation of a precarious labor force,' benefiting private Lebanese interests 'in need of cheap and docile Syrian labor force,' and a new source of insecurity for the refugees (Turner, 2015, p. 398), not to mention highly unsustainable land-use and servicing patterns (involving tree-cutting, contamination of agricultural land, soil and water (Fawaz et al., 2014)).

Furthermore, the Syrian refugees who have settled in urban areas have had to contend with additional risks. Urban areas represent a uniquely challenging environment, due to the close intermingling of the multiple actors and audiences involved in the refugee response (see also Sanyal, 2014). These include not only the host and refugee communities, but also the clustering of governmental, civil society, humanitarian, and development actors – be they local or international – with competing mandates and jurisdictions. The most vulnerable refugees in the city face similar restrictions as do refugees settling in rural areas, at least in terms of lack of access to employment. They also experience heightened insecurity in the form of predatory relationships emerging between them and property owners, realtors, and slumlords who, in the absence of housing regulations and of any housing rights enforcement on the part of the authorities, are setting rents at unrealistic levels (Fawaz et al., 2014). Observers note that 'many evictions [are] characterized by repeated threats and harassment and, in some cases, backed by the implicit or explicit threat of force by armed militia, or even the police force' (Fawaz et al., 2014, p. 7).

In the Lebanese case, the perspective from the host communities is twofold. On the one hand, the insecurities plaguing the Syrians refugees have spread to the local urban poor and the Palestinian refugees, who now must compete with Syrians in increasingly congested sub-standard housing and labor markets (Figure 3.4). On the other hand, the Lebanese society as a whole has benefitted from the increased demand, in multiple sectors and services, generated by the Syrian refugees, as well as from the foreign aid that has been directed to Lebanon in response to the crisis (roughly US$2.2 billion since 2012) (UNDP, 2015). In their assessment of the impact that the humanitarian aid has had on the Lebanese economy, the United Nations Development Program (UNDP) estimate that Lebanon's GDP

Figure 3.4 Refugee housing, Jordan

grew by an additional 1.3% in 2014 alone (UNDP, 2015, p. 7). It is interesting to contrast this figure with the official rhetoric employed by political leaders and which casts refugees as an enormous burden on the local infrastructures and economy. While not entirely inaccurate, the accusations made against the refugees have often been overblown, mainly for political purposes, as they point to internal dysfunctions that far predate the crisis.[9]

Beyond economic measures, one might begin to discern signs of positive resilience by looking at the situation from the perspectives of the municipalities and the international humanitarian organizations (IHOs). In the absence of a strong central government, it is no surprise that the municipalities have often been the first responders to the influx of refugees (Fawaz et al., 2014). It is equally encouraging that the IHOs have channeled many of their operations through such municipal entities, providing support, capacity building, and expertise. As they 'go urban,' IHOs are fundamentally adjusting their practices and expertise, moving away from activities aimed at 'reducing individual vulnerability,' in favor of operations that enhance community resilience (Reid-Henry and Sending, 2014; Potvin, 2019).

Beyond territorial location: The role of housing tenure in mediating risk and buttressing resilience

This focus on how Lebanon and Jordan have dealt with refugee crisis suggests that resilience must be understood as unfolding at several territorial scales simultaneously – community, city, nation, and globe. These cases also revealed the multiple and overlapping forms of risk associated with settlement location, such as the different experiences in camps versus the city (see for example Campbell, 2006; Crisp, 2004). Such distinctions support further efforts to call into question any policy or action strategy for refugees that is built around a single territorial or scalar framing of the problem. Actions intended to reduce risk and strengthen resilience at one territorial scale can impact capacities for resilience at another, particularly when refugees are considered intrusive outsiders whose rights to stay put are guaranteed more by international humanitarian agencies than local and national authorities. To push this idea further, it is instructive to share some evidence about how efforts to accommodate displaced and refugee populations might be unfolding in a different spatial, territorial, and sovereignty context, drawing from conditions in Colombia. The status of refugees in Colombia differs from the Jordanian and Lebanese situation, in that those displaced from its longstanding civil war are considered internally displaced peoples or IDPs, not externally displaced refugees. As such, refugee resettlement programs and policies have been developed by local and national authorities who are expected to respond to displaced peoples because of the formal citizenship status of the displaced – a situation that differs from the two prior cases. The question is whether citizenship not only affects the locational strategies of resettlement, but also the means by which displaced peoples respond to the challenges of resettlement, and whether they demonstrate the same patterns of risk and resilience.

Although many of Colombia's IDPs migrate to large metropolises like Bogota, evidence from Granada, a smaller city in the province of Meta, allows us to understand how IDPs confront various options for resettlement in places that are rural as much as urban – i.e. locations that straddle the territorial conditions of city and camp discussed above. Many IDPs ended up in Granada precisely because they prefer to live in a smaller town with easier access to agricultural lands, working under the assumption that securing daily subsistence would be easier in an 'urban' settlement than in a large city.[10] Granada offers both agricultural employment and food cultivation opportunities, and the fact that the far majority of Colombian IDPs come from rural setting further explains this preference. In this sense, refugees/IDPs setting in Granada already have shown some proactive agency by calculating which location might be better for minimizing risk and enhancing resilience, a strategy that is reinforced by government programs to help foster resettlement.

Granada is a city of less than 150,000 inhabitants that has received more than 25,000 displaced persons between 2000 and 2015.[11] By 2015, Granada had more than 300 ex-combatants and more than 20,000 displaced persons, thus suggesting that the challenges of resettlement and re-integration in Granada also carried with them the possibility of ongoing conflict and tension between groups that are often seen as enemies. With more than one-third of Granada's population formally listed as 'victims' of the armed conflict, public authorities of Granada also have been facing critical challenges in both accommodating IDP housing demands, managing population growth, and facilitating the social, economic, and spatial integration of these vulnerable populations. How IDPs and the state navigate this complex relationship is very much impacted by the territorial and land-use context.

Much of what we know about IDP resettlement in Colombia (Ibáñez, 2003; Atuesta, 2019) focuses on the exclusion and marginality experienced by IDPs in large metropolitan areas (e.g. spatial segregation, limited access to key city services and amenities, mismatch in the quality of life and life expectations, and stigmatization of populations and places), all factors that can exacerbate risk and limit resilience. Yet in 'urban' locations like Granada, there is evidence that although space matters in fostering risk and resilience, refugee conditions are determined by much more than just the physical isolation and place-based stigmatization of IDPs. The nature of property ownership and land-use management also matter. Through examination of Granada's refugee resettlement patterns at a more granular spatial scale, and by comparing three different neighborhoods on the outskirts of the same city, we see considerable variation in coping strategies as well as in degrees of risk and resilience experienced by IDPs across settlement types. This variation owes to differences in how IDPs acquired housing as well as how these processes impacted both quality of life and degrees of collective social mobilization.

One settlement, called Makatoa, is a horizontal (i.e. superblock) urbanization of 180 public housing units, built with support from national authorities under laws designated to guarantee housing to displaced populations of Colombia's internal armed conflict.[12] Residents in this settlement have received subsidies from the government to acquire apartments, meaning the IDPs in this site have formal tenure rights and their settlement is sanctioned by the state (Figure 3.5). The second neighborhood, called Paraiso, is an irregular urbanization of self-built houses on a large parcel of private land subdivided by a single landowner who sold more than 6,000 lots, but where resident IDPs live without legal license. That is, in this neighborhood the displaced do not have formal rights to land, or tenure in a legal sense, in contrast to those living in Makatoa. But, they do have homes and are informally 'protected' from eviction by the landlord who illegally sold the plots and who is politically powerful enough to have done so without local government approval (Figure 3.6). The third settlement, called Villas, is an illegal land invasion on a private property that had already been zoned and designated for other purposes, but whose development stalled. It now is comprised of more than 1500 illegal lots, where the displaced live without formal recognition either by the state or an intermediary landowner (Figure 3.7).[13]

With many IDPs living in close residential proximity to each other on the outskirts of Granada one might expect to see a high level of social organization capable of supporting displaced people's collective efforts better

Figure 3.5 Formal housing in Grenada, 2019

their situation, perhaps similar to what happens in the camp settings discussed earlier. Yet instead of working together for the long-term benefits of all victims, the IDPs in Granada developed relations to each other and to authorities based primarily on land tenure arrangements, rather than displaced status *per se*. In Paraiso, for example, they relied on clientelistic relationships with a local landlord, thus making IDPs less likely to push for bettering conditions through collective action. Dependence on an individual strongman for a better future might be seen as a form of negative resilience.

Figure 3.6 Irregular settlement, Paraiso, 2019

This contrasts to the case of Makatoa, where residents have regular meetings with city authorities, owing to their legal status and the fact that housing acquisition was sanctioned through formal government programs. These meetings take place because local authorities are responsible for supervising and providing social support to the beneficiaries of the national government's Free Housing projects for ten years; and with municipal authorities in Granada acknowledging their statutory role in accommodating victims of the civil war, residents feel free make claims and define themselves as a vulnerable community of displaced peoples.

Figure 3.7 Informal settlement, Villas de Granada, 2019

This experience contrasts greatly with the other two neighborhoods. Although lots in Paraiso were initially acquired without legal license in 2009, by 2011 the city council approved a modification of its land-use plan, which then designated Paraiso as a recognized area of urban expansion.[14] When Paraiso became an official neighborhood of Granada in 2016, its residents still lacked services, although some new settlers did receive individual property titles (justified by legislation that designated the area for affordable housing). This co-existence of various distinct property arrangements affected the ways in which Paraiso residents interacted with each other and with the state, primarily because it produced divisions by class, social status, and housing tenure. Originally, low prices have made lots accessible to poor households, and among them many IDPs.[15] Yet since being formally recognized as an area of urban expansion, Paraiso began to attract land speculators, with non-IDPs coming from Bogota to capitalize on the low land costs for the purposes of constructing rental housing.[16]

One consequence is that the area is upgrading, with some residents building two to three floor houses and establishing large businesses such as hardware stores or warehouses. This has led to a complex mix of classes and statuses, with IDPs living side by side with longstanding local populations or other non-IDP rural migrants. These differences not only limit collective solidarity; the irregularity of housing acquisition (some still without title yet recognized and protected) does not produce collective mobilization – among either land speculators or low-income buyers – because both of these constituencies rely on the landlord's social and power to achieve their aims, whether this involves purchasing more illegal lots or protecting themselves against eviction. Complicating matters, even without legal status, both groups feel free to resell the lots to future interested buyers, thus preventing many in the settlement from challenging the arrangement. As such, unlike the other two settlements discussed here, the quasi-formal/informal (i.e. irregular) situation both makes it possible for IDPs to survive by acquiring housing, a form of resilience, but does not lead them to challenge the status quo of precarity associated with having a single 'strongman' at the helm of the settlement's irregular urbanization. Because he holds the power over their housing futures, at best we might characterize this situation as equilibrium rather than positive resilience.

Of the three neighborhoods examined here, the one that hosted the most collective organization and active claim-making was illegal settlement of Villas, and for this reason it also might be identified as the most resilient (Figure 3.8). The neighborhoods high degree of social organization owed to the fact that the mere act of illegal land invasion required a certain degree of cooperation and leadership. Purposeful occupation empowered a cadre of local, grassroots leaders who supervised land grabs, informed newcomers about available lots, and controlled settlers and potential land speculators alike by enforcing a norm of one lot per household. They also built alliances with local authorities in order to defend residents against evictions. With a large portion of the settlers in Villas identifying as IDPs, local leaders tapped into various humanitarian organizations to defend the many displaced households as well as to acquire resources for building infrastructure. For example, the community sued the regional energy provider for refusing to bring the necessary electricity to the neighborhood and for sabotaging the illegal connections they had originally established. Finally, community members made claims on local and national authorities to invest in

Figure 3.8 Meeting between villa residents and land-use plan contractors, Villas de Granada, 2019

the neighborhood, doing so by building on their alliances with local and national NGOs committed to attending the needs of displaced and other sorts of victim populations.

Paradoxes of resilience: The Janus-faced dynamics of IDP status

The story of Villas suggests that an illegal settlement of IDPs may be better positioned to foster collective mobilization than a formal settlement with clear property rights or an irregular settlement with a politically powerful leader calling the shots. We might consider it is the paradox of resilience, in that we are seeing strong mobilization bringing tangible and positive outcomes among populations considered to be highly precarious, at least in property rights terms. One way to explain the high degree of resilience among illegal land occupiers in Villas is to recognize the extent to which collective action is driven by dire conditions of scarcity where, as Solnit (2009) notes, deep human bonds of caring and mutual support are established because that is the only way forward to acquire housing and secure necessary services. However, there is another important element to the story of this neighborhood's collective mobilization, having to do with displacement status itself. Although residents of Villas included both groups of

IDPs that are considered mortal enemies (Steele, 2018) – the demobilized (perpetrators) and the displaced (victims) – such divisions did not prevent them from working together to make claims on local authorities for a better future. This contrasts to the irregular and formal settlements holding both displaced and demobilized peoples in Granada, where we see less collective mobilization and civic engagement, with social leaders in those communities becoming, in some instances, dividers rather than agents of unity and engagement. Such observations raise questions about the subjectivity of refugees and displaced peoples, and the fact that social statuses and identities may also matter in determining degrees of resilience.

Indeed, despite the fact that Paraiso also hosted IDPs coming from different backgrounds who lived in residential proximity to each other, residents there did not frequently use the categorical identity of displaced person as a means for mobilization or claim-making. Residents there were more likely to declare that although they know that 'victims' do live in their settlement, they do not really know who exactly is a 'victim' and where that family lives.[17] More specifically, in Paraiso the language of displacement or victimhood was not viewed as necessary, or useful, to defend households against eviction or bring domiciliary services to the neighborhood. Both tasks were understood to be better mediated through clientelistic engagement with the powerful landowner who sold them the lots and was thus influential among public authorities. As such, categories of victimhood or forced displacement were useful in social interaction at the scale the neighborhood, and, in fact, it could be divisive. In Makatoa, likewise, residents did not see a 'need' to use the language of displacement to defend the legality of land titles and homeownership, precisely because they already had access to permanent and formal housing, owing to the government programs formally designated for the displaced. Because of this, moreover households were just learning to live together under the specific regulations and urbanistic designs, imposed on them by public authorities; they did not yet have a track record of unifying together to make insurgent claims.

In Villas, however, the category of displacement was used strategically as a unifying factor during the whole process of land occupation. The language of displacement became a means to defend households against eviction, to call for NGOs to invest in the neighborhood, and to make claims to defend their right of homeownership. And in the process, residents were able to establish strong relationships with local authorities, in no small part because Villas was already officially recognized as a neighborhood of Granada despite the fact that its occupation was illegal. Residents did not have to wait and work for the legalization of the neighborhood in order to be officially included in the different urban development plans and programs of the municipal and provincial administrations. Residents were thus able to defend their right to possession against eviction, a process of claim-making that strengthened collective efficacy and cooperative action. Indeed, current residents who participated in the original process of land occupation recall those moments as a time of fellowship and deep commitment and support among neighbors. Neighbors organized regular potlucks, for hundreds of people, where families shared bowls of soup and hot chocolate on safeguarded open spaces of the neighborhood. Everyone contributed with labor or whatever means they had available to support the construction of sewage infrastructure. Together, residents created a communitarian flood warning mechanism, and the most highly exposed households could count on squads of neighbors to help them bring mattresses and household appliances to higher grounds before flooding. The influx of IDPs into the neighborhood was an important mobilizing tool in all these events, up to and including the capacity to raise capital from national organizations working on reintegration and reconciliation to support their efforts.[18]

Conclusion: Horizontal and vertical connections in refugee resilience

In this chapter, we have tried to shed light on the relationship between refugee status, resettlement, and resilience by highlighting the role of institutional and policy actors as well as the displaced themselves, and by understanding each of their actions on various scalar dimensions. After all, forced migration is a deeply dynamic process, both spatially and temporally, and it triggers a wide array of responses from both the refugees and the communities or cities they are entering. To address the array of interconnections that characterize migration, resilience must be cultivated and embodied in cross-scalar efforts that strengthen networks, reconcile antagonistic relationships, and mediate between diverse interests.

In the cases of Jordan and Lebanon, we have seen that greater risks or resilience on one scale (i.e. within a camp) can produce problems at other scales (i.e. for the host country), or vice versa. In Jordan, encampment meant isolation from jobs and other opportunities for bettering livelihoods among the refugees, although the formalization of camp conditions brought funds to the Jordanian government. In the case of Lebanon, where refugees lacked formal recognition and were distributed across the city rather than isolated in camps, we see that acts of resilience among refugees had both negative and positive repercussions on other scales, the former with respect to producing risk for other vulnerable native populations in the same city, and the latter in the form of resources and labor that benefited the urban economy as a whole. With this final set of observations about Colombia, we do not see a scalar trade-off in positive versus negative resilience, or even risk, so much as a territorial disaggregation of conditions at a scale smaller than the city, with some neighborhoods more resilient than others.

These variations can be explained by the various land tenure arrangements through which housing and shelter have been provided or implemented. Thus, our main takeaway is that all shelter/settlement/housing arrangements for displaced populations are not alike, and the difference is not merely based on location in cities vs. camps, or even because of the conditions of isolation versus integration that accompany these locations produce risk to resilience, but because the *process through which housing is acquired* also matters.

In our cases and generally when it comes to studying refugees and displaced populations, housing or shelter possibilities are frequently enabled by government policies and programs as well as the involvement of humanitarian actors. Decisions made by authorities lay the framework for coping and adaptation strategies of refugees. As such, we want to understand or facilitate the likelihood of resilience, as pay attention to the policy environment. But, the Colombian case also brings home the message that housing acquisition is determined by the actions and orientations of the refugees or displaced populations themselves. In some locational settings, there is less room for maneuver of refugees to pursue multiple paths of housing acquisition. When options are determined only by the policy contours set by the state, strategies are limited. Indeed, the ways in which Syrian refugees landed in Jordanian camps may in fact be similar to the ways in which Colombian IDPs ended up in formal housing in Granada, and the formal programs that governed housing distribution did little to empower their recipients. What sets the Colombian case apart is the wide variety of strategies for acquiring housing in Granada, and the fact that the most irregular and least formalize means for acquiring housing also led to the greatest resilience; and this resilience was strengthened by the horizontal and vertical relationships of reciprocity established within and between residents and authorities through the process of housing acquisition.

These findings suggest that it might be important to 'de-exceptionalize' the study of refugees, resettlement, and resilience, and to look for parallels in research undertaken on other vulnerabilities, including urban violence. In some of my own writings on this topic, I have examined the role that urban design plays in enhancing security and fostering resilience. One of the key parallel findings has been that not all urban design strategies will produce the same outcomes in all times and places; rather, socio-spatial context, community power structures, and stakeholder assemblages also matter, as does the nature and source of risk (Davis, 2019). Stated differently, interventions in physical space are more likely to produce positive results when they reinforce principles of community engagement, when they are scaled towards the neighborhood, and when they take historically specific urban conditions into account.[19]

Such lessons might be readily applied to efforts to accommodate or securitize the conditions of refugees, forced migrants, and other displaced peoples as well as to help them recover from the trauma and disaster that have accompanied their vulnerable status. Framing these lessons within this book's focus on recovery also allows us to move beyond the far too trendy appropriation of the concept of resilience. After all, resilience is a tricky word, veering into the ideological and building on the assumption that any form of coping or adapting by a vulnerable individual will constitute a form of resilience. Yet as we have seen with our overview here, strategies for generating resilience and reducing risk, just like the notion of recovery itself, involve collective action as well as engagement with local authorities, not merely individual volition. In that sense, both resilience and recovery should be framed less as technical fixes and more as means for enabling socially engaged projects that produce strong horizontal and vertical relationships or institutionalized connections within and between citizens and authorities. It is with these new relationships that citizens and practitioners will be best able to lay the foundations for meaningful and effective push back against insecure, unjust, or distressing conditions, thus paving the way for both individual and collective recovery after disaster, manmade or otherwise.

Notes

1. Much of the material on Jordan and Syria presented in this chapter draws on work that I have conducted with Marianne Potvin, a former doctoral student at Harvard, with whom I co-authored a piece on resilience and spaces of migrant refugees for a *Handbook of Security Studies*. The section on Colombia draws heavily on dissertation research by Maria Atuesta, a current GSD doctoral student conducting fieldwork on demobilized and displaced peoples in the province of Granada. This chapter would not have been possible without the assistance, intellectual collaboration, and ongoing independent research of both students.
2. Lebanon is now host to over 1.8 million refugees; and recent statistics suggest that Syrian refugees, alone, represent one-fourth of the Lebanese population. While Palestinian refugees have lived in long-standing refugee camps, embedded in the local urban fabric for several decades, the more recent Syrian, Palestine refugees from Syria (PRS) and Iraqi refugees have settled in the innumerous informal settlements that have recently sprung up in rural and low-income urban neighborhoods, following the Lebanese Government decision not to build any new camps. (These figures come from UNRWA, 2016 and UNDP, 2015.)
3. In my role as co-director of a Master in Design Studies (MDes) program on Risk and Resilience, I have taken a very critical stance vis-à-vis the concept of resilience, focusing instead on risk. In this program at the GSD, we take as our starting point the fact that world faces unpredictable challenges at increasing intensities – natural disasters, ecological uncertainty, public health crises, extreme social inequity, and rising violence. While the program is grounded in the physical and tectonic realities of location, the social and political conflicts often emerge as sites of investigation. We then ask our students to think about confronting risk at these sites through acts of resilience, and to do so emboldened by design thinking. Design thinking, for its part,

must start from an understanding of this complexity of embedded systems of risk and resilience, and proceed with the acknowledgement that any single design project or intervention will have implications far beyond its targeted scope, both in scalar and sectoral terms. We thus start from the assumption that any effort to enable constructive action in context of multiplying and inter-connected vulnerabilities unfolding at multiple scales one must also embrace the idea of risk, and build, design, and plan for risk rather than merely resilience.

4. It could be that this number is on the rise precisely because of tightening restrictions on the crossing of international borders, as well as the increase in prevention and containment practices (Isotalo, 2009, p. 77).
5. One must recall that the very definition of 'refugee' hinges on a specific definition of risk, in this case, on an 'existential threat' that one fears in one's country of origin. The 1951 Refugee Convention defines the refugee as someone who: owing to a well-founded fear of being persecuted for reasons of race, religion, nationality, membership of a particular social group or political opinion, is outside the country of his nationality, and is unable to, or owing to such fear, is unwilling to avail himself of the protection of that country.
6. The distinction between positive and negative resilience that I develop in this chapter derives from my earlier work for a USAID study titled Urban Resilience in Situations of Chronic Violence. Through that research, focused on seven cities facing chronic violence around the world, we developed the distinction between positive, negative, and equilibrium resilience. For more, see: web.mit.edu/cis/urban_resilience.html.
7. UNHCR data, Syrian Regional Refugee Response – Inter-agency Information Sharing Portal, http://data.unhcr.org/syrianrefugees/regional.php, accessed March 15, 2016.
8. UNHCR data, Syrian Regional Refugee Response – Inter-agency Information Sharing Portal, http://data.unhcr.org/syrianrefugees/regional.php, accessed March 15, 2016.
9. Interviews with INGOs in Beirut, August 2015, conducted by Marianne Potvin.
10. For example, in Colombia, Bogota and Medellin – the biggest Colombian cities by population size – have received the largest amounts of victims of the internal armed conflict since 1984 – each city has received more than 480,000 victims. However, in total for the whole nation, the municipalities above 150,000 inhabitants have received 3,407,299 cases of displacement since 1984, whereas municipalities below that number have received 4,445,858 cases. Aggregately, small municipalities are receiving a larger bulk of the population, without getting equal attention by the literature.
11. With an estimated population of 44,130 in 2000, Granada since then has grown by at least half its population size, with most of its population growth explained by conflict-driven migration.
12. The 2011 administration of President Juan Manuel Santos passed the Victims and Land Restitution Law, but it was criticized for not including provisions to facilitate housing access to victims of the armed conflict. One year later, his administration launched the 100,000 Free Housing Program. Makatoa was built under the auspices of this program.
13. Fieldwork and interviews in these three settlements were conducted by Maria Atuesta, and form part of her dissertation research at Harvard's Graduate School of Design. For more details on the housing acquisition strategies and spatial integration of displaced peoples in Colombia, see Atuesta (2021).

14. Although originally Paraiso was located just outside the urban growth boundary established in the 2004 land-use plan, in 2012, minister Vargas' passed a national law regulating the implementation of his Free Housing program, which established that any land located just outside an urban perimeter, could be automatically included in the urban perimeter of the given municipality (just once), if demonstrated that the land was adequate for the development of low-income housing. Subsequently, Paraiso was sanctioned as an affordable housing development within Granada's urban perimeter, even though it actually did count with proper access to any domiciliary services.
15. In 2009, lots in Paraiso were sold for 200 dollars, while today each lot costs about 3,000 dollars.
16. There is a dynamic secondary market for lots in Paraiso, regulated by El Gurre's office directly, where the interested parties (often land speculators from Bogota) pay a fee to be able to transfer an original purchase receipt over to a new buyer. If needed, each interested buyer can get into a flexible payment scheme with the landowner's office, distributing the whole cost of the lot in monthly payments. Buyers are just not allowed to move and build any infrastructure on the lot before they have paid the whole amount. The informality of individual land titles was not considered to be a problem for either land speculators or low-income buyers, who trust in the landlord's power to protect them against evictions, and who even without legal status continue to resell the lots to any interested buyers.
17. Interviews with residents in all three neighborhoods were conducted by Maria Atuesta as part of her dissertation research. The results presented in this and the next paragraph are built around general impressions drawn from interviews identified through snowball sampling, not via a systematic survey.
18. In exchange for resident participation in a series of workshops held between displaced populations and the demobilized, these national organizations provided the money to build a community center and a soccer court. Today, several of the demobilized persons who participated in this process continue to be active in neighborhood community organizations and have become members of the neighborhood community board.
19. In such settings, the mere re-design of street lighting, commitment to better garbage collection, or construction of straighter pathways may be welcome ways of reducing risk, but they may be unable to fundamentally challenge the local political conditions that empower the perpetrators of violence.

References

Agier, M. (2011). *Managing the Undesirables: Refugee Camps and Humanitarian Government.* Cambridge; Malden, MA: Polity.

Atuesta, M. (2019). Progressive Politics and Inclusive Social Housing: Enablers and Barriers to Transformative Change in Bogota. Working Paper, Joint Center for Housing Studies, Harvard University.

Atuesta, M. (2021). *Forced Migration and Neighborhood Formation*: How Communities of Internally Displaced Persons Find Residential Stability in an Unstable World. Ph.D. Dissertation, Harvard Graduate School of Design.

Beehner, L. (2015). Are Syria's Do-It-Yourself Refugees Outliers or Examples of a New Norm? *Journal of International Affairs*, *68*(2), 157–XII.

Betts, A. (2013). *Survival Migration: Failed Governance and the Crisis of Displacement*. Ithaca, NY: Cornell University Press.

Bigo, D. (2002). Security and Immigration: Toward a Critique of the Governmentality of Unease. *Alternatives: Global, Local, Political*, *27*(1), 63–92.

Black, R., Adher, W.N., Arnell, N.W., Dercon, S., Geddes, A., and Thomas, D. (2011). The Effect of Environmental Change on Human Migration. *Global Environmental Change*, *21*, S3–S13.

Bourbeau, P. (2013). Resiliencism: Premises and Promises in Securitisation Research. *Resilience*, *1*(1), 3–17.

Bourbeau, P. (2015). Migration, Resilience and Security: Responses to New Inflows of Asylum Seekers and Migrants. *Journal of Ethnic and Migration Studies*, *41*(12), 1958–1977.

Bourbeau, P. (ed.). (2017). *Handbook of Migration and Security*. Cheltenham, UK: Edward Elgar Press.

Brown, W. (2010). *Walled States, Waning Sovereignty*. New York: Zone Books; Cambridge, MA.

Campbell, E.H. (2006). Urban Refugees in Nairobi: Problems of Protection, Mechanisms of Survival, and Possibilities for Integration. *Journal of Refugee Studies*, *19*(3), 396–413.

Cavelty, M.D., Kaufmann, M., and Kristensen, K.S. (2015). Resilience and (In) security: Practices, Subjects, Temporalities. *Security Dialogue*, *46*(1), 3–14.

Chandler, D. (2012). Resilience and Human Security: The Post-Interventionist Paradigm. *Security Dialogue*, *43*(3), 213–229.

Crisp, J. (2004). *The Local Integration and Local Settlement of Refugees: A Conceptual and Historical Analysis*. Geneva: UNHCR Evaluation and Policy Analysis Unit.

Davis, D.E. (2012). *Urban Resilience in Situations of Chronic Violence*. Final report for USAID. Available at: web.mit.edu/cis/urban_resilience.html.

Davis, D.E. (2019). Reducing Vulnerabilities through Urban Design: Strategic and Tactical Interventions for Generating Resilience in Communities at Risk, in T. Banerjee and A. Loukaitou-Sideris (eds.), *The New Companion to Urban Design*. Oxfordshire, UK: Taylor and Francis.

Fawaz, M., Saghiyeh, N., and Nammour, K. (2014). *Housing, Land and Property Issues in Lebanon. Implications of the Syrian Refugee Crisis*. Beirut: United Nations High Commissioner for Refugees and UN-Habitat.

Gemenne, F., and Blocher, J. (2016). How Can Migration Support Adaptation? Different Options to Test the Migration–adaptation Nexus. International Organization for Migration Working Paper Series. DOI: 10.13140/RG.2.1.2010.6008.

Goodwin-Gill, G.S., and McAdam, J. (2007). *The Refugee in International Law*. Oxford and New York: Oxford University Press.

Hammerstad, A. (2000). Whose Security? UNHCR, Refugee Protection and State Security after the Cold War. *Security Dialogue*, *31*(4), 391–403.

Hammerstad, A. (2012). Securitisation from Below: The Relationship between Immigration and Foreign Policy in South Africa's Approach to the Zimbabwe Crisis. *Conflict, Security & Development*, *12*(1), 1–30.

Herz, M. (2011). South America Regional Order: The Growth of Brazil as a regional power, in T.J. Volgy, R. Corbetta, K.A. Gran, and R.G. Baird (eds.), *Major Powers and the Quest for Status in International Politics*. London: Palgrave Macmillan, 256.

Ibáñez, A. (2003). *Instrumentos de atención de la población desplazada en Colombia: una distribución desigual de las responsabilidades municipales.* Documento 03–37 (Universidad de los Andes (Bogotá, Colombia). Centro de Estudios sobre Desarrollo Económico). Bogotá: Centro de Estudios sobre Desarrollo Económico Facultad de Economía.

IDMC. (2015). *Global Estimates 2015: People displaced by disasters.* Geneva: Internal Displacement Monitoring Centre. https://www.internal-displacement.org/sites/default/files/inline-files/20150713-global-estimates-2015-en-v1.pdf.

Isotalo, R. (2009). Politicizing the Transnational: On Implications for Migrants, Refugees, and Scholarship. *Social Analysis*, *53*(3), 60–84.

Jacobsen, K. (2006). Refugees and Asylum Seekers in Urban Areas: A Livelihoods Perspective. *Journal of Refugee Studies*, *19*(3), 273–286.

Kaldor, M. (1999). *Old and New Wars: Organized Violence in a Global Era.* Stanford, CA: Stanford University Press.

Kaufmann, M. (2016). Exercising Emergencies: Resilience, Affect and Acting out Security. *Security Dialogue*, *47*(2), 99–116.

Leaning J. (2009) Health and Human Security in the 21st Century, in H.G., Brauch, N.C., Behera, P., Kameri-Mbote, J., Grin, Ú., Oswald Spring, B., Chourou, C., Mesjasz, and H. Krummenacher (eds.), *Facing Global Environmental Change. Hexagon Series on Human and Environmental Security and Peace*, vol 4. Berlin, Heidelberg: Springer, 541–551.

Lischer, S.K. (2005). *Dangerous Sanctuaries: Refugee Camps, Civil War, and the Dilemmas of Humanitarian Aid.* Ithaca: Cornell University Press.

Loescher, G., and Milner, J. (2005). Security Implications of Protracted Refugee Situations. *The Adelphi Papers*, *45*(375), 23–34.

Methmann, C., and Oels, A. (2015). From 'Fearing' to 'Empowering' Climate Refugees: Governing Climate-Induced Migration in the Name of Resilience. *Security Dialogue*, *46*(1), 51–68.

Mountz, A., and Hiemstra, N. (2014). Chaos and Crisis: Dissecting the Spatiotemporal Logics of Contemporary Migrations and State Practices. *Annals of the Association of American Geographers*, *104*(2), 382–390.

Nayel, M.A. (2014). On the Struggle of Syrian refugees in Lebanon. *Jadaliyya.* https://www.jadaliyya.com/Details/30530.

Newhouse, L.S. (2015). More than Mere Survival: Violence, Humanitarian Governance, and Practical Material Politics in a Kenyan Refugee Camp. *Environment and Planning A*, *47*(11), 2292–2307.

Potvin, M. (2019). Humanitarian Urbanism: Cities, Technology, and the Hybrid Practices of Humanitarian Planners. Doctoral Dissertation, Faculty of Arts and Sciences, Harvard University.

Reid-Henry, S., and Sending, O.J. (2014). The 'Humanitarianization' of Urban Violence. *Environment and Urbanization*, *26*(2), 427–442.

Sanyal, R. (2014). Urbanizing Refuge: Interrogating Spaces of Displacement. *International Journal of Urban and Regional Research*, *38*(2), 558–572.

Solnit, R. (2009). *A Paradise Built in Hell: The Extraordinary Communities that Arise in Disaster.* New York: Viking.

Steele, A. (2018). IDP Resettlement and Collective Targeting During Civil Wars: Evidence from Colombia. *Journal of Peace Research*, *55*(6), 810–824

Turner, L. (2015). Explaining the (Non-)Encampment of Syrian Refugees: Security, Class and the Labour Market in Lebanon and Jordan. *Mediterranean Politics*, *20*(3), 386–404.

UNDP. (2015). *Impact of Humanitarian Aid on the Lebanese Economy.* Beirut: United Nations Development Program.

UNHCR. (2016). *Global Report 2016.* Geneva: UNHCR. http://reporting.unhcr.org/sites/default/files/gr2016/pdf/Book_GR_2016_ENGLISH_complete.pdf.

Walker, J., and Cooper, M. (2011). Genealogies of Resilience from Systems Ecology to the Political Economy of Crisis Adaptation. *Security Dialogue*, *42*(2), 143–160.

Weinstein, L. (2018). *The Durable Slum: Dharavi and the Right to Stay Put in Globalizing Mumbai.* Minneapolis: University of Minnesota Press.

Weizman, E. (2011). *The Least of All Possible Evils: Humanitarian Violence from Arendt to Gaza.* London and New York: Verso.

4 Spatial patterns, gray spacing, and planning policy implications

The urbanization of forced population displacement in Lebanon

Mona Fawaz, Mona Harb, and Carla Al-Hage

Introduction

Over the past two decades, population displacements have followed the 'urban turn' (Fábos and Kibreab, 2007; UNHCR, 2010; Landau, 2014). Of the total global refugee population of 25.9 million people (UNHCR, 2018), almost two-thirds live in cities.[1]

As researchers looking into the determinants of the urbanization of forced population displacement, they found that in line with nationals, refugees prefer cities because of promises of economic mobility, social support, and physical freedom (Landau, 2014). In addition, research has shown that in cities, many refugees find the possibilities for the long-term improvements that they aspire to secure. These include better education for their children (Campbell, 2005, 2006; Hovil, 2007; Jacobsen, 2006; Landau and Jacobsen, 2004; Macchiavello, 2004; Sommers, 1999, 2001), better access to social services, such as healthcare, or better access to recreational and intellectual activities (Macchiavello, 2004; Women's Commission for Refugee Women and Children, 2002). Others have found that refugees are attracted to higher connectivity in cities (e.g. internet), which allows them to maintain communication with relatives elsewhere, but also with better opportunities for social integration (Horst, 2002; Jacobsen, 2006; Kibreab, 1996; Landau and Jacobsen, 2004; Macchiavello, 2004; Women's Commission for Refugee Women and Children, 2002). Urban settlements are also chosen for relative improvements in personal safety and security when compared to camps. Thus, the anonymity and invisibility provided by the city are attractive prospects to refugees who are often stigmatized as 'outsiders' (Swerts, 2017; Horst, 2002; Jacobsen, 2006; Macchiavello, 2004; Sommers, 1999).

Studies investigating the outcomes of refugees settling in cities have confirmed that their preference for urban areas is well founded. Thus, urban settlement was found to be conducive to better integration for refugee and host communities (Bolzoni et al., 2015). Unlike camps, where refugees are constrained to a temporality and a geography *outside the city*, 'impinging upon but never fully integrating with the city' (Ramadan, 2013, p. 74), urban settlement, scholars have argued, allows refugees to claim a certain normalcy and permanency in their settlement. Consequently, the UNHCR has endorsed urban displacement as a way to facilitate refugee integration (Rokem et al., 2017). Along the same line, resettlement strategies in Europe and Canada have been critiqued when they do not resettle refugees in cities. This is because confinement to rural areas exacerbates refugees' isolation and exposes them to more discriminatory practices as well as higher levels of urban poverty (Brun, 2015; Bolzoni et al., 2015; Andersen et al., 2013; Carter and Osborne, 2009), in addition to potential threats of disease outbreak, as experienced during the Covid-19 pandemic in shelters in Canada.

While everyone concurs that refugee settlement is globally urbanizing, few have explored the spatial patterns through which this urbanization happens. To many, the camp remains the main anchor or reference on which refugee studies are based. Others have looked at the spatial relations between camps and cities, showing how distinctions between camps and cities are increasingly blurred. Sanyal (2017), Doraï (2010), and Martin (2015) have demonstrated the merging of Palestinian refugee camps within Beirut's urban fabric, rendering these neighborhoods indiscernible from immediately adjacent neighborhoods. Conversely, Agier (2011) and others (Hyndman and Giles, 2011) have shown that encampments (particularly across the African continent) now function like cities, with internal economies and socio-economic patterns of organization. Similarly, in Palestine, camps like Dheisheh have grown into full fledge cities (Abourahme and Hilal, 2009).

Outside camps, the literature covering the 'urbanization' of displacement too often assumes a uniform model in which refugees arrive in pre-existing urban fabrics where they settle seamlessly. As a result, very few scholars have documented or analyzed the spatial forms that the urbanization of forced displacement has taken.[2] As such, parallel readings of the city offer, on the one hand, the findings of relief agencies and solidarity groups and, on the other hand, the forecasts of planning agencies and municipal authorities. Meanwhile, cities grow in intertwined forms, bringing together displaced, refugees, migrants, and other vulnerable groups that are poorly understood and, consequently, inadequately served.

In this chapter, we seek to relocate the urban production of space at the center of our analysis, using the built environment as a window to our study of forced population displacements and their implications for today's cities. We specifically look at the production and reorganization of urban quarters that have accommodated large flows of refugees in Lebanon's secondary cities. We ask: What are the forms that the urbanization of forced population displacement takes? How are these forms produced? Who are the actors involved? How do they influence ongoing urban transformations? What role do planning frameworks play in these processes?

Our first argument is that far from a seamless inscription in pre-existing neighborhoods, refugee displacement is profoundly reshaping the organization of cities, requiring us to look at the modalities in which it is occurring and the forms it is taking. In Lebanon, we found this urbanization to be heavily driven by individual, profit-driven initiatives instigated by impoverished local populations. Additionally, we found this urbanization to be determined by a range of territorial considerations that shape the locality's urban governance, and by deficient and outdated planning systems that turn a blind eye to the presence of refugees. As such, we argue that in line with other contexts (Martin, 2015), refugee migration is expanding spaces of exclusion in numerous cities, areas tagged by Yiftachel (2009, 2015) as 'gray spaces' (also see Datta, 2012) where relatively higher levels of informal management and sharp environmental deterioration have occurred, where residents navigate an increasingly precarious and criminalized presence, where they are, at best, tolerated as 'inferior' others.

In closing, we reflect on the imperative for decision-makers to engage city planners in their response to the management of forced population displacement, which has become an integral element of contemporary city making. We also advocate that planners review refugee response policies in ways that integrate tools and approaches of urban governance that address the aggravating spatial and environmental challenges and curtail the expansion of gray spaces in cities.

Our research methods relied on in-depth investigations of three secondary cities in Lebanon, Zahle, Saida, and Halba during 2016–2017. Each city includes a substantial number of Syrian refugees since the outbreak of the war in 2011 (Table 4.1). By extending our analysis to secondary cities and selecting comparatively localities where urbanization displays different geometries and patterns, we sought to disentangle the multiple settlement scenarios lumped under the label of 'urban refugees.'

Data gathering was conducted as part of a larger investigation of the Syrian refugee settlement in Lebanon's secondary cities, analyzing the intersection of community and individual agency with the modes of local governance that responds to their presence. This investigation was funded by IDRC and conducted by the Beirut Urban Lab, the Lebanese Center for Policy Studies (LCPS), and the Syrian Center for Policy Research (SCPR). Over 100 qualitative, in-depth interviews were conducted with refugees, members of the host communities, and local stakeholders in Halba, Saida, and Zahle in order to explore the modalities through which the production of space is happening,

Table 4.1 Number of Syrian refugees and Lebanese individuals in the cities of Saida, Zahle, and Halba in 2018

	Number of Syrian refugees	*Number Lebanese individuals*
Halba	7,979	5,833
Saida	12,242	78,450
Zahle	33,313	53,580

focusing on the strategies deployed by local governments and other stakeholders in each city. In addition, a large-scale survey of Living Condition of Refugees and Host Communities in Lebanon (LCSRHCL) was conducted in 2018 by LCPS and SCPR. The survey covered 1,556 Syrian and Lebanese households (7,208 individuals: 2,879 Lebanese and 4,326 Syrians) and inquired about social networks, refugees' profile, legal status, economic and financial status, relationship with the local authorities and aid agencies, access to urban services, access to shelter, type of shelter, and shelter conditions.

For the purpose of this study, we additionally looked for the principles that guided the expansion of existing urban settlements to accommodate refugee settlement and investigated how it was intersecting with earlier patterns of urbanization. We recorded new building typologies and conditions and, when possible, collected available documents from public authorities (e.g., municipality, local DGU offices) to understand how this urbanization was being regulated. Additionally, refugee settlement was mapped, documented, and analyzed in order to classify the form that this urbanization has taken and unravel its organizing mechanisms.

The chapter is divided in three sections. We begin by describing the three identified forms in which the urbanization of forced population displacement is occurring across the cities we examined. We identify typologies, production processes, building conditions, and levels of services in each. We show how the term 'urbanization' actually masks rather distinctive modalities of settlement. The chapter then links these processes to social patterns and practices before moving to an analysis of their outcomes, using the notion of gray spacing in urban settings. The chapter concludes by pleading for a different role for urban planning, one that responds to the historical proposition of the profession to foster more inclusive and just cities.

Spatial patterns of urbanization

In order to analyze the spatial organization of the urbanization produced by forced population displacement, we began by documenting the multiplicity of forms that the urbanization of forced population displacements has taken in Lebanon since 2011. Given the largely informal nature of this production, we extended the methodologies developed by earlier studies of informal settlement expansion to trace the physical development (Léna, 2012; Yassine et al., 2019). These methodologies map the formation of new buildings and complexes within existing lot geometries, identifying the position and size of developments in relation to pre-existing land lots. They also point to the organization and geometries of left-over spaces, and their relation to nearby infrastructure networks such as main roads and/or sewer networks.

In each city, we sought to unravel the determinants of this spatial production: Who was adding spaces and what was motivating these additions? Our aim was to apply frameworks widely used in the study of urban development, but rarely applied to conditions of displacement. These include morphological analysis as well as investigations about the context of urban regulations and their impact on cities' development. We identified three distinctive modalities of varying densities and consolidation levels that were present in the cities of Halba, Zahle, and Saida. Each of the modalities we identified was also present in other contexts in Lebanon. As a result, our description provides a general overview of the conditions of settlement as a way to transcend the specificity of each city and reflect, more generally, on the urbanization of displacement. The three modalities are:

1 *Neighborhood Densification:* In deteriorated old cities (e.g., Saida) or existing urban 'slums' (e.g., refugee camps in Saida, Beirut, or Tyre, informal settlements in Saida or Beirut), rent is a valuable source of income for impoverished landlords who have strived to expand the stock of housing each of them commands. With or without the support of slumlords, these landlords have taken the lead in the production of additional housing units, often by subdividing existing apartments, and/or adding rooms and floors. They thus densify the urban fabric to respond to the housing demand. They rent out these units to refugees (informally), sometimes in exchange of labor, often for direly needed cash. Building expansions often benefit from the support of local and international non-profit organizations, particularly in the form of building materials or direct subsidies. They typically include transactions with the local police force or powerful parties who turn a blind eye to these unregulated additions—as we further discuss in the next section. Additions also typically follow a number of locally upheld constraints such as respect of mutual privacy, although dire conditions sometimes lead to infractions.[3] In districts where political control is an important consideration, settlement is filtered through the presence of strong men or party members who may approve—or not—the presence of specific families or individuals.[4]

In Saida, where the LCSRHCL survey went in depth to look at this urban typology, housing additions were particularly concentrated in the old city and in the poor urban peripheries (Haret Saida). As Figure 4.1 shows, it is virtually impossible to locate the additions in the top view of the Old City, as these additions are nested in the urban fabric and reflected in the increments of the building's elevations. As expected, the survey showed that the spatial conditions of these units were old relatively bad, with users complaining about the small size and poor finishes of the buildings.[5] These findings echo earlier documentations of refugees living in this typology (Fawaz, 2017; Affordable Housing Institute, 2017), which showed these dwellings to typically suffer from water leaks, overcrowding, lack of heating/cooling, and damp walls. Our surveys in the old city showed that services are typically secured through ad-hoc, informal providers who also ensure connections to other, typically vulnerable, population groups also living in the neighborhood.

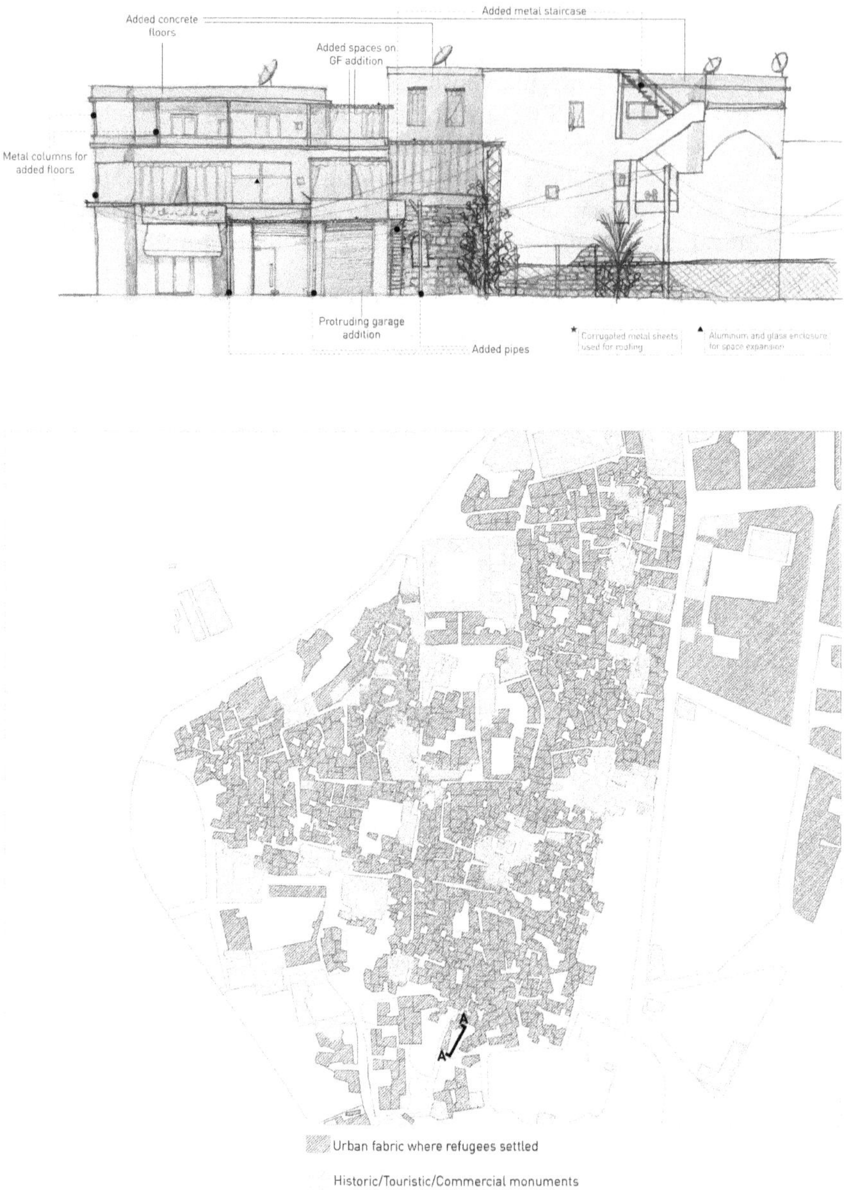

Figure 4.1 (a) Densification of existing fabric in the Old City of Saida, sketch of a street elevation. (b) Densification of existing fabric in the Old City of Saida. Additions are invisible in the top view. Legends: Urban residential fabric, monuments, and building fabric

The quality of services is often dismal, and further aggravated by high densities: Electricity and water outages are frequent, and waste rarely flows through adequate pipes. All in all, the urbanization amounts to an increase in existing densities, and is visible to neighborhood residents or visitors because of its typically unfinished and dilapidated quality.

2 *Housing Compounds:* Tagged as 'mujamma'at el Suriyyeen' (Syrian Compounds), large-scale compounds present a unique geometry of rapid refugee urbanization in several cities of North Lebanon (Affordable Housing Institute, 2017). These compounds extend typically along the main roads, developed on large agricultural tracts where they replicate the longitudinal form of land lots in rural areas. Each includes around 300 apartments, mostly rented out to refugees by Lebanese landlords, in a relatively distributive form.

Our in-depth analysis of one of the cases (Halba, see Figure 4.2) showed that the city and its vicinities witnessed a building development boom through the production of affordable housing units in large-scale

Figure 4.2 Location of recently built housing compounds along the main roads in Halba. Legends: Urban residential fabric, constructed housing compounds

housing compounds since 2011, through informal business partnerships between developers, landowners, as well as public agencies and/or municipalities. These developments were initially fueled by a massive influx of capital in the country, and they targeted a middle-class population of Lebanese public sector employees and expatriates. Financing was sustained though the supply of state-subsidized loans targeting Lebanese Army soldiers, Lebanese expatriates, and others who benefited from subsidized bank loans during the 2011–2018 period. The arrival of refugees in the city (and others) modified this process of housing production, providing an opportunity for landlords to rent out units in exchange for an income that would cover mortgages costs. The lucrative opportunity of renting out units to refugees encouraged further development, pushing the building of denser and larger compounds that reflect the current typology. However, as the resident population included more refugees, building quality deteriorated: Building facades were left unfinished, unpainted, and only plastered. Inside apartments, roofs leak, walls are damp, units are overcrowded, heating is inadequate, and doors and windows are unsealed. Moreover, compounds suffer from poor levels of services. They rely on illegal hook-ups for electricity, dig illegal wells for water, and channel their sewers often without proper networks to nearby locations, to an open pit in at least one location. Those who can afford it also pay to informal providers who insure better provision of water and electricity.

The analysis of the production processes of these compounds reveals that despite a seeming adherence to planning regulations, numerous informal institutional arrangements sustain their making. In many cases, low-level public agents have partnered with developers in the building development process, suggesting that some informal transactions must have facilitated it—as we discuss further below. In addition, landlords who benefited from state-subsidized loans are renting out the units to refugees, despite the explicit legal prohibition to do so. Hence, ironically, the rent and financial subsidies provided by the UNHCR to refugees are being used to pay the monthly rent to Lebanese landlords who use the money to settle their mortgage installments.

3 *Tented settlements:* In several urban contexts, particularly when refugee presence is deemed undesirable as in Zahle (Figure 4.3), the ‘urbanization’ of settlement has taken the form of vast ‘tented settlements’ in the immediate peripheries of cities. Given the dwindling value of agricultural land, landowners find lucrative opportunities in renting out land lots to refugees. They hence transform large agricultural tracts into the so-called ‘informal tented settlements’ (ITSs), subdivided into smaller lots, each occupied by an individual household who builds their tent.

Syrian households typically rent their access to land/tents through the mediation of a Syrian strongman, a *shaweesh*, who is also often entrusted

Figure 4.3 Agricultural tracts in Zahle and location of informal tented settlements. Legends: Agricultural tracts, ITSs

with organizing access to labor. The *shaweesh* may manage the camp alone or with his wife, and they often also organize access to labor and secure profit from the mediation with the Lebanese landlord, and the array of service providers who service the camp (UN-agencies, local and international NGOs, and religious organizations).

Living conditions in ITSs are consistently reported as the worst among all refugee settlements (VASYR, 2017): Overcrowding, leaking, lack of heating, lack of proper lighting, and small size of dwelling. Rehabilitation projects, when they happen, occur with the support of INGOs and the UNHCR, and they usually rely on self-help interventions. Refugees typically access electricity through informal measures that depend on the context where they are located: In Zahle, for example, they are legally hooked to the electricity meters (Electricité de Zahle, EDZ) while in numerous other areas they buy electricity from informal suppliers. Other services such as sewage networks are usually provided through aid agencies generating hybrid systems of service provision that cause severe environmental harm (Makki, 2018).

Discussion

The forms and geometries of this urbanization are clear indications that refugee displacement is profoundly reshaping the organization of cities in Lebanon and transforming their spatial characteristics. These modalities, it is worth noting, often co-exist within a same city, although there are sensible variations in their distribution. Indeed, in the three studied cities, most of the urbanization in Saida is occurring through neighborhood densification spread across the multiple quarters of the city (i.e., historic core, early modern city, and peripheral incremental development). Conversely, the majority of refugees in Zahle is concentrated in the city's peripheries where they are mostly confined to informal tented settlements. Halba offers a greater diversity—with a distinctive twist of large-scale complexes that were only documented in North Lebanon.

All in all, we can speak of two processes of urbanization (Figures 4.4 and 4.5): (1) densification in incremental dimensions within historical urban cores and low-income informal settlements, as well as (2) expansions of urban peripheries, typically along main roads, either through the form of consolidated housing compounds or precarious informal tented settlements. Furthermore, and looking comparatively at the three patterns of urbanization, we note that the term urbanization conceals critical differences in livability across contexts. Thus, refugees in peripheral tented settlements or collective shelters tend to cope with substantially poorer living conditions, heavily suffering from overcrowding and lack of privacy. Accordingly, we also identified varying levels of integration: Refugees in compounds and ITSs expressed more frequently in the survey their inability to easily access

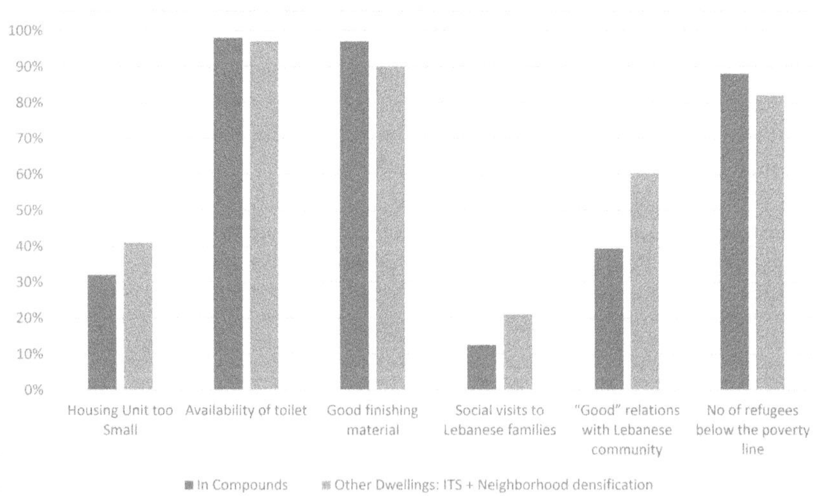

Figure 4.4 Housing quality in compounds and other forms of housing in Halba

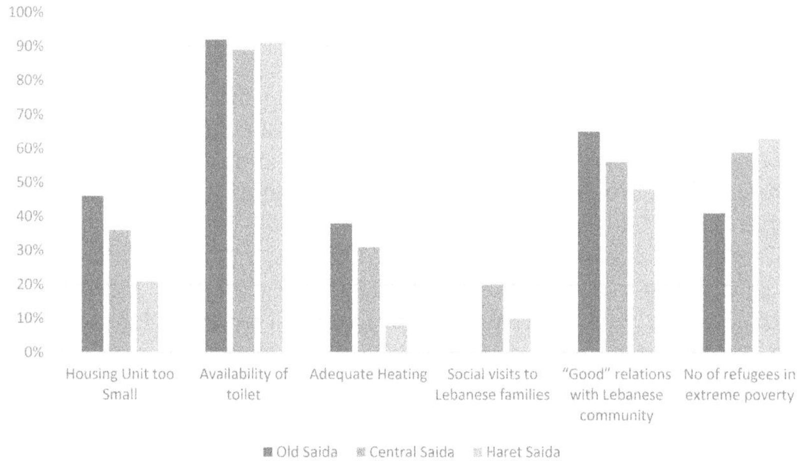

Figure 4.5 Comparative Housing Indicators across three locations in Saida

facilities and schools, and noted higher movement restrictions due to the lack of access to means of transportation—than their peers settled in urban neighborhoods.

An analysis of the urbanization of displacement

How are these modalities inscribed in cities, and to what extent do they fulfill the promises of refugee urbanization, as described in the development literature, notably in terms of better integration and increased opportunities (Crisp et al., 2012). Relocating the documented spatial patterns within the pre-existing urbanization of the three cities where they occur, we examine in this section their processes of production. We find that, while levels of integration and poverty vary, urbanization amounts everywhere to an expansion of urban 'gray spaces'—namely zones in which residents, activities, and territories are only partially incorporated in cities, to paraphrase Yiftachel (2009, 2015). Typically associated with informal processes of urban production and exchange,[6] gray spaces are best understood as a mode of negotiated urban production that produces neighborhoods or urban pockets (e.g., individual buildings and empty lots) that are neither fully integrated in cities, nor openly rejected.

Spatially, gray areas are characterized by the precarious conditions and poor level of urban services that exacerbate the vulnerability of their often-criminalized populations (Holston, 2009; Miraftab and Wills, 2005; Yiftachel, 2015). While other sections of the city may be more rigidly regulated and better serviced, these areas stand to benefit. They include, in our case, deteriorated historical cores, sprawling large-scale compounds in urban suburbs, failed large-scale developments, impoverished agricultural tracts, and informal urban extensions in urban peripheries. They benefit from their 'under the radar' position to expand and provide typically substandard shelter without the necessary urban amenities (Yiftachel, 2015). Building amendments are rarely authorized by municipal authorities. Such expansion occurs largely via ad-hoc incremental additions to housing units. More visible additions, such as the housing compounds developed in the peripheries of Halba and Minieh in North Lebanon, are made by numerous legal trespasses, many construction irregularities, and no compliance with official permits and regulations, maintaining instead what scholars have called a 'tolerated ceiling' (Fawaz, 2009; Durand-Lasserve, 1998). Similarly, tents are established and consolidated without municipal authorization. They rely on concerted decision-making among refugees, *shaweesh* and property owners, with the support of international organizations.

In what follows, we illustrate how gray spacing is sustained by bottom-up and top-down dynamics of spatial production.

1 *Gray spacing from below:* Processes of urban production/expansion are sustained by an array of informal economic transactions that involve four categories of actors. The first includes *those actors who hold claims over the spaces* used as shelters, be they property owners (e.g., farmers, landowners, and apartment owners) or illegal occupiers (e.g., squats in the old city of Saida and in numerous informal settlements), and who are hence in a position to rent them out to refugees for income.[7] This applies to the informal tented settlements described above, where the promise of a lucrative rent encouraged the transformation of agricultural land, many of which was abandoned urban peripheries, sometimes divided and rendered unproductive by road development and/or the lack of irrigation water, into informal tented settlements.[8]

   The second group of actors includes *those who facilitate the transactions*, such as realtors, slumlords, and community organizers—be they Syrian (such as the *shaweesh*) or Lebanese, or small-scale public agents (e.g., mukhtar). Among them, we find the religious and non-religious, as well as local and international relief organizations that provide building materials, building fixtures, and temporary relief that can be either directly used by dwellers, or turned into cash (through sale) to finance the production of this urbanization. The *informal service providers* form the third group: They either provide the 'service' of illegal hook-ups or organize make-shift forms of service provision as in electricity generation, water cisterns, etc.

   A fourth category of actors is *the police force* whose men turn a blind eye to incremental constructions that violate urban regulations in exchange of a small kickback. Through these negotiations, everything has a price, including legality and/or turning a blind eye to the law as we eloquently documented in the cases of housing compounds in Halba where additional floors and/or rooms are secured through the intervention of public actors from the local planning agency (Fawaz et al., forthcoming).

In this 'gray' economy, all transactions are person to person, unrecorded, undeclared, and tacit. This informality applies to contractual agreements regulating rental transactions. Irrespective of the shelter form (e.g., piece of land, apartment, and garage), all contracts are oral, and agreements are often sketchy and open ended, leaving both the landlord and the tenant the option to change terms if conditions were to change. These agreements do not reference state regulations, nor do they rely on public institutions for their organization. Rents are not registered in municipal offices, as legally mandated. Recourse in cases of default or the threat of eviction is very limited, sometimes allowing a few local figures to mediate a delay in payment, often not.

The production and maintenance of gray spaces relies heavily on social networks. Refugees recognize such networks as critical for their survival, as reflected in several surveys that showed 'proximity to refugees' to be the most determining factor in the localization and consolidation of refugee settlement.[9] By clustering, researchers have shown refugees to build on networks of solidarity, what some have labeled an 'infrastructure of care' through which they help each other secure jobs, healthcare, and more generally information to navigate the new environments where they dwell (Yassine et al., 2019; Fiddian-Qasmiyeh, 2015, 2016a, 2016b; Simone, 2004; Neis et al., 2018; Rokem et al., 2017). These networks are, however, not always clearly distinguishable from the 'host' communities since numerous cross-marriages and the historical continuity of the terrain and the families that occupied it, particularly in North Lebanon, mean that relations are often intertwined.

Despite sharing the destitution, all the dwellers of gray spaces are not necessarily equal in their conditions. As tenants, temporary residents, or visitors, refugees share these spaces with other city dwellers, including lower-income Lebanese households, migrant workers, and earlier waves of inhabitants. A comparative analysis of tenants and property owners in all three cities reflected a much higher level of poverty for tenants, when compared to property owners, irrespective of nationality. Still, refugees remained considerably poorer across local populations, irrespective of housing status. Having to navigate their criminalized presence, without their social networks, often restricted from working, refugees have to endure antagonism from the local community and lower security because they have no recourse if they get to be evicted from their homes.

Transactions and negotiations in gray spaces are not merely economic or social. In a country where political leadership rests on the provision of favors that often take the form of protection against the law, negotiations also involve numerous permissions to circumvent urban regulations, particularly during the election season. At other times, such permissions remain discrete, nuanced, but amount to calculations in which political figures maintain control over territories. This is done through the exchange of scattered 'covers,' extended through the local strongmen of the party or the *za'im*, empowering the developers of large-scale compounds or small-scale property owners in historical cores or other urban districts. They subdivide a housing unit into several smaller homes, add a room or a floor, or turn a garage into a makeshift shelter.

2 *Gray spacing from above:* In addition to being enabled from below by these variety of actors, gray spacing is also the outcome of top-down actions through which international non-governmental organizations (INGOs), public actors (ministries, municipal authorities, and planning agencies), as well as political and religious stakeholders determine the modalities through which refugee urbanization occurs.

A key factor of urbanization form is the institutional set-up though which the refugee arrival is regulated and managed, and access to shelter and urban services are organized. The absence of a clear national territorial strategy to address the influx of refugees led, as of 2012, to a variety of responses through which refugee presence was managed at the local level. After a phase of 'no policy,' a set of national-level regulations were issued, at different points in time, to regulate refugees predominantly as laborers in specific sectors—construction, agriculture, and solid waste collection (Fawaz et al., 2018). Characterized by their ambiguity, these regulations portrayed Syrians as criminalized others who needed to remain as invisible as possible in the city. They also enabled their policing and control, endowing local governments with the tasks of monitoring their numbers, their status, and their movements. An apt description of this mode of public governance is manufacturing vulnerability, the process through which the legal framework is designed to maintain the conditions of refugees precarious, and consequently affecting their livability (Saghieh, 2015). This process helps informalize the public authorities' response, and keep their commitments and responsibilities to a minimum, thus allowing them the flexibility to displace and/or evict refugees, as they see fit, evoking security or, more recently health concerns.

In each city, the municipality became the main public actor managing the refugee presence. They handled this responsibility variably, depending on their administrative and financial resources, but also depending on the territorial politics they are an integral part of. As such, in localities with strong municipal councils (e.g., Saida and Zahle), the municipality appears as an important player, albeit with a role limited to policing and containment. Indeed, municipalities tolerate refugee urbanization as long as it occurs in specific districts—namely away from the city center, in areas that are already informal and/or poor, or on peripheral agricultural lands, away from sight. They keep regular counts of the population, and police, or even evict, those who do not conform to 'invisibility' requirements—which could be more or less tacitly expressed, and which could also change over time. Conversely, in localities where the municipal institution is weaker, such as Halba, refugee governance is overtaken by other groups of local actors as discussed above.

In either case, with a strong or weak municipality, municipal governance is mostly tactical, responding to the territorial politics that distinguish the locality itself—which are often sectarian-based—helping gray spacing in various ways. In Halba, a predominantly Sunni city which used to be aligned, prior to 2015, onto Syrian tutelage, the local strongmen and elites have been struggling to identify a new patron that would lend them protection and benefits, in exchange of loyalty. They tried to appeal to the parliament representatives of the North close to Saad Hariri—the most powerful Sunni national patron, without much success. When Syrians came to Halba, these strongmen and elites quickly grasped that this was a great opportunity to capture some of the resources that come along with any refugee crisis (Jacobsen, 2006). Given Halba had attracted most of the Syrian opposition groups, which were receiving funding from Saudi Arabia and Qatar, the Sunni religious organizations in the city positioned themselves as key mediators and seized most of the refugee aid. They helped distribute the aid in multiple forms that contribute to an urbanization of displacement densifying the neighborhoods of the central city, and expanding gray spaces in its outskirts—where the compounds are currently located.

In Saida, refugees' resources were also channeled to the large network of Sunni religious organizations in the city, but under the close supervision of the municipality, where the coordinator of said organizations serves as a municipal councilor. Thus, refugees' resources strengthened the alliance of religious organizations in Saida with the municipality and, hence, with Bahia al-Hariri, the aunt of Saad, which is the main city patron. Their redistribution was done in ways that did not disrupt the dominant socio-economic distribution of populations in Saida, thus densifying the old city center, its immediate vicinity, and some of its agricultural peripheries. Conversely, in Zahle, given the mayor is a vehement opponent to the presence of Syrians in the city, claiming it is a conspiracy that seeks to disrupt the demographic balance of the country and that threatens Christian minorities, refugees resources bypassed the city, and became concentrated in its fertile valley, in the ITSs. Indeed, the municipal policy of exclusion tipped the shelter form towards more precarious typologies, and delegated the governance of the refugee crisis to the hybrid groups of actors we identified, with a key role to INGOs, albeit keeping the prerogative of legitimate violence.

Besides municipalities and religious organizations, planning agencies and ministries also play a significant role in the form of refugee urbanization. The changes in the zoning law that increased the exploitation ratios in Halba in 2012 incentivized developers to build extensively, and encouraged the addition of floors to already constructed compounds. Similarly, encroachments spiked when a 2017 informal decree issued by the Minister of Interior allowed the illegal construction of 150 m^2 on every lot in the months preceding national elections.

In sum, every detail of the urbanization of forced displacement is governed and negotiated among a multiplicity of actors, from above and from below. Furthermore, much of this governance and these negotiations extend over long durations, include interruptions, and involve multiple calculations of current and future benefits (Simone, 2005, 2011, 2013). Documented elsewhere as quiet encroachments (Bayat, 2010) or celebrated as claims for substantive citizenship (Holston, 2009; Miraftab, 2009), these practices intensify the tolerated, under the radar practices of the poor, and expand the cities' gray spaces. They are sustained by maintaining a semblance of sketchiness, temporariness, and precarity that secure the tacit tolerance of city authorities and/or powerful elites, without granting their agents their due rights of political societies (Chatterjee, 2004; Roy, 2009).

Implications for cities and planning policy

Relocating refugee spaces within cities in Lebanon, the chapter has found that the urbanization of forced displacement has amounted to an expansion of gray spacing—areas produced and managed through informal, ad-hoc actions. Refugees occupy 'gray spaces' between legality and illegality in urban environments (Yiftachel, 2009), but their struggle is shared with many of the vulnerable host populations who also negotiate daily formal and informal practices in the city (Darling, 2017). We further showed that this urbanization is the result of the combined actions of public and other stakeholders' who capture refugees' resources but deliberately exclude and criminalize them, and the cumulative negotiations of an array of actors including brokers, landowners, service providers, who can reap profit from its production and servicing.

Our findings are well in line with what researchers have argued about cities of the North where the inclusivity and transparency of the urban housing market proved to be essential in the positive settlement of refugees. In its absence, refugee settlement countered the hopeful assumption that urbanization opens possibilities for refugees to improve living conditions, as shown by researchers in the contexts of Winnipeg, Turin, Berlin, Leipzig, and Stockholm among others (Andersen et al., 2013; Carter and Osborne, 2009; Bolzoni et al., 2015; Murdie and Borgegard, 1998; Sherrell et al., 2007).

In order to secure the promises of urbanization, it is imperative for cities to recognize the transformative impacts of refugee settlement and initiate strategic plans through which they account for the severe transformations imposed by rapid population increases on their already crumbling infrastructures. In most contexts, the association of refugee settlement with rural contexts and camps has kept city planning at bay. Leaving refugee hosting to solidarity networks, political and religious authorities, administrative personnel, or relief agencies has meant that their settlement has been organized through ad-hoc, spontaneous measures, rather than actual planning. The chapter discussed how the absence of a national plan through which refugee response was regulated and governed allowed municipalities and other stakeholders to enable the formation and expansion of territories of neglect housing refugees and impoverished host populations. We also underscored how this expansion amounts to a criminalized presence that ultimately undermines the potentials of refugee urbanization and furthers its gray spacing instead.

Similarly, numerous European cities have found that the expected positive externalities of urban settlement were not materializing; refugees are isolated, impoverished, and failing to benefit from the opportunities predicted by the scholarly and grey literatures with which we opened the chapter (Eckardt, 2018; Fontanari, 2015; Seethaler-Wari, 2018; Werner et al., 2018). Ironically, when planners were called upon to respond to the refugee crisis, like in the case of the Tempohomes in Berlin, their work amounted to a mere spatial reorganization of refugee camps. Referencing a 'village' rhetoric, planners developed a grid like plan for Berlin, with aesthetically pleasant-looking neighborhoods. The main characteristics in the process was to east the management of the camp, their legibility, hence reinforcing high levels of control and facilitating disciplinary practices (Dalal et al., 2018).

In order to respond to refugee urbanization, cities need to acknowledge the reality of this form of urban expansion and integrate it within a renewed, inclusive planning practice. This is all the more urgent as the spread of diseases as Covid-19 show that cities are as vulnerable as their most vulnerable sections, creating the momentum for change in urban governance practice. As such, it is imperative that public agencies, local governments, and international organizations recognize the transformative impacts of the refugee settlement on their cities and infrastructures, the rising demand for more services, and the sprawling expansions of their peripheries. It is further urgent that these stakeholders initiate the necessary planning processes to integrate these settlements within their territories. It is further urgent that local governments are given the financial means and independence to conduct these interventions, while fostering an inclusive national discourse that supports these initiatives. We propose, by way of conclusion, three types of planning interventions, all borrowing from traditional and widespread planning approaches that aim to include refugee populations as integral components of communities in urban areas:

1 *Strategic Spatial Planning:* The practice of articulating flexible, performative planning frameworks to guide urban development, is a widely used planning tool particularly in European countries. Departing from the rigid land-use plans that require frequent revisions and dictate constraining options, strategic plans can be developed in participatory

form involving various groups of stakeholders who share the same urban or regional spaces. Participants articulate shared visions of the long-term development of the town in which trade-offs are defined and concerted decisions are taken. These shared visions can then translate into actual project interventions that are endorsed and applied with the input of stakeholders.[10]

Given the diversity of actors involved in the ongoing urbanization of refugee responses in Lebanon's cities, the high level of informality in which the production of space is happening, and its individual form, it seems imperative that local governments are given incentives to identify the current patterns of growth. They can then launch planning initiatives that mitigate their impacts and control future expansion. Furthermore, given that much of the above-described urban changes reflect pre-existing patterns of urban growth in which the most vulnerable population already dwells, local governments in particular should initiate consultation processes. This is to bring together members of these vulnerable communities (i.e., host and refugees) to discuss the organization and management of shared spaces, the connection to urban infrastructures, the expansion of settlements, the sharing of costs, and the respective burdens each is bringing. Urban planners and designers favoring participatory area-based approaches and place-making interventions can be key actors leading such processes. International agencies should thus consider supporting local governments to engage in such a practice, placing the urban at the heart of the refugee policy response. Accordingly, the role of local governments would be up-scaled from coordinating policing and humanitarian responses to managing development and planning. Then, local governments will play a leading role in allowing for an integrated management of territories and all the populations dwelling in them, irrespective of their nationalities and/or voters' status.

2 *Land-Use Planning at Multiple Scales*: The development of informal settlements as a form of urban expansion associated to the individual decisions of landowners to sacrifice the agricultural value of their land has demonstrated sizable long-term negative effects in terms of loss of food security and ecological degradation. Years after settlement, the interruption of natural waterways, the pollution of underground water tables, the erratic accumulation of solid waste, and the disruption of fragile ecosystems generate irrecoverable costs (Ministry of Environment in Lebanon – European Union – UNDP, 2014). Studies overlapping current ITSs settlement with flood plains and water resources have shown that settlement have neither accounted for precious agricultural land, nor have they respected waterways and protected resources. In addition, ITSs in the periphery of the city of Zahle extend within flood plains, exposing refugees to severe threats (Makki, 2018), as happened in Fall 2019 when severe rains drained some settlements in the peripheries of the Beqaa valley.

To respond to such challenges, land-use planning should be adopted at multiple scales. On the one hand, national authorities should reactivate the national land-use plan adopted in 2009 (SDATL), where water resources, forests, and agricultural fertile zones are delineated, as are areas of severe seismic risk and flood plains. Indeed, given that planning agencies have failed to translate the directives of the national land-use plan in mandatory planning documents, the general vision of the national master plan is still not translated into actual planning directives. Such directives should include guidelines for relocating refugee settlements currently built within the SDATL's protected zones, while paying special attention to the needs of their dwellers.

On the other hand, cities and their peripheries need to re-evaluate their land-use plans, earmarking areas of urban expansion where housing compounds or temporary settlements can be developed and serviced without extensive costs. In Zahle, for example, Makki (2018) showed that a large number of informal tented settlements followed existing road and sewer networks' lines, making it possible to develop connections to the city's infrastructural system and consequently improve the livelihoods of refugees, while planning for environmental protection. In Halba, El Merehbi (2018) showed how the sprawl of large-scale complexes pushed urbanization to far urban peripheries, triggering additional building development. Such land-use plans can translate into locally adapted planning regulations, allowing local governments more flexibility, such as canceling ad-hoc additions, prompted by the permission to add 150 m^2 without permit introduced by the Ministry of Interior in 2017, mentioned earlier.

3 *On-site settlement upgrading* is an area-based intervention through which city authorities target specific low-income, dilapidated neighborhoods with an integrated upgrading intervention. Building on discourses of urban justice, the right to housing, or simply health imperatives (as heard recently with the Covid-19 pandemic), authorities intervene in neighborhoods developed in violation of urban and/or building regulations (e.g., squatters and self-help) where they extend services and expand the shared infrastructures. They may also provide loans and/or direct material help to upgrade houses and introduce regulatory mechanisms to support existing rental markets. These interventions build on decades of 'informal settlement upgrading' approaches of the type advocated by UN-Habitat, the World Bank, and other agencies in the late 1970s (Fernandes and Varley, 1998). Prioritizing the right-to-housing over the exchange value of land, such interventions have come a long way in unbundling property ownership from shelter acquisition. In Brazil, for example, neighborhoods with contested property rights and high density of precarious populations can be earmarked as *zones of special social interest* where property rights are suspended and the right to shelter is decoupled from the right to sell and/or bequeath a piece

of land (Santoro, 2015; Rolnik, 2014). Public authorities consequently develop land reserves for affordable housing where they can provide services without seeing housing prices increase, gentrification take over, and their attempts to build affordable housing reserves undermined by market forces. More recent interventions have further integrated principles of place-making in the upgrading of shared spaces—interventions that are direly needed in Lebanon's cities, particularly in inner neighborhoods and dense slums (Samaha, 2015).

At first glance, there seems to be a world of difference separating the language of planning at the end of this chapter with the three spatial patterns of refugee urbanization we documented and analyzed. Yet, we see that each of pattern incorporates the ingredients that can help in expanding, rather than gray spaces, the definitions of actors, institutions, and planning visions beyond the narrow imagination currently dominating the country's sectarian political system and its bureaucracy. Indeed, we are appealing to the historical proposition of planning as a profession committed to inclusive and livable urbanization and to its role as an organizational system that provides the tools necessary for the expansion of our political imagination, in order to advance social and spatial integration in cities.

Acknowledgments

We are grateful to Dunia Ezzeddine and Diala Makki who contributed to the field research towards this work.

Notes

1. The approximate proportion of urban refugees was estimated at 61% globally. The data coverage was limited to around 56% of the refugee population. Given that the coverage was the lowest in high-income countries and in order not to bias the results towards lower-income, it was assumed that refugees in high-income countries would be urban (UNHCR, 2018).
2. Al-Harithy (2019/2020) and Doraï (2010) have looked at the incremental development of individual shelter. Yassine et al. (2019) have also investigated the development of settlement strategies where refugees illegally settle in clusters of families, eventually forming dense neighborhoods.
3. Samhan (2008) had documented such practices in the Palestinian Camp of Burj el Barajneh South of Beirut where she was able to map an informal regulatory code guiding urban development. Although our study did not map the same level of details, our observations in the Old City of Saida reflects similar practices, as illustrated in Figure 4.1.
4. Fawaz et al. (forthcoming) have investigated the materialization of municipal refugee responses into territorial practices of governance and filtration. This policing is particularly intensive in areas under control of Hezbollah in Lebanon as in the city of Hermel, the southern and eastern peripheral districts of Beirut, numerous villages in South Lebanon. It is common in these districts to represent refugees as a potential refugee threat and to filter them according to political and religious affiliation.

5. A comparison of the conditions mapped in Saida by the LCSRHCL 2018, LCPS-SCPR nonetheless show better conditions than earlier research in the peripheries of Beirut (Fawaz et al., 2014; Fawaz, 2017).
6. In reference to the multiple violations of property and/or urban and building codes that support the production of spaces, as well as the unstable and tacit agreements that sustain their organization.
7. The LCPS/ SCPR Living Conditions Survey shows that in all three cities, these rental units are distributed whereby an individual typically claims 1–2 units.
8. Although the typology may suggest that we are dealing with a 'camp', the process of housing acquisition, is one based on rental agreements. However, instead of the apartment, the housing product is a land lot and its transformation into shelter relies heavily on aid agencies who provide the material needed to build and consolidate tents into shelters.
9. Several studies have shown that refugee households prioritize social networks in housing decisions, choosing to move near relatives and/or neighbors (Fawaz et al., 2018).
10. In Lebanon, strategic spatial planning came in vogue during the 2007–2015 period, during which several international organizations supported the development of such interventions in Lebanon's regions.

References

Abourahme, N., and Hilal, S. (2009). Intervention: (Self) Urbanization and the Contours of Political Space in Dheisheh Refugee Camp. *Jerusalem Quarterly, 8*, pp. 42–45.

Affordable Housing Institute. (2017). *Strategic Housing Solutions for Vulnerable Populations in Lebanon*, research report submitted to the Norwegian Refugee Council in Lebanon.

Agier, M. (2011). *Managing the Undesirables*. Malden, MA: Polity.

Andersen, H.S., Turner, L.M., and Søholt, S. (2013). The Special Importance of Housing Policy for Ethnic Minorities: Evidence from Comparison of Four Nordic Countries. *International Journal of Housing Policy, 13*(1), 17–36, DOI: 10.1080/14616718.2013.764659.

Bayat, A. (2010). *Life as Politics: How Ordinary People Change the Middle East*. Palo Alto: Stanford University Press, DOI: 10.5117/9789053569115.

Bolzoni, M., Gargiulo, E., and Manocchi, M. (2015). The Social Consequences of the Denied Access to Housing for Refugees in Urban Settings: The Case of Turin, Italy. *International Journal of Housing Policy, 15*(4), 400–417, DOI: 10.1080/14616718.2015.1053337.

Brun, C. (2015). Active Waiting and Changing Hopes: Toward a Time Perspective on Protracted Displacement. *Social Analysis, 59*, 19–37, DOI: 10.3167/sa.2015.590102.

Campbell, E. H. (2005). Formalizing the Informal Economy: Somali Refugee and Migrant Trade Networks in Nairobi. Global Commission on International Migration, 47. Available online at: https://www.refworld.org/docid/435f87254.html

Campbell, E. H. (2006). Urban Refugees in Nairobi: Problems of Protection, Mechanisms of Survival. *Journal of Refugee Studies, 19*(3), 396–413, DOI: 10.1093/jrs/fel011.

Carter, T. S., and Osborne, J. (2009). Housing and Neighbourhood Challenges of Refugee Resettlement in Declining Inner City Neighbourhoods: A Winnipeg Case Study. *Journal of Immigrant and Refugee Studies, 7*(3), 308–327. DOI: 10.1080/15562940903150097.

Chatterjee, P. (2004). *The Politics of the Governed: Reflections on Popular Politics in Most of the World.* New York: Columbia University Press.

Crisp, J., Morris, T., and Refstie, H. (2012). Displacement in Urban Areas: New Challenges, New Partnerships. *Disasters*, *36*, S23–S42, DOI: 10.1111/j.1467-7717.2012.01284.x.

D'Auria, V., Daher, R., and Rohde, K. (2018). From Integration to Solidarity: Insights from Civil Society Organisations in Three European Cities. *Urban Planning*, *3*(4), 79–90, DOI: 10.17645/up.v3i4.1688.

Dalal, A., Darweesh, A., Steigemann, A., and Misselwitz, P. (2018). Planning the Ideal Refugee Camp? A Critical Interrogation of Recent Planning Innovations in Jordan and Germany. *Urban Planning*, *3*(4), 64–78, DOI: 10.17645/up.v3i4.1726.

Darling, J. (2017). Forced Migration and the City: Irregularity, Informality, and the Politics of Presence. *Progress in Human Geography*, *41*(2), 178–198, DOI: 10.1177/0309132516629004.

Datta, A. (2012). *The Illegal City: Space, Law and Gender in a Delhi Squatter Settlement.* Ashgate: Aldershot.

Doraï, M. K. (2010). *From Camp Dwellers to Urban Refugees? Urbanization and Marginalization of Refugee* Camps *in Lebanon.* The Lived Reality of Palestinian Refugees in Lebanon, Institute for Palestine Studies. Available online at: https://halshs.archives-ouvertes.fr/halshs-00495742

Durand-Lasserve, A. (1998). Law and Urban Change in Developing Countries: Trends and Issues, in *Illegal Cities – Law and Urban Change in Developing Countries*, edited by E. Fernandes and A. Varley. London and New York: Zed Books. 233–257.

Eckardt, F. (2018). European Cities Planning for Asylum. *Urban Planning*, *3*(4), 61–63, DOI: 10.17645/up.v3i4.1834.

El Merehbi, S. (2018). Actors, Governance and Modalities of Electricity Supply: The Case of Low-Income Neighborhoods and Refugee Compounds in Halba. Thesis submitted for the degree of Master of Urban Planning and Policy: American University of Beirut, Beirut, Lebanon.

Fábos, A., and Kibreab, G. (2007). Urban Refugees: Introduction. *Refuge: Canada's Periodical on Refugees*, *24*(1), 1–19.

Fawaz, M. (2017). Planning and the Refugee Crisis: Informality as a Framework of Analysis and Reflection. *Planning Theory*, *16*(1), 99–115.

Fawaz, M. (2009). The State and the Production of Illegal Housing: Public Practices in Hayy el-Sellom, Beirut-Lebanon, in *Comparing Cities: The Middle-East and South Asia*, edited by K. Ali and M. Rieker, London: Oxford University Press, 197–220.

Fawaz, M., Al-Hage, C. Harb, M. (forthcoming). *Urban Housing for Refugees: Three Case Studies from Lebanon.* Manuscript submitted for review.

Fawaz, M., Nammour, K., and Saghieh, N. (2018). *You Get what you pay for, Housing, Land and Property Issues of Syrian Refugees from Homs City in Lebanon.* Report submitted to UN-Habitat and UNHCR.

Fawaz, M., Saghieh, N., and Nammour, K. (2014) *Housing, Land, and Property Issues in Lebanon. Implications of the Syrian Refugee Crisis.* Beirut: UN-Habitat and UNHCR.

Fernandes, E., and Varley, A. (Eds.). (1998). *Illegal Cities: Law and Urban Change in Developing Countries.* London: Zed Books.

Fiddian-Qasmiyeh, E. (2015). Refugees Helping Refugees: How a Palestinian Camp in Lebanon Is Welcoming Syrians. *The Conversation,* Nov. 4. Available online at: https://theconversation.com/refugees-helping-refugees-how-a-palestinian-camp-in-lebanon-is-welcoming-syrians-48056

Fiddian-Qasmiyeh, E. (2016a). Refugee–Refugee Relations in Contexts of Overlapping Displacement. *International Journal of Urban and Regional Research.* Available online at: https://www.ijurr.org/spotlight-on/the-urban-refugee-crisis-reflections-on-cities-citizenship-and-the-displaced/refugee-refugee-relations-in-contexts-of-overlapping-displacement/

Fiddian-Qasmiyeh, E. (2016b). Refugees Hosting Refugees. *Local Communities First and Last Providers of Protection FMR*, *53*, 25–27.

Fontanari, E. (2015). Confined to the Threshold. The Experiences of Asylum Seekers in Germany. *City*, *5*(19), 714–726, DOI: 10.1080/13604813.2015.1071112.

Holston, J. (2009). Insurgent Citizenship in an Era of Global Urban Peripheries. *City and Society*, *21*(2), 245–267, DOI: 10.1111/j.1548-744X.2009.01024.x.

Horst, C. (2002). Vital Links in Social Security: Somali Refugees in the Dadaab Camps, Kenya. *Refugee Survey Quarterly*, *21*(1 and 2), DOI: 10.1093/rsq/21.1_and_2.242.

Hovil, L. (2007). Self-Settled Refugees in Uganda: An Alternative Approach to Displacement? *Journal of Refugee Studies, 20*, No. 4, DOI: 10.1093/jrs/fem035.

Hyndman, J., and Giles, W. (2011). Waiting for What? The Feminization of Asylum in Protracted Situations. *Gender, Place & Culture*, *18*(3), 361–379.

Jacobsen, K. (2006). Refugees and Asylum Seekers in Urban Areas: A Livelihoods Perspective. *Journal of Refugee Studies*, *19*(3), DOI: 10.1093/jrs/fel017.

Kibreab, G. (1996). Eritrean and Ethiopian Urban Refugees in Khartoum: What the Eye Refuses to See. *African Studies Review*, *39*(3), 131–178, DOI: 10.2307/524946.

Landau, L. B. (2014). Urban Refugees and IDPs. *The Oxford Handbook of Refugee and Forced Migration Studies*, vol. 1. UK: Oxford University Press, pp. 139–150.

Landau, L., and Jacobsen K. (2004). Refugees in the New Johannesburg. *Forced Migration Review*, *19*, 44–46.

Léna, E. (2012). Mukhalafat in Damascus: The Form of an Informal Settlement. *Popular Housing and Urban Land Tenure in the Middle East.* Cairo and New York: The American University in Cairo Press.

Macchiavello, M. (2004). Livelihoods Strategies of Urban Refugees in Kampala. *Forced Migration Review*, *20*.

Makki, D. M.-A. (2018). *Actors, Governance and Modalities of Sanitation Services: Informal Tented Settlements in Zahleh (Lebanon).* Unpublished Master's Thesis in Urban Planning and Policy: American University of Beirut, Beirut, Lebanon.

Martin, D. (2015). From Spaces of Exception to 'Campscapes': Palestinian Refugee Camps and Informal Settlements in Beirut. *Political Geography*, *44*, 9–18.

Ministry of Environment in Lebanon – European Union – UNDP. (2014). Lebanon Environmental Assessment of the Syrian Conflict & Priority Interventions. Available online at: https://www.aub.edu.lb/facilities/ehsrm/Documents/EASC-ExecutiveSummaryEnglish.pdf

Miraftab, F. (2009). Insurgent Planning: Situating Radical Planning in the Global South. *Planning Theory*, *8*(1), 32–50, DOI: 10.1177/1473095208099297.

Miraftab, F., and Wills, S. (2005). Insurgency and Spaces of Active Citizenship: The Story of Western Cape Anti-Eviction Campaign in South Africa. *Journal of Planning Education and Research*, *25*(2), 200–217, DOI: 10.1177/0739456X05282182.

Murdie, R. A., and Borgegard, L. E. (1998). Immigration, Spatial Segregation and Housing Segmentation of Immigrants in Metropolitan Stockholm, 1960–95. *Urban Studies*, *35*(10), 1869–1888, DOI: 10.1080/0042098984196.

Neis, H. J., Meier, B., and Furukawazono, T. (2018). Welcome City: Refugees in Three German Cities. *Urban Planning*, *3*(4), 101–115, DOI: 10.17645/up.v3i4.1668.

Ramadan, A. (2013). Spatialising the Refugee Camp. *Transactions of the Institute of British Geographers*, *38*(1), 65–77, DOI: 10.1111/j.1475-5661.2012.00509.x.

Rokem, J., Fregonese, S., Ramadan, A., Pascucci, E. A., Rosen, G., Charney, I., and Sidaway, J. D. (2017). Interventions in Urban Geopolitics. *Political Geography*, *61*, 253–262.

Rolnik, R. (2014). Place, Inhabitance and Citizenship: The Right to Housing and the Right to the City in the Contemporary Urban World. *International Journal of Housing Policy*, *14*(3), 293–300 DOI: 10.1080/14616718.2014.936178.

Roy, A. (2009). Why India Cannot Plan its Cities? *Planning Theory*, *8*(1), 76–87.

Saghieh, N. (2015). Manufacturing Vulnerability in Lebanon: Legal Policies and Efficient Tools of Discrimination, in *The Legal Agenda*. Available online at: http://legal-agenda.com/en/article. php?id=690andfolder=articlesandlang=en

Samaha, P. K. (2015). *Informal Streetscapes: Weaving the Patches and Patterns of Contested Public Spaces in Nabaa (Bourj Hammoud-Beirut)*. Master's dissertation, Department of Architecture and Design: American University of Beirut, Beirut, Lebanon.

Samhan, H. M. K. (2008). *Housing Alternatives: The Case of the Bourj Al-Barajneh Refugee Camp*. Master's dissertation, Department of Architecture and Design: American University of Beirut, Beirut, Lebanon.

Santoro, P. F. (2015). Urban Planning Instruments for Promoting Social Interest Housing: From Zoning to Obligatory Percentages in São Paulo, Brazil, in Dialog with Bogotá, Colombia. *Revista Brasileira de Estudos Urbanos e Regionais*, *17*(2), 99–117, DOI: 10.22296/2317-1529.2015v17n2p99.

Sanyal, R. (2017). A No-camp Policy: Interrogating Informal Settlements in Lebanon. *Geoforum*, *84*, 117–125.

Seethaler-Wari, S. (2018). Urban Planning for the Integration of Refugees: The Importance of Local Factors. *Urban Planning*, *3*(4), 141–155, DOI: 10.17645/up.v3i4.1696.

Sherrell, K., D'Addario, S., and Hiebert, D. (2007). On the Outside Looking In: The Precarious Housing Situations of Successful Refugee Claimants in the GVRD. *Refuge: Canada's Journal on Refugees*, *24*, 64–75.

Simone, A. (2004). People as Infrastructure: Intersecting Fragments in Johannesburg. *Public Culture*, *16*(3), 407.

Simone, A. (2005). Urban Circulation and the Everyday Politics of African Urban Youth: The Case of Douala, Cameroon. *International Journal of Urban and Regional Research*, *29*(3), 516–532, DOI: 10.1111/j.1468-2427.2005.00603.x.

Simone, A. (2011). The Surfacing of Urban Life. *City*, *15*(3-4), 355–364, DOI: 10.1080/13604813.2011.595108.

Simone, A. (2013). Cities of Uncertainty: Jakarta, the Urban Majority, and Inventive Political Technologies. *Theory, Culture and Society*, *30*(7–8), 243–263, DOI: 10.1177/0263276413501872.

Sommers, M. (1999). Urbanization and its Discontents: Urban Refugees in Tanzania. *Forced Migration Review*, (4) 22–24.

Sommers, M. (2001). Young, Male and Pentecostal: Urban Refugees in Dar es Salaam, Tanzania. *Journal of Refugee Studies, 14*(4), DOI: 10.1093/jrs/14.4.347.

Swerts, T. (2017). Creating Space for Citizenship: The Liminal Politics of Undocumented Activism. *International Journal of Urban and Regional Research, 41*(3), 379–395, DOI: 10.1111/1468-2427.12480.

UNHCR. (2010). *2009 Global Trends: Refugees, Asylum-seekers, Returnees, Internally Displaced and Stateless Persons.* Geneva UNHCR. Available online at: https://www.unhcr.org/statistics/country/4c11f0be9/2009-global-trends-refugees-asylum-seekers-returnees-internally-displaced.html

UNHCR. (2018). Global Trends: Forced Displacement 2018. Available online at: https://www.unhcr.org/globaltrends2018/

VASYR. (2017). *Vulnerability Assessment of Syrian Refugees in Lebanon.* Beirut: UNHCR.

Werner, F., Haase, A., Renner, N., Rink, D., Rottwinkel, M., and Schmidt, A. (2018). The Local Governance of Arrival in Leipzig: Housing of Asylum-Seeking Persons as a Contested Field. *Urban Planning, 3*(4), 116–128, DOI: 10.17645/up.v3i4.1708.

Women's Commission for Refugee Women and Children. (2002). Fending for Themselves: Afghan refugee Children and Adolescents Working in Urban Pakistan. Available online at: https://www.google.com/url?sa=t&rct=j&q=&esrc=s&source=web&cd=&ved=2ahUKEwiI34nly7rvAhVh8OAKHWorAJEQFjAAegQIAxAD&url=https%3A%2F%2Fwww.refworld.org%2Fdocid%2F48aa82dc0.html&usg=AOvVaw36tMRsco_LsNxY3z8_Dg7r

Yassine, B., Al-Harithy, H., and Boano, C. (2019). Refugees Hosting Other Refugees: Endurance and Maintenance of Care in Ouzaii (Lebanon). *Journal of Refugee Studies*, DOI: 10.1093/jrs/fez098.

Yiftachel, O. (2009). Critical Theory and 'Gray Space': Mobilization of the Colonized. *City, 13*(2-3), 246–263, DOI: 10.1080/13604810902982227.

Yiftachel, O. (2015). Epilogue—from 'Gray Space' to Equal 'Metrozenship'? Reflections on Urban Citizenship. *International Journal of Urban and Regional Research, 39*(4), 726–737, DOI: 10.1111/1468-2427.12263.

5 Governing displaced cities

Calibrating reconstruction amidst instability

Loren Landau

Introduction

Scholars of forced migration often present displacement as a disruptive event that fragments if not shatters people's socio-economic foundations. Underlying that is a presumption of loss: not only of basic rights and physical security, but also of membership in normatively sedentary social worlds (Gibney, 2015, p. 13). Such framing engenders a perspective on reconstruction founded on restoring, replicating, or at least approximating the *status quo ante.* Much of the global humanitarian apparatus and humanitarian scholarship is oriented towards this goal; considering their responsibilities over when stability is achieved (see Black and Koser, 1999; Long, 2014). Indeed, even while recognizing that displacement, like other movements, catalyses irreversible change (see Lubkemann, 2010), discussions of durable solutions nonetheless tend to evoke a sense of permanent resolution: a return to predictability, solidity and solidarity.[1] This chapter argues that this may be a fundamental mistake.

The pursuit of solidarity is evident in both definitions of 'durability' and humanitarian efforts to 'reconstruct' and achieve sustainable solutions. In some ways, this is not surprising as inclusive urban development programmes often presume spatially stable populations. The emphasis on reunification, a return to families' mechanical solidarity whether in sites of origins, sanctuary or elsewhere, is perhaps the most visible illustration of an effort to realize this status (see UNHCR, 1983). Beyond households, resettlement, return or local integration offer complementary endpoints rooted in the organic solidarity of exchange, labour markets and generalized interdependence. Such solidarities will be differentiated and unequal with disparities persisting for generations (see Portes and Zhou, 1993). Nonetheless, the understandings of durability embedded in the 1951 *Geneva Convention Related to the Status of Refugees* evoke metaphors of relatively stable, almost organic formations in which the various parts interact along systems of shared rules and mutual recognition (see Malkki, 1995). Debates on the nature of family and the specific mechanisms and meaning of integration continues, yet most scholarship—and almost all policy interventions—still conceives displacement as a temporally defined arc beginning with tragedy and ending with durable forms of solidarity fostered by return or reconstruction. This is echoed in many local government approaches to governing displacement which speaks about integration, stability or social cohesion.

This chapter questions the desirability of governing for solidary durability. It argues that due to changes in the global economy, coupled with the disruptions of displacement, often undermine the potential or desirability of building lives in singular locations. It accepts solidarity and sedentariness remain strong normative guides for many displaced people, humanitarians and local government officials. However, the structures of global capitalism and urbanization means that displacement to or within southern cities shifts the basis on which people chart life courses, not only for the displaced but also for other full- or part-time urban residents. For some at least, there can be little sense of site-specific durability, but rather a future imagined as continual mobility accompanied by tactical connections and disengagements. Effective reconstruction must recognize these multiple trajectories.

In an era in which finding economic security increasingly means spreading family and individual risk across multiple spaces, displacement may create new opportunities and demands within existing and emerging translocal relationships and diasporas. Moreover, given the challenges of rebuilding 'sending' communities, access to external resources and remittances may become central to individual and familial economic security. In some instances, individual survival or success are dependent on shunning solidarity and going at it alone for abbreviated or extended periods. Under such conditions, the most durable solution may well be the opportunity to continue moving within or across borders unencumbered by social responsibilities. If so, what first appears as the (self) preservation of social marginalization and vulnerability in the form of alienation from family, kin and community represents tactical agency: a quest for protection and durability amidst the structural uncertainty of contemporary urbanism. While such behaviours may well be tactical—a response of the generally disempowered to structurally oppressive conditions—they nonetheless represent an effort to build lives (cf. de Certeau, 1984). But in this case, long-term welfare is premised on the ability to avoid what we typically understand as 'durable solutions'. That is, relative long-term security comes from keeping things unsettled.

The remainder of this chapter proceeds through four primary sections. The first comments on my approach and the empirical inquiries underlying my analysis. The second embeds questions of urban humanitarianism and reconstruction within broader sets of 'urban questions'. It draws particular attention to the precarious nature of African urban life and how stability and fixity can work against residents' ability to accumulate the social and material capital necessary to thrive or survive. The third illustrates means by which refugees and other migrants (the two are often legally and sociologically indistinguishable (see Landau and Duponchel, 2011)) forge socialities that allow them to negotiate the urban condition while avoiding precisely the kind of solidarities typically presumed to shape durable solutions. The last section—more schematic than substantive—points to a range of practical, ethical and epistemological concerns stemming from the empirical discussion. Apart from problematizing teleologies of solidarity embedded in international refugee law, it speaks to how knowledge is constructed within 'refugee and forced migration studies'.

Methods and approach

This chapter explores the meanings of durability and agency amidst urban precarity and fluidity. In doing do so, it makes no claims to represent the totality of people's diverse intentions, aims or experiences but instead points to avenues for further exploration and the possibility that disconnection and social alienation—shunning solidarity—may be critical to achieving lasting welfare and *de facto* protection. While potentially uncomfortable and normatively unappealing, it is premised on the position that only by first understanding the nature and necessities of urban life can we then consider their potential practical and ethical implications.[2] It also raises fundamental questions for the governance of urban spaces where populations may remain mobile, disconnected and disinvested.

Although this is largely a conceptual inquiry, I draw on an ecumenical set of data in support of my position. The data also enable me to shift from more generalized assessments to individualized stories and accounts. Most of the information used here stems from migration-related research in Southern and Eastern Africa—beginning with Johannesburg and expanding to Nairobi, Maputo and Lubumbashi—undertaken between 2002 and 2015. Each of the cities included in these studies are destinations and transit points for domestic and international migration including significant numbers of displaced persons. Together, they represent a range of social, economic and political characteristics that allows for modest generalizations about trends and possibilities about African cities and, potentially, the rapidly expanding metropolises of the global South.

The survey data used intermittently throughout stem from interviews with 2,211 people in three of the cities: Johannesburg, Maputo and Nairobi. These data do not fully represent either the migrant or host populations in any of the sites, let alone the experience of migration and displacement elsewhere on the continent.[3] Rather, data collection targeted particular groups of foreigners categorized by nationality. With the exception of Mozambicans included in the Johannesburg survey, data were collected on selected groups—Somalis, Rwandans, Sudanese and Congolese—that straddle the line between economic migrants and those who might be considered (in substance, if not in law), forced migrants or displaced persons. Such sampling does not cover the cities' full geographic footprints as doing so would have generated too thin a sample. Instead, sampling concentrated on areas on gateway areas and 'estuarial zones' of high traffic and diversity (see Landau, 2014; Saunders, 2011; Singer *et al.*, 2008). Although statistics do not feature prominently in this chapter, they inform prior works (cited throughout) which inform it. Instead, the text relies more heavily on illustrations from secondary sources and qualitative fieldwork conducted in collaboration with doctoral students in Johannesburg and Nairobi, two African cities attracting significant numbers of displaced persons (see Campbell *et al.*, 2011; Krause-Vilmar and Chaffi, 2011).

The emergence of urban humanitarianism

It is for good reason that urban refugees have received a great deal of attention of late.[4] As the UNHCR notes on its 'urban refugees' webpage, 'only one-third of the world's 10.5 million refugees now live in camps' (http://www.unhcr.org/pages/4b0e4cba6.html – accessed July 2017). Across Africa, there is a growing awareness that the majority of refugees and other displaced people—in some places the vast majority of them—live and seek protection in what are often deeply impoverished, unequal and under-capacitated but increasingly global urban centres (see UN-Habitat, 2014).

Be it Kabul, Khartoum or a dozen other cities proximate to seemingly interminable conflicts, refugees and the internally displaced significantly contribute to urban population growth and position in regional and global flows of information and material (Beall and Esser, 2005, p. 6; Gow, 2005; also Segbers, 2007). One need only glance at the cities of Jordan and Lebanon to know that displacement from Syria's civil war has reshaped their internal composition and global position (see World Bank, 2015). Even where the displaced are proportionately less, their presence can rapidly reconfigure social and economic life. Elsewhere, the displaced move almost invisibly into cities, disappearing among longer-term residents who may share class, language, religious or other commonalities. Given that the alternatives to urban settlement include decades in camps, administrative detention or another 'protracted refugee situation', it is hardly surprising that the displaced increasingly find their ways to population centres (see Kamal, 2016). Although the urban displaced may not find golden paved streets, cities nonetheless offer at least faint promises of upward economic mobility and physical freedom.

Although refugees' presence in global cities is not as novel as many suggest—authors have flagged urban refugees' presence for decades (see Cooper, 1992; Malkki, 1995; Rogge and Akol, 1989)—strong normative, political and financial motivations have recently fixed the humanitarian gaze on what the eye had previously 'refused to see' (Kibreab, 1996). Rather than crediting a rise growth in the urban, the emergence of 'urban refugees' as an object of study extends from broader efforts to 'visibilise displacement': to identify and expose the vulnerability of varied groups and define them in terms that make them suitable objects of humanitarian action (see Lubkemann, 2010; Polzer and Hammond, 2008). Even if driven by a valuable humanitarian impulse, this result is a particular kind of analytical blindness that leaves presumptions unchallenged. In particular, a continued focus on vulnerabilities associated with displacement—an orientation due to researchers' close connection to humanitarian action and organizations—often limits the field's engagement with broader questions of urbanism, precarity and resilience. That characterizing social realities in rapidly transforming urban centres is difficult without the added challenge of hunting down new arrivals who may seek invisibility (Jacobsen and Furst, 2011) only exacerbates this isolation. If nothing else, this chapter suggests there are compelling normative and intellectual attention for sustained attention to both cities and the people seeking protection within them. However, such inquiries' potential will only be achieved through a substantial redefinition of the modes through which we 'see' and understand displacement and humanitarian intervention.

As noted earlier, the focus here is on what the presence of displaced people in cities—particularly those in the global south—means for understanding displacement and, particularly, what it means to 'reconstruct' urban space in ways that are inclusive and durable. It also hints at the inverse of the equation by asking what displacement may mean for studies of global urbanism. Doing so contributes to a better understanding of an increasingly diversifying pool of cities proximate to conflicts and other sources of displacement. It also asks us to reconsider the forms of protection displaced persons seek, the means through which should protection is achieved and the meaning of durable solutions. At the heart of this discussion is how nature of the South's conflicts and global cities—precarity crossed with potential; fluidity and resistance—creates forms of almost perpetual displacement and mobility. In an era in which finding economic security increasingly means spreading family and individual risk across multiple spaces (Halliday, 2006), displacement may create new opportunities within existing and emerging translocal relationships and diasporas. Moreover, given the challenges of rebuilding 'sending' communities, access to external resources and remittances may become central to families' economic security (see Deshingkar and Farrington, 2009). Without discounting the possibility or desirability of return, local integration or third country resettlement, it argues that durable solution may well be premised on shunning the solidarity and stability often underlying a language of 'durable solutions'. While the focus here is explicitly on the highly informal and infrastructurally frail cities of Southern and Eastern Africa, there are lessons in this analysis for displaced people in politically marginal but deeply networked urban spaces worldwide.

Protection in flux

Understanding the meaning of solutions forged within urban spaces demands moving beyond immediate efforts to improve service delivery and protection in urban spaces. We must instead delve more deeply—empirically and philosophically—in to the nature of contemporary urban life. Doing so reveals extraordinarily high levels of economic and physical precarity associated with living in Africa's urban centres and residents' tactical and strategic responses to these conditions. (For the most part, such risks face all poor urban residents but they can be potentially exacerbated by a history of displacement.)[5]

This is not the place for a thorough review of Africa's 'urban condition' let alone cities across the global south. Indeed, a piece of this length cannot accurately reflect the diversity of opportunities and obstacles present within even a single city (for such approaches see De Boeck and Plissart, 2004; Meyers, 2011). Rather, it summarizes a condition perhaps best characterized as 'possibility amidst precarity'.[6] The precarity of Southern city life is a familiar strain in the Malthusian narratives underlying much academic and policy work. Many of these accounts note the tremendous growth rates (UNDESA, 2014; UN-Habitat, 2014) and that as cities swell, new arrivals (and long-term residents) find few new jobs. Others note that rather than embracing their population while planning for future growth, urban governance systems are instead captured by narrow interests or crumbling from lack of financial support, technical capacity and *de facto* political centralization. With a forthcoming 'youth bulge', cities will again host growing numbers with little concomitant growth in formal economic opportunities (UN-Habitat, 2014).

Apart from those profiting from the poor, the elite—political and economic—are largely disengaged from the cities' majority, often physically separating themselves in gated communities or new cities (Kermeliotis, 2013). When possible, they are retreating to physically removed, purpose built urban sites. In Kihato and Muyeba's (2015, p. 3) words, 'the dominant images that Africa's cities conjure are ones where the landscapes are rolling shantytowns, where toilets 'fly' and rivulets of sewerage snake through cobbled-together homes. Africa's cities it seems, tell a story of failure—the failure of modernization, capitalism and the African state'.

Yet for all of their evident shortcomings and risks, Africa's globalizing cities and their counterparts elsewhere nonetheless allure with their often elusive potential for profit, protection or passage elsewhere. Ideal by almost no imaginable developmental metric, these sites nonetheless offer relative access to varied opportunities and possibilities and urbanization is correlated with increased human development indices, a better quality of life and access to services and the possibility to earn (UNDESA, 2014). Even slum dwellers often live longer and healthier lives than similar folks living beyond the city. Indeed, the potential for investment and work in rapidly expanding urban markets offer comparative advantage in terms of trade and economies of scale when compared to Africa's relatively sparsely populated rural hinterlands.[7]

For present purposes, the question is what people—migrants, IDPs, refugees and even long-term residents—do to capitalize on cities' opportunities while avoiding the pitfalls of economic and physical precarity. With few possibilities for return to communities of origin or organized resettlement to a wealthy western country, *de facto* durable solutions for millions of displaced persons will be forged in these urban sites. In the absence of elaborate states or humanitarian interventions, these solutions, whatever their form, will be designed and realized by urban residents working alone, together or with connections living beyond the urban edge. This is a form of tactical agency that is often overlooked or dismissed as survivalism in terms that analytical disconnect it from residents' potential longer-term objectives.

In describing the necessities of urban life, Simone (2010) suggests that across much of the 'global south', daily existence is a process of constant breaking down and recomposition. People living in the city constantly contend with current conditions while also imagining and preparing for a different future. And as movement comes to characterize the city, the possibility of fixing people in place becomes more difficult. So even as authorities seek to create a stable and ordered mode of living—an impulse that Scott (1996) ascribes to the genetics of stateness—people's every day necessities and imaginations help to undermine the very possibility of doing so.

Going further, Simone argues that through their knowledge and relationships, people are cities' most significant infrastructure. Due to the fluid nature of employment combined with perennial material precarity, urban success demand people regularly reconfigure and if possible upgrade the infrastructure to which they are connected. In his words (Simone, 2010, p. 126):

> Where people are the important infrastructure, the value of an individual existence rests less in the elaboration of a 'meaningful' life or a coherent story about a person's character. What is important is individuals' ability to be 'hooked in' to different daily scenarios, dramas, networks, and affiliations that provide a constant set of alternatives for how to put bread on the table or how to become a person that can be taken seriously.

In an earlier work (Simone, 2006, p. 358), he similarly posits, 'Cities everywhere exert pulls on each other in a force field in which the maintenance of localized coherence becomes increasingly problematic'. Part of this desire for constant restructuring and ongoing motion is rooted in the need to maintain an ever expanding or diversifying set of 'weak ties' (Granovetter, 1973). This may be accomplished from a fixed point or it may demand movement within cities or between them.

Reciprocal obligations along with the economic precarity of the post-industrial (or perhaps more accurately, the neo-liberal, non-industrial) era further encourage an impulse to decouple from local and distant social relations and the corollary connections to space. Take, for example, Worby's (2010) discussion of Zimbabwean refugees/migrants living in Johannesburg.

In his account, the desire for disconnection is rooted less in the nature of the city, *per se*, although he recognizes the risks from police, employers and others typically associated with urban life. Rather, the need for consistent connection and reconnection stems from the elaborate and occasionally oppressive social networks extending among vulnerable populations migrant and otherwise. These take the form of demands from family and sending communities to send money, food, durable goods and clothing to their places of origin (see also Dzingirai *et al.,* 2014; Kihato, 2013; Madsen, 2004). Perhaps more immediately, these demands may also manifest in requests to physically host and support relatives and friends—a flexible, often expansive category of people—in continual succession or even simultaneously. The result being a kind of dependency pyramid in which a refugee pioneer ends up hosting close relatives followed by ever more distant connections, frequently under cramped and unsustainable conditions: self-built houses, storage rooms, sub-let and sub-divided apartments, or even occupied offices.

Rather than offering comfort or social security, such relationships introduce not only material obligations and inconvenience but an additional source of uncertainty. In Worby's (2010, p. 417) words:

> The open-ended condition of displacement brought about by violence, insecurity and the present political impasse within Zimbabwe lends itself to a particularly vexing moral economy for Zimbabweans abroad—one in which the temporal horizon for reconciling credits and debts, present capacities and future dependencies, is unknowable.

For people living in the fluid, economically marginal neighbourhoods of southern cities, denying others support when even marginally more resource endowed is almost unimaginable. Those with the courage to so risk acute costs: shaming oneself and even one's parents who may be accused of raising a disrespectful child. For migrants, betraying expectations not only means exiting from potential reciprocal support at some indeterminate date, but may elicit explicit countermeasures ranging from reputational costs to curses, theft or physical threats. In these cases, the strong social networks and sets of overlapping reciprocal obligations—social formations scholars and practitioners often celebrate for providing protection and resilience (see Aguilera and Massey, 2003; Deshingkar and Farrington, 2009; Kanaiapuni *et al.*, 2005; Landau and Duponchel, 2011; Munshi, 2003; World Bank, 2005)—become burdensome and counterproductive. The failure to maintain them during periods of crisis can lead to what Kankonde (2010) terms, 'social death': one's indefinite social exclusion and stigmatization.

Confronted with the likely failure to meet social obligations or the sometimes onerous costs of doing so—not only short-term material costs, but comfort and capital-accumulation possibilities—refugees hoping to make their lives in fluid, global cities often seek forms of social invisibility. For many, desiring disconnection and invisibility stems from the seemingly shameful activities one undertakes to make it in a new place (see Kankonde, 2010; Kihato, 2013; Malauene, 2004; for example). For those involved in sex work or other criminalized behaviours, the source of stigma is perhaps obvious even if many take pride in supporting their families (see Oliveira and Vearey, 2015). For others the downgrading of skills—from accountant to taxi driver; from teacher to day labourer—may be what they hope to shroud. Even where the work is 'honourable', maintaining local and translocal connections comes with costs: observational disciplines and networks that risk revealing less laudable truths, spreading rumours and making unreciprocated demands.

Even if desires for disconnection are not universal, they nonetheless capture an important aspect of contemporary life in global cities of the South. Economic precarity, predatory policy and states, and reciprocal obligations are giving rise to a desire for fluidity and ethics of disconnection that furthers what is perhaps a generic condition of city life. In one of sociology's foundational works, *The Metropolis and Mental Life*, Simmel (1997, p. 179) remarks on a tendency to social distance among metropolitans, a reserve that shades, 'not only into indifference, but, more often than we are aware ... a slight aversion, a mutual strangeness and repulsion'. Going further, Jacobs (1964, p. 403) speaks of cities as places in which one finds freedom about 'countless numbers of people', a freedom in part that comes from the ability to be disengaged from those physically closest to you. As Tonkiss (2003, p. 298) notes, 'While a language of community has been important for articulating various politics of difference, I suggest that forms of *indifference* also afford certain rights to and freedoms in the city'. Young (1990, p. 241) goes further in arguing for an ethics of indifference related to the capacity to be unseen, to be unexceptional and to be impersonal in a social field where 'differences remain unassimilated' (see also Wirth, 1995, p. 70).

Reconstruction and amidst mobility and disconnection

In cities where rooting is either undesirable or effectively impossible, residents—including the displaced—treat the sites in which they may live for decades as sites in which to gather the resources needed to further onward movement or build status and security beyond the city's boundaries (Geschiere, 2009). Doing so often means directly countering community and the forms of place based solidarity—organic or mechanical—evoked by talk of 'durable solutions' and informing programmes attempting to rebuild. Or reconstruct. Proximate and translocal rooting may be important, but so too are regular shifts between rural (or peri-urban) and urban areas, within urban areas or between cities. For many, urban sites are 'places of flows' (Castells, 1996) or 'nowherevilles' (Bauman, 2000) where rooting and local representation is not the goal and the burdens and binding that connections and political participation offer are often something to be avoided (Kankonde, 2010; Madsen, 2004). Given the insecurity of land tenure, the possibility of violence and ongoing economic deprivation, a durable solution means actively maintaining feet in multiple sites without firmly rooting themselves in any (Freemantle, 2010). As such, migrants—including refugees and other displaced persons—are turning cities (or parts of them) into stations, rather than destinations on part of ongoing journeys. This helps generate a kind of permanent temporariness in which tactical agency is visible in actively resisting incorporation (Kihato, 2013; Landau, 2014; Malauene, 2004).

In such environments, formal citizenship and stability may have value for some, but with only limited enforcement capacity and a minimal reliance on state-provided services—schools, clinics, jobs—it is safe to say that documentation and legal status often do little. At the most practical level, they are poor predictors of people's welfare (Landau and Duponchel, 2011). Even in South Africa, arguably the continent's 'strongest' state, these processes are negotiated on the ground through a panoply of rationalities and calculations, sometimes involving laws and state actors but not always in predictable ways (Hansen and Stepputat, 2010). In such environments, finding space and resources within a city involves an ongoing process of forming and abandoning horizontal and vertical connections and constant, often constrained, calculations.

The forms of belonging and solidarity that matter for people increasingly are themselves fluid, syncretic and translocal. Even churches—often seen as instruments of local integration, community formation or stable transnational mobilization (see Cadge and Ecklund, 2007)—are now sites for 'tactical cosmopolitanism' and other forms of fluidity and liminality (see Garbin, 2012; Landau and Freemantle, 2010, 2016; Schiller *et al.*, 2006). Within them and similar bodies, people find ways of maintaining the levels of social engagement and recognition necessary to negotiate everyday life, but without the kind of place based fixity often associated with durable solutions. The forms of solidarity forged through these bodies are often inherently transient, translocal and even post-terrestrial. Not only do people frequently shift liturgical allegiances, but the churches themselves foster a kind of disconnection from those beyond the ever more fragmented congregations (Kankonde, 2016).

Conclusions: Ethical and epistemological reflections

By using such strategies of partial inclusion and rights claiming, international migrants' illegality and continued mobility can prevent capture by the state or communities—local or diasporic—enabling a kind of invisibility that, though precarious, allows them to elude obligations and, occasionally, exploit the city for resources to which they are not legally or socially entitled. Instead of transplantation and legibility to the society and political systems under which they ostensibly live, urban refugees may well be striving for a kind of 'usufruct' rights—a form of exclusion that is at least partially compatible with social and political marginalization. Rather than integrating or assimilating, they exploit their position as the permanent outsiders in ways that 'distances him or her from all connections and commitments' (Said, 2001, p. 183). They may use a variety of tactics to claim rights in their sites of residence, but we must not presume this reflects a desire for the forms of solidarity that typically inform externally designated durable solutions.

Their behaviours raise a number of important questions. Perhaps most obviously, they raise questions about what an 'emic' understanding of durability or reconstruction might look like within the highly fluid and fragmented spaces that characterize global cities. In African cities—as elsewhere—inclusion is something more than claiming a 'right to the city' or becoming part of a stable, urban community. We must avoid assuming the existence of such communities but also recognize that for many domestic and (especially) international migrants and the displaced among them, moving to city whatever the reason, is also a step into a global 'imaginary' and spaces beyond bounded place. For them, the goal of reconstruction may be more than finding a place to stay, work or find physical protection, but also entrée into global culture, new universal urban lifestyles (however understood), or, more concretely, opportunities for onward journeys. They may never realize these ambitions, but the city is nevertheless a space of possibility with access to trading and travel opportunities unavailable in villages or camps. Where integration or inclusion into a city of residence is either impossible or undesirable, membership into these decentred, socially regulated, globalized networks may represent a far more significant form of belonging and source of rights and recognition. Alternatively, they may pursue almost total alienation or loose, easily reformed networks that provide information and guidance without observation and obligation.

Urban refugees' potential desire for continued instability raises important ethical and policy issues that we can only grapple with by better understanding the southern 'urban condition', which they are helping to create. The rapid expansion of urban populations—and the concentration of migration and urban growth in particular urban gateway neighbourhoods—questions the use of the term 'local' or 'host' to talk about the destination areas (see Landau, 2014). It also raises concerns over what reconstruction might look like in *practical* terms and, more importantly, what it *should* look like. Reconstruction efforts often imagines a future of stability and place bound solidarity, but at what point do these models start to recognize the crisis-like, fluid modes of everyday urban life? What, under such conditions, would humanitarian benchmarking or urban development look like? The SPHERE standards may help provide guidance on material conditions, but within contemporary urban life almost no one—refugee or 'local'—lives at such standards. Moreover, achieving these standards requires not just new modes of urban governance, but a fundamental shift in global supply chains and domestic politics. In all cases, effective interventions may best be understood as expanding agency which may afford opportunities for both fixed settlement and ongoing mobility and disconnection.

Apart from fundamental questions of policy intervention, the trends outlined above raise epistemological questions for scholars seeing to understand displaced persons' agency under constrained circumstances. Arguing that striving for ongoing flux and mobility is part of a durable solution and reconstruction raises significant methodological challenges to the observer. It is, after all, difficult to know if someone is in constant movement because they are unable to settle or because they have read the environment in which they operate and are actively choosing to remain unrooted and potentially in motion—what I have termed 'tactical agency'. To be sure, the fact that churning is part of what contemporary urban life requires suggests a degree of vulnerability and structural constraints that all residents—particularly those who are relatively poor—experience. But this is no more so then those who remain trapped in place, limiting their ability to capitalize on the potential benefits of migration (which, after all, may include basic physical security and the ability to maintain even bare life) (see Lubkemann, 2008). In a world where economic precarity can be thrust on individuals by local policies, global structures, and even one's family and kin, the desire to move and the ability to do so is of enormous tactical importance. In this regard, we must take the lead from critiques of the human trafficking and sex work literature, which ask us to problematize questions of victimization. Yet as Sen (1993) and others note, choice under highly constrained situations is itself a problematic concept and people's desires are often shaped as much by perceived possibilities and objective or consistent moral principles and commitments.

In closing, it is worth noting that whether one's primary interests are scholarly or practical, there are benefits of recalibrating the nature of 'refugee and forced migration studies': there is much to gain from embedding our approaches to displacement within a broader scholarship on the places and processes in which refugees participate. In this case, an explicit engagement with both urban studies and African urbanism helps reframe the meaning of durability and its potential desirability in the ways it has typically been understood. It also helps us recode actions that may in part appear to represent signs of constraint to ones of empowerment or, more mildly, agency. Beyond thematic broadening, there are further gains from following Comaroff and Comaroff's (2012) suggestion to begin more substantially theorizing from the South.

As Connell (2007) notes in describing the sociology of globalization, we tend to look towards the south as a source of data to flesh out models formed in the backyards of the more powerful knowledge producers. A number of urbanists have begun this process (see Mbembe and Nuttall, 2004; Parnell and Robinson, 2013; Watson, 2009), and there is much to learn from them. However, given that the humanitarian enterprise and the headquarters of most global humanitarian agencies are rooted primarily in 'the north' and operate through knowledge forged there (see Fassin, 2011; Malkki, 1995, 1996), it is often difficult for 'refugee studies' to shift fundamental frames. Indeed, if we take seriously the question of agency, we must then also take seriously the question of how we theorize about the agency of the displaced. Beginning from a position that 'solutions' are compatible with state objectives and looking at how refugees and displaced people might ultimately be accommodated within them, still grants an unnecessarily privileged position to global and national regulatory frameworks. If instead we start from the perspective of what is required—or what is perceived to be so—among refugees, our understanding of 'solutions'—durable or otherwise—begin to look substantially different. As scholars we must not allow our desire for relevance and policy influence to foreclose the possibilities and critiques emerging from authors generating insights from a wide range of geographic and intellectual sites and traditions.

Notes

1. On the quest for stability and solidarity, see various contributions to Black and Koser (1999) and Allen and Morsink (1994); for Durkheimian solidarity, see Lukes (1973).
2. This responds to calls by Skrbis *et al.* (2004, p. 132); also Beck (2009).
3. For more details on the survey data employed here, see Madhavan and Landau (2011).
4. See, for example, *Forced Migration Review's* 2010 issue dedicated to Urban Refugees along with a range of other studies conducted in Africa, Asia, the Middle East and Latin America by or on behalf of humanitarian organizations (Zetter and Deikun, 2010) and a series of studies by such high-profile organizations as the Women's Refugee Commission (2011a,b) and the associated International Rescue Committee (Lyytinen and Kullenberg, 2013). More recent debates around inclusive cities and the sustainable development goals together with the global compacts on refugees and migrants all recognize the presence of displaced people in urban areas.
5. Earlier research indicates that people who claim to have moved due to persecution and conflict were on aggregate no more vulnerable than those who moved for other reasons (see Landau and Duponchel, 2011). However, in situations where cities are proximate to active conflicts—Gulu in Northern Uganda, for example—or under situations of mass and rapid displacement, there may be close correlations between people's reasons for movement and their socio-economic position.
6. While my use of 'precarity' is influenced by the work of Standing (2011) and others, I am not speaking of the economic uncertainty associated with declining industrial production, the weakening welfare state, or crumbling unionization. These processes are largely foreign to Africa outside the more industrialized Southern cone. Instead, precarity refers to a generalized condition of economic uncertainty and unpredictability.
7. For more on the socio-economics of African urban development, see (Meyers, 2011; Potts, 2010; UN-Habitat, 2014).

References

Aguilera, M.B., and Massey D. (2003). Social Capital and the Wages of Mexican Migrants: New Hypotheses and Tests. *Social Forces*, *82*(2), 671–701.

Allen, T., and Morsink, H. (eds.). (1994). *When Refugees Go Home*. Oxford: James Currey.

Bauman, Z. (2000). *Globalization: The Human Consequences*. New York: Columbia University Press.

Beall, J., and Esser, D. (2005). *Shaping Urban Futures: Challenges to Governing and Managing Afghan Cities*. Issues Paper Series. Kabul: Afghanistan Research and Evaluation Unit.

Beck, U. (2009). Imagined Communities of Global Risk, *Lecture for the Risk Conference in Shanghai*. First Draft: Uncorrected Version, 2.

Black, R., and Koser, K. (1999). The End of the Refugee Cycle? in *The End of the Refugee Cycle? Refugee Repatriation and Reconstruction*, edited by R. Black and K. Koser. Oxford: Berghahn Books.

Cadge, W., and Ecklund, E.H. (2007). Immigration and Religion. *Annual Review of Sociology*, *33*, 359–379.

Campbell, E., Crisp J., and Kiragu, E. (2011). *Navigating Nairobi: A Review of the Implementation of UNHCR's Urban Refugee Policy in Kenya's Capital City*. Geneva: UNHCR, Policy Development and Evaluation Service.

Castells, M. (1996). The Space of Flows, in *The Castells Reader on Cities and Social Theory*, edited by I. Susser. Oxford: Blackwell, 314–365.

Comaroff, J., and Comaroff, J. (2012). *Theory from the South, Or, How Euro-America is Evolving toward Africa*. New York: Paradigm.

Connell, R. (2007). The Northern Theory of Globalization. *Sociological Theory*, *25*, 368–385.

Cooper, D. (1992). *Urban Refugees: Ethiopians and Eritreans in Cairo*, *15*(2). Cairo: American University in Cairo Press.

De Boeck, P., and Plissart, M. (2004). *Kinshasa: Tales of the Invisible City*. Brussels, Belgium: Ludion.

de Certeau, M. (1984). *The Practice of Everyday Life*, (Trans. S. Rendall). Berkeley: University of California Press.

Deshingkar, P., and Farrington, F. (eds.). (2009). *Circular Migration and Multilocational Livelihood Strategies in Rural India*. Oxford: Oxford University Press.

Dzingirai, V., Mutopo, P., and Landau, L.B. (2014). *Confirmations, Coffins and Corn: Kinship, Social Networks and Remittances from South Africa to Zimbabwe*. Migrating Out of Poverty Research Consortium Working Paper 18, Sussex: University of Sussex.

Fassin, D. (2011). *Humanitarian Reason: A Moral History of the Present*. Berkeley: University of California Press.

Freemantle, I. (2010). You Can Only Claim Your Yard and Not a Country: Exploring Contexts, Discourse and Practices of Quotidian Cosmopolitanism amongst African Migrants in Johannesburg. Doctor of Philosophy, University of the Witwatersrand.

Garbin, D. (2012). Marching for God in the Global City: Public Space, Religion and Diasporic Identities in a Transnational African Church. *Culture and Religion*, *13*(4), 425–447.

Geschiere, P. (2009). *The Perils of Belonging: Autochthony, Citizenship, and Exclusion in Africa and Europe*. Chicago: University of Chicago Press.
Gibney, M.J. (2015). Refugees and Justice Between States. *European Journal of Political Theory*, *14*, 448–463.
Granovetter, M. (1973). The Strength of Weak Ties. *American Journal of Sociology*, *78*, 1360–1380.
Gow, G. (2005). Rubbing Shoulders in the Global City: Refugees, Citizenship and Multicultural Alliances in Fairfield, Sydney. *Ethnicities*, *5*(3), 386–405.
Halliday, T. (2006). Migration, Risk, and Liquidity Constraints in El Salvador. *Economic Development and Cultural Change*, *54*(4), 893–925.
Hansen, T.B., and Stepputat, F. (eds.). (2010). *States of Imagination: Ethnographic Explorations of the Postcolonial State*. Durham: Duke University Press.
Jacobs, J. (1964[1961]). *The Death and Life of Great American Cities: The Failure of Town Planning*. Harmondsworth: Penguin.
Jacobsen, K., and Furst, R. (2011). *Developing a Profiling Methodology for Displaced People in Urban Areas*. Medford: Feinstein International Center.
Kamal, B. (2016). Now 1 in 2 World's Refugees Live in Urban Areas'. *Inter Press News Service* (19 May 2016): http://www.ipsnews.net/2016/05/now-1-in-2-worlds-refugees-live-in-urban-area/
Kanaiapuni, S., Donato, K., Thompson-Colon, T., and Stainback, M. (2005). Counting on Lin: Social Networks, Social Support, and Child Health Status. *Social Forces*, *83*(3), 1137–1164.
Kankonde, P. (2010). Transnational Family Ties, Remittance Motives, and Social Death among Congolese Migrants: A Socio-Anthropological Analysis. *Journal of Comparative Family Studies*, *41*, 225–244.
Kankonde, P. (2016). *Taking Roots in the Name of God? Super Diversity and Migrant Pentecostal Churches' Legitimation and Social Integration in Post-Apartheid South Africa*. Doctoral Dissertation, University of Göttingen.
Kermeliotis, T. (2013). Africa's "New Cities:" Urban Future or Utopian Fantasies, CNN Online (30 May 2013): http://edition.cnn.com/2013/05/30/business/africa-new-cities-konza-eko/
Kibreab, G. (1996). Eritrean and Ethiopian Urban Refugees in Khartoum: What the Eye Refuses to See. *African Studies Review*, *39*(3), 131–178.
Kihato, C.W. (2013). *Migrant Women of Johannesburg: Everyday Life in an In-Between City*. London: Palgrave.
Kihato, C.W., and Muyeba, S. (2015). The Challenges and Prospects of African Urbanization: Forging Africa's Economic Growth Through Sustainable Urban Policies. *Unpublished Report for the Cities Alliance*.
Krause-Vilmar, J., and Chaffi, J. (2011). *No Place to Go but Up: Urban Refugees in Johannesburg, South Africa*. New York: Women's Refugee Commission.
Landau, L.B. (2014). Conviviality, Rights and Conflict in Africa's Urban Estuaries. *Politics & Society*, *14*(3), 359–380.
Landau, L.B., and M. Duponchel (2011). Laws, Policies, or Social Position? Capabilities and the Determinants of Effective Protection in Four African Cities. *Journal of Refugee Studies*, *24*(1), 1–22.
Landau, L.B., and I. Freemantle. (2010). Tactical Cosmopolitanism and Idioms of Belonging: Insertion and Self-Exclusion in Johannesburg. *Journal of Ethnic and Migration Studies*, *36*(3), 375–390.

Landau, L.B., and I. Freemantle. (2016). Beggaring Belonging in Africa's No-Man's Lands: Diversity, Usufruct and the Ethics of Accommodation. *Journal for Ethnic and Migration Studies*, *42*(6), 933–951.

Long, K. (2014). Rethinking Durable Solutions, in *The Oxford Handbook of Refugee and Forced Migration Studies*, edited by E. Fiddian-Qismeyeh, G. Loescher, K. Long, and N. Sigona. Oxford: Oxford University Press.

Lubkemann, S. (2008). Involuntary Immobility: On a Theoretical Invisibility in Forced Migration Studies. *Journal of Refugee Studies*, *21*(4), 454–475.

Lubkemann, S. (2010). *Culture in Chaos: An Anthropology of the Social Condition in War*. Chicago: University of Chicago Press.

Lukes, S. (1973). *Emile Durkheim: His life and Work*. London: Allen Lane.

Lyytinen, E. and J. Kullenberg. (2013). *Urban Refugee Research: An Analytical Report*. New York: International Rescue Committee.

Madhavan, S., and Landau L.B. (2011). Bridges to Nowhere: Hosts, Migrants and the Chimera of Social Capital in Three African Cities. *Population and Development Review*, *37*(3), 473–497.

Madsen, M.L. (2004). Living for Home: Policing Immorality among Undocumented Migrants in Johannesburg. *African Studies*, *63*, 173–192.

Malauene, D. (2004). *The Impact of the Congolese Forced Migrants' "'Permanent Transit" Condition on their Relations with Mozambique and its People*. MA thesis, University of the Witwatersrand.

Malkki, L.H. (1995). Refugees and Exile: From "Refugee Studies" to the National Order of Things. *Annual Review of Anthropology*, *24*, 495–523.

Malkki, L. (1996). Speechless Emissaries: Refugees, Humanitarianism, and Dehistoricization. *Cultural Anthropology*, *11*(3), 377–404.

Mbembe, A., and Nuttall, S. (2004). Writing the World from an African Metropolis. *Public Culture*, *16*(3), 347–372.

Meyers, G. (2011). *African Cities: Alternative Visions of Urban Theory and Practice*. Chicago: University of Chicago Press.

Munshi, K. (2003). Networks in the Modern Economy: Mexican Migrants in the U.S. Labor Market. *The Quarterly Journal of Economics*, *118*, 549–599.

Oliveira, E., and Vearey, J. (2015). Images of Place: Visuals from Migrant Women Sex Workers in South Africa. *Medical Anthropology: Cross-Cultural Studies in Health and Illness*, *34*(4), 305–318.

Parnell, S., and Robinson, J. (2013). (Re)theorizing Cities from the Global South: Looking Beyond Neoliberalism. *Urban Geography*, *33*(4), 593–617.

Polzer, T., and Hammond, L. (2008). Invisible Displacement. *Journal of Refugee Studies*, *21*(4), 417–431.

Portes, A., and Zhou M. (1993). The New Second Generation: Segmented Assimilation and its Variants. *Annals of the American Academy of Political and Social Science*, *530*, 74–96.

Potts, D. (2010). *Circular Migration in Zimbabwe and Contemporary Sub-Saharan Africa*. Oxford: James Currey.

Rogge, J.R., and Akol, J.O. (1989). Repatriation: Its Role in Resolving Africa's Refugee Dilemma. *International Migration Review*, *23*(2), 184–200.

Said, E. (2001). *Reflections on Exile and Other Essays*. Cambridge, MA: Harvard University Press.

Saunders, D. (2011). *Arrival City: The Final Migration and our Next World*. New York: Knopf.

Schiller, N.G., Caglar, A., and Guldbrandsen, T.C. (2006). Beyond the Ethnic Lens: Locality, Globality, and Born-again Incorporation. *American Ethnologist*, *33*(4), 612–633.

Scott, J. (1996). *Seeing Like a State*. New Haven: Yale University Press.

Segbers, K. (ed.). (2007). *The Making of Global City Regions: Johannesburg, Mumbai/Bombay, São Paulo, and Shanghai*. Baltimore: Johns Hopkins University Press.

Sen, A. (1993). Capability as Well-Being, in *The Quality of Life*, edited by M. Nussbaum and A. Sen. Oxford: Oxford University Press.

Simmel, G. (1997). The Philosophy of Fashion, in *Simmel on Culture: Selected Writings*, edited by M. Featherstone and D.P. Frisby. London: Sage Publications, 187–205.

Simone, A.M. (2006). Pirate Towns: Reworking Social and Symbolic Infrastructure in Johannesburg and Doula. *Urban Studies*, *43*(2), 357–370.

Simone, A.M. (2010). *City Life from Jakarta to Dakar: Movements at the Crossroads*. New York: Routledge.

Singer, A., Brettell, C.B., and Hardwick S.W. (eds). (2008). *Twenty-First Century Gateways: Immigrant Incorporation in Suburban America*. Washington: Brookings Institution Press.

Skrbis, Z., Kendall G., and Woodward, I. (2004). Locating Cosmopolitanism: Between Humanist Ideal and Grounded Social Category. *Theory, Culture and Society*, *21*(6), 115–136.

Standing, G. (2011). *The Precariat: The New Dangerous Class*. London: Bloomsbury Academic.

Tonkiss, F. (2003). The Ethics of Indifference: Community and Solitude in the City. *International Journal of Cultural Studies*, *6*(3), 297–311.

United Nations Department of Economic and Social Affairs Population Division (UNDESA). (2014). *World Urbanisation Prospects: The 2014 Revision: Highlights*. New York: UNDESA.

United Nations High Commissioner for Refugees (UNHCR). (1993). *UNHCR Guidelines on Reunification of Refugee Families* (July 1983). Geneva: UNHCR.

UN-HABITAT. (2014). *African Cities 2014: Re-Imagining Sustainable Urban Transitions*. Nairobi: UN-HABITAT (http://unhabitat.org/the-state-of-african-cities-2014).

Watson, V. (2009). Seeing from the South: Refocusing Urban Planning on the Globe's Central Urban Issues. *Urban Studies*, *46*, 2259–2275.

Wirth, L. (1995[1938]). Urbanism as a Way of Life, in *Metropolis: Centre and Symbol of Our Time*, edited by P. Kasinitz. Basingstoke: Macmillan, 58–82.

Women's Refugee Commission. (2011a), *Bright Lights, Big City: Urban Refugees Struggles to Make a Living in New Delhi*. New York: Women's Refugee Commission.

Women's Refugee Commission. (2011b), *No Place to Go but Up: Urban Refugees in Johannesburg: South Africa*. New York: Women's Refugee Commission.

Worby, E. (2010). Address Unknown: The Temporality of Displacement and the Ethics of Disconnection among Zimbabwean Migrants in Johannesburg. *Journal of Southern African Studies*. *36*(2), 417–431.

World Bank. (2005). *Global Economic Prospects 2006: Economic Implications of Remittances and Migration*. Washington DC: World Bank

World Bank. (2015). *The Welfare of Syrian Refugees: Evidence from Jordan and Lebanon*. Washington DC: World Bank.

Young, I.M. (1990). *Justice and the Politics of Difference*. Princeton: Princeton University Press.

Zetter, R., and Deikun, G. (2010). Meeting Humanitarian Challenges in Urban Areas. *Forced Migration Review*. 34: 5–7.

6 City Development Frame

The City as a unit of the countryside recovery (Azaz City in Syria)

Ghiath Al Jebawi

I received an email in 2016 from a German company requesting my advice on the development of a masterplan in Syria. I was, at the time, engaged in work in Azaz. We faced obstacles in establishing a common understanding of how architects, engineers, and urban planners engaged in the project define technical terms such as *planning* and *values*. Experts from Azaz initially told us that they already have a masterplan in place, considered as a trusted social contract with people.

The population of north Aleppo where Azaz is located was estimated at one million in 2019. Around half of the residents are Internally Displaced People (IDPs) with at least 200,000 living in IDP camps. Azaz alone had around 90,000 IDPs living in formal and informal camps near the city, and 75,000 live within the city itself (URC, 2018b). North Aleppo was opened to Afrin in 2018 where a massive demographic change happened, and later to Idlib with a population of at least three million residents. At least 40 percent of these residents are IDPs (ACU, 2019). This complexity is a part of a national crisis in Syria where around half of its population in 2010, estimated at 20.7 million, are now displaced (The World Bank, 2017). While 6.2 million are estimated to be internally displaced, 5.6 million were registered as refugees in neighboring countries in 2019 (UNHCR, 2019). In 2015 and 2016, around 800,000 Syrians sought asylum in Europe, mainly in Germany.

Additionally, there is a general belief that international money will fund reconstruction efforts. While it is difficult to disprove these beliefs, it is important to survey the use of available land in Syria and its resources. This includes the land rich with resources in the northwest province. Even before the conflict, much of the land in the countryside was abandoned. The percentage of people living in urban contexts from 60 years before the conflict increased from 33 percent to 56 percent (The World Bank, 2017; Aita, 2017). Nahr Aisheh in South Damascus is an example of early informalities caused by rural migration in Syria. While other types of informality are primarily attributed to political factors, such as the spread of Palestinian refugee camps across the region, a third type of informality is emerging in independent settlement IDP camps in the northwest region. Along this continuous uprooting and informal urbanization, several sustainable building and production practices on a household scale were abandoned.

This abandonment questions the neglect of valid construction techniques and fields of agricultural land. It also questions the role of urbanists, architects, and engineers. In the governance process, engineering consultants in iNGOs provided infrastructure for IDP camps without the flexibility to address fundamental questions about how the camp and its communities will evolve in 5 to 50 years. There was also a need to investigate the changes to land layouts and harness knowledge on local construction techniques. Investors and building contractors are often urbanists, architects, and engineers who operate under syndicates that control the building process. There is a need to reform regulations and accessibility to available land and resources for communities to access urgent housing and resettlement needs, as well as enhance the local economy.

The five-year planning period failed to deal with the abandonment of the countryside throughout Syria and proved incapable of leading to localized plans in urban areas. To respond to these shortcomings, my local network in Azaz and I started our work on the premise that another masterplan will not solve problems. It also became clear why the strategic plan of the local council which suggested devising a masterplan as early as 2016 did not function properly.

Our analysis of plans and actors on country, city, and regional scales provided evidence that the recovery of Azaz is possible. The City Development Frame (CDF) that we proposed included cultural, social, environmental, economic, legal, and institutional aspects for recovery. While the CDF works on the local community level, it addresses issues of affordable housing, local value chains, and urban wellbeing. It sees the management of land through knowledge and information systems fundamental to other physical and communal layers of the city.

Cities as nodes in a network

Even before the conflict in Syria, urbanization devastated the production sector. After being only at 33 percent in 1950, 56 percent of the Syrian population in 2011 has lived in urban areas, with 37 percent of the urban population in Damascus and Aleppo cities alone (20 percent of the country's population) (The World Bank, 2017; Aita, 2017). This led to transformations in socioeconomic factors and a reduced dependence on production and local value chains and broke the balance of urban settlements into congested central or abandoned smaller cities. To address recovery in Syria, it is essential to consider this transformation as a root cause and prioritize middle-sized cities as accelerators of local recovery and development. This requires better resource management and capacity building (Figure 6.1).

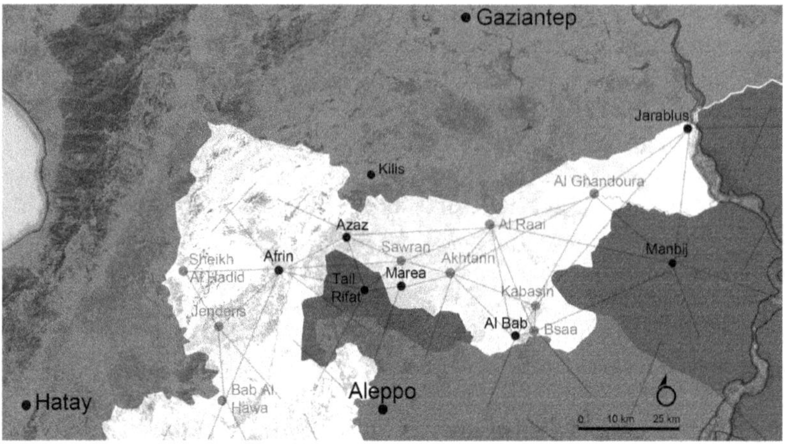

Figure 6.1 Cities network in north Aleppo

In north Aleppo, conflict positioned Azaz, Jarablus, and Al-Bab as major cities that assumed new political roles since 2011. With this massive flux of IDPs, the population of Azaz and Al-Bab was estimated between 150,000 and 200,000 each. The population of cities such as Jarablus, Marae, and Akhtarin with more than 50,000 people and developed operative local councils. North Aleppo also became connected to Afrin and Idlib, resulting in the northwest region with common characteristics in terms of percentage of IDPs, at around 50 percent, and the dependency of people on aid and import rather than on local production that has always characterized the area. This requires the right scales of governance, production, and accessibility to land. Equilibrium must be based on a pattern of urban nodes within adequate networks. The scale of each node should facilitate the governance and management of information and the measurement of its local resources and economy. It should also be sensitive to the needs of rural landscapes. Networks of coordination and mobility between the nodes are encouraged to exchange resources, production, and knowledge and decrease pressure on large urban centers. Ten cities in the region are transformed into centers of new administrative areas in north Aleppo. Along other considerations, they were identified by the most operative local councils in the area. The population of these cities ranges between 200,000 and slightly less than 50,000 residents.

The CDF allows cities to become aware of their resources and connect with other places in the northeast and beyond. The network is based on the self-sufficiency of each node (city) identified in relation to human and natural resources and local economy, as well as other technical rather than political considerations. Therefore, it is neither constrained, nor an opponent of administrative or other divisions. It is autonomous by nature. It is still important to analyze and plan sustainable resources and risk with other areas through channels for horizontal cooperation.

Internal components of the cities

Activities and resources in Azaz are scarce in comparison to small cities and are easily manageable, which makes it an adequate accelerator for local recovery. The risks and governance complexities are moderately balanced in comparison to those in large cities. Azaz has the potential to be a node among a wider system of nodes, while maintaining itself as an autonomous entity and developing the correct codes of exchange with peer cities. Cities with populations ranging between 50,000 and 200,000 seem to be able to play this role as they have major resource risk management.

As resources, these cities have a large working age population in different sectors. Furthermore, there are many natural resources such as available land that allows local agricultural production to minimize imports as well as quarries for locally sourced construction materials. The high need for housing is a catalyst in the job market, creating high employment rates. Furthermore, the agility in local governance and flexibility in legal frameworks could lead to a margin of flexibility to propose and find solutions, especially with the presence of community actors that have the capacity to participate in the development process.

However, these cities face problems and risks in different sectors. They depend on import and trade rather than the production economy. The economic elite impact the big trade activities and speculate on land value leading to unaffordable housing. The legal and institutional frameworks allow for such practices, while hindering efficient construction. They do not prevent the continuous informal urbanization – peripheral sprawl and in IDP camps – which has social and environmental impacts and causes urban segregation and posed social risks within the city.

City Development Frame – six sectors

Focusing on urban development processes, the projects faced many issues that were addressed in the six sectors of the CDF. For example, modify the land use of residential complex and a public school can be problematic. It requires coordination with the Technical Office in the local council, which is responsible for the management of the current masterplan which no longer corresponds to the pre-2011 administrative and legal conditions. Modifying the land use of a land plot of a school project must be done through the Directorate of Education, which operates on the provincial level. The directorate is no longer administratively connected to Azaz.

The HLP Registry Administration was also not functional, which added complexity to non-registered ownerships since land dealings were done through non-registered deeds. This poses legal and institutional problems that do not allow for the secure transformation of property ownership.

The digital map of the city had displacements in its basic layers; plans were no longer precise to allow for adequate urban planning. We must involve the Survey Plans, some of which are lost, and others are in Azaz or in inaccessible areas in the rest of the Aleppo Governorate due to the political divisions. Similarly, partitioning or transforming a private land into a public property required land appropriation; the necessary institutional frameworks for land appropriation were absent.

These examples highlight issues related to governance and HLP Registry, among other local registries, based on the analysis of processes on a small scale of the city. The analysis however uncovered other issues related to masterplanning and urban development. For example, facing the high need of housing, there is an economic recession manifested in unemployment and low purchasing power. Furthermore, expensive modern construction materials are important even though local construction materials are available in the region. In addition to this, land values are high despite the low purchase power and demand. Legal and institutional frames and current practices are among the drivers of these problems.

An economic condition based on import as opposed to production is prevent even if surrounding agricultural lands are available, and community actors are excluded from the process. Furthermore, estimates of current or projected population counts are not available, including IDPs numbers, which identify housing needs and economic potentials. When all these issues were categorized, they identified the six sectors of the CDF: Governance, local resources and housing, local economy, HLP and civil registries, civil society and community involvement, and research and planning.

Constitution of the process

The CDF focused on highlighting risks, resources, problems, and needs. As a result, two main tools are proposed as part of the CDF: The template tool and the Elements Map tool. The function of the map tool is to categorize content into sectors and issues, measure them, and then highlight processes and sequences across them. The Elements Map tool focuses on interaction between actors, protocols, and events that feed into the template tool. Both will be explained later in three modes of function of the CDF.

Research and analysis tools

Template Tool: The process of the city's recovery is categorized through the template tool into six sectors. It is the planning tool on which sequences of projects, goals, plans, solutions, experiments, assessments, legislations, and actions are identified. These sectors are divided under primary issues, and a list of secondary issues in each sector. Then the secondary issues are grouped by links across sectors to visualize the possible processes which lead towards further goals. It is not a strategic plan, but rather a dynamic tool that is synched with the Elements Map.

Elements Map Tool: Through the Elements Map tool, the CDF deals with three components: Actors, events, and protocols, and feeds into the template tool. It detects relations, identifies sequences, and delineates smaller systems all necessary for the whole process of planning. In this way, it categorizes and understands actors in depth, and arranges protocols with their different categories as legal, administrative, and knowledge content. In addition, it identifies events such as activities and meetings. Through identifiable sequences of actors, events, and protocols, it helps construct the template tool by suggesting new emerging topics in each of the six sectors. The upper two tools help keep track of the project and provide easy access to all its information.

Experts Meetings Analysis: Facts, opinions, and recommendations that are collected in meetings are analyzed by counting the duplication of keywords and ideas, while giving more relevance to the specialization and weights of the source person. Accordingly, they are analyzed, and outcomes and recommendations are produced. Meetings with experts are the main methods to understand the function of governance and legal frames, which has influenced the core of the work from spatial masterplanning to the CDF and its six sectors.

Systems Analysis: Systems are broken down into their main components such as actors, technical tools, and micro-processes. An example of the System Analysis is the HLP system complexity analysis, described later in the chapter.

Spatial Analysis: The CDF uses overlapping multi-layered classic vector maps and pixel pictures to detect spatial trends such as urban sprawl or masterplan infringes. However, it is more efficient to use GIS programs and link the spatial analysis with data sets, other used systems in the city such as HLP and Civil Registries, and a 3d model of the city, which is accessible by the community (Figure 6.2).

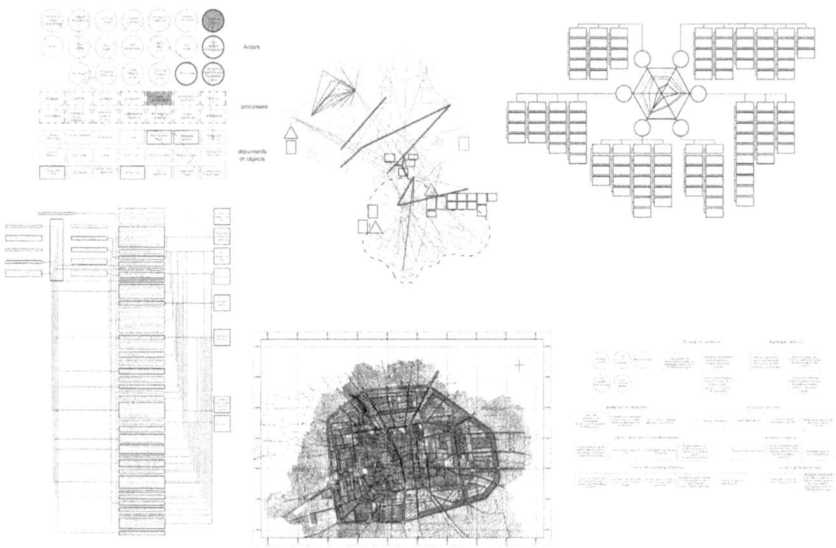

Figure 6.2 Research and analysis tools

A System of Protocols: The emerging scene of iNGOs as donors and employers after 2011 has resulted in competition between local actors to engage with external donors. Due to the limited resources of the project, a commitment to work on a voluntary basis with no profit was established by team members in Azaz. It was based on the evolving trust of shared understanding to the technical values that all team members agreed upon.

The team established trust with different local actors. Based on this and taking into consideration the experimental nature of the project, the CDF tea, established a declaration of trust with the local council to test the possibility of facilitating its mission.

This declaration states that the team is the advisor for the local council for its strategic and spatial planning, while it ceases to engage in any profit-making. The declaration of trust values sincerity in the evaluation of plans upon a clear vision of an open contemporary city with self-awareness, dependence on local resources primarily for the interest of the city's people, and secondarily with its local council. The declaration included the application of urban management models as a culture and site-sensible tool paying attention to HLP justice, land prices, social inclusive, and environment. The team committed to knowledge exchange with local and international cooperation actors in appliance with the core values and practice of sustainability and resilience in the management of opportunities, local resources, and risks. The team committed not to receive any material benefits or form any conflicts of interests with the local council or other local or external actors, prioritizing the values of knowledge and mutual trust.

By doing so, the team released itself from the burdens of bureaucratic engagements and aimed at creating a consensus through similar written protocols between actors based on trust and commitment. This allowed the team to stay neutral, having its presence between actors as a generator of knowledge and networking, without making any monetary benefits. It is not yet clear where this would lead, but it has facilitated the team's engagement on a local level, while at the same time failing to engage with other important actors or potential partners in Azaz.

Spatial/Multidisciplinary Analysis and Planning: In the case of Azaz, spatial management has failed to cope with the pre-2011 war conditions and solve the housing crisis that the city currently faces. Understanding the urban development process as a mere spatial issue that ends with a finalized masterplan has failed. Masterplans are often static products that fail at implementation due to bureaucratic, legal complexities, and behavioral issues, and uncontrolled construction.

Analyzing these problems and risks together with the assessment of the 2008 masterplan and its application has led to issues related to the stagnation of the urban management system. Stagnation in land dealings prevailing since 2011 because of general economic stagnation. In addition to this, this condition is attributed in Azaz to the lack of trust in the legal and institutional frame and is manifested in informal land ownerships and inflation of land values. Due to the suspension of the HLP Registry in 2012 and 2019, land dealings were conducted through informal contracts circulated between people without being verified through legal procedures. Inflation of land value is attributed to safety conditions and general belief in land as an asset to preserve values in crisis, transforming other properties and holdings into land and contributing to the inflation of its values. In 2017, it resulted in a high land value of US$70–100 per square meter in average within the areas inside the masterplan zone.

Urban management systems are problematic in Azaz due to the failure of its 2008 masterplan, manifested in urban sprawl and informal construction. The sprawl is caused by the absence of local administration, leaving construction and infrastructure provision based on the priorities and financial capacities the sole responsibility of landowners and investors (Figure 6.3).

Informal construction did not follow the masterplan in place. Less than 30 percent of the buildings in Azaz informally acquired a building permit. Some constructions exceeded the density limits for monetary profit, while

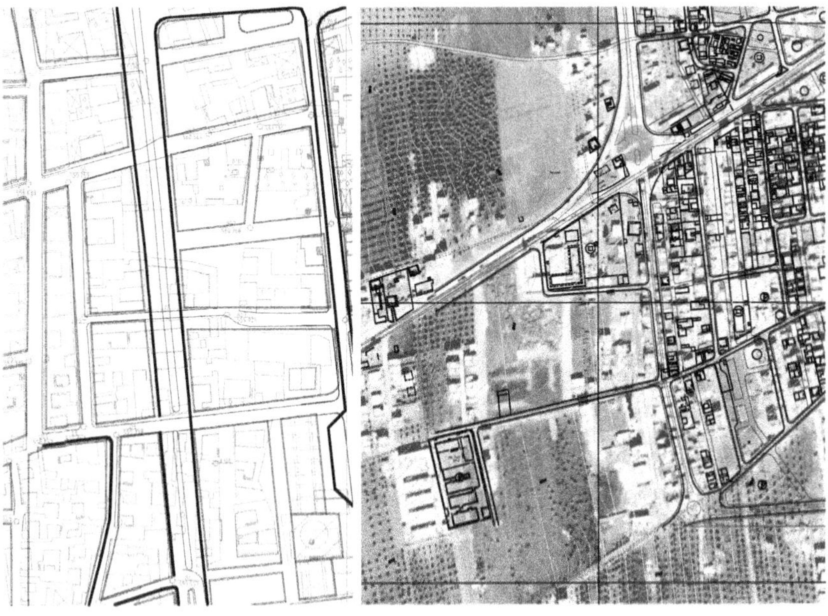

Figure 6.3 Informal building and urban sprawl - houses built on planned streets and houses built outside the zone of 2008 masterplan

an estimate of 15 percent of new buildings[1] follow the legal building typology framework.[2] Furthermore, there were buildings constructed in the footprints of public spaces. This is partially the result of the non-satisfactory, non-timely, and often unfair land appropriation. Additionally, the absence of an institutional and legal frame for appropriation since 2011 accelerated the construction process before land appropriation is completely suspended.

The Land Registry Administration re-opened again in Azaz in February 2019 with a functioning temporary Land Registry that increased the trust in land dealings. This, together with the influx of IDPs, has led the region to a slight revival of the construction market, which meant again another wave of building infraction and increase in prices.

In the face of increasing prices and building infractions, spatial solutions on the masterplan level will only partly function. An expansion in the masterplan will temporarily reduce land prices but will not solve problems on the long term. In fact, land prices outside of the masterplan area have increase. Experts and urban planners in the city believe that this expansion should be made, outskirts should emerge, and satellite cities should be constructed to provide more land for construction and decrease the price of lands in the city. It is also highly believed constructing high-rises will solve the current crisis; new small apartment typologies with non-spatial will be discussed more in the urban governance section.

If we discuss the current practices and perceived solutions, a typical physical expansion will reduce inflation in land values and stagnation, since the expansion will make more land available, and thus reduce prices and revive land dealings. Enthusiasts and local investors are likely to succeed in generating an active urban expansion, especially in the first period, which may also allow large project developments to succeed. Nevertheless, there is no guarantee to the continuity of the process.

Apartments in new buildings will likely be sold, but the cost may be much higher than the average purchase capacity of the middle-class family. The process may face stagnation again before housing those in need. Another risk is the return to inflation if richer communities from around the northwest region return, resulting in another form of urban segregation and doubling the inflation of land values in Azaz and restricting affordable housing for the middle class. These are all factors that regenerate conflict. At the same time, such solutions are not viable without revising pre-2011 and current urban problems related to implementation of legislations.

Therefore, to guarantee the success of the spatial planning and future expansions, a development plan should be coordinated within other sectors. A flexible plan with spatial and non-spatial aspects must be placed and revised periodically in relation to the other sectors within the CDF.

Modifying land use is not possible on a large scale as is the case in classic masterplans in Syria. This is due to the complexity and length of the process of modifying land uses in pre-2011 condition, and the absence of legal and institutional frames for it in Azaz after 2011. Thus, spatial aspects must set

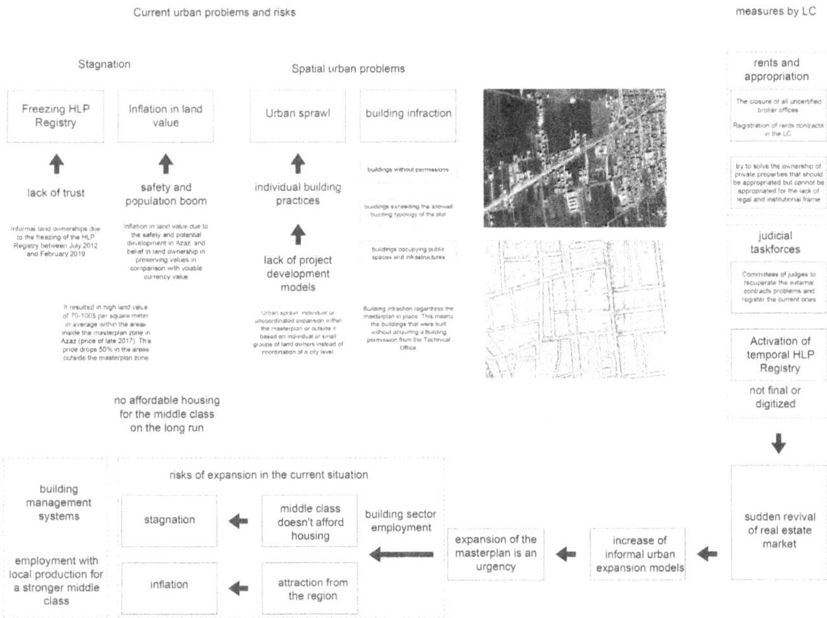

Figure 6.4 Azaz urban challenges

in strategic projects on the city level to activate socioeconomic centers. This is accompanied by specifying locations of public services and infrastructures within the masterplan zone.

Middle-class communities, as well vulnerable, need to be involved in the planning process to generate value based on investment in local resources and production and access to affordable housing. Thus, it must be accompanied by the revision and development of legal and implementation frames ensuring that the same spatial urban problems do not occur again while the recovery process is ongoing. In this case, all the CDF sectors are involved in the wider process of masterplanning (Figure 6.4).

Local resources and housing

Housing, local resources, and the local economy are connected, nevertheless, housing, and local resources are grouped under a single sector because of the availability of local resources for construction.

Housing challenges

After 9 years of war, destruction and temporary habitation still impact the housing sector in Syria. According to the 2010 Syrian census data, 4.12 million conventional dwellings existed across Syria (The World Bank, 2017).

Until 2017, at least 650,000 housing units were destroyed in addition to another 1 million damaged units. These numbers reflect a total loss of 30 percent of the value of the housing stock until 2017, a percentage accumulated for 18 years (Urban Housing and Question of Property Rights in Syria, 2017). The northwest region alone suffered the destruction of almost 20 percent of its housing stock (ACU, 2017).

Still, there is a need for housing for IDPs. In 2019, an estimate of 75,000 IDPs lived in Azaz and around 90,000 in neighboring refugee camps (URC, 2018a). Jarablus and Al-Bab in the same region house around the same number of IDPs, while around half a million live in refugee camps or in tents in the Idlib region. Households treat their shelters individually according to their needs and capacities. They expand their tents or repair them with cement or concrete according to climate conditions and others dig informal sanitation holes.

Inflation of land prices in property markets is a massive challenge with land speculators active in buying land around the city using international sources of money, further contributing to the crisis. Additionally, communities are unable to afford the construction process itself, regardless of land prices, because of imported construction materials and fuel. The abandonment of the countryside since the mid-last century was accompanied by the abandonment of sustainable and affordable traditional construction practices. Abandoning local construction methods and materials caused the reliance on lower standard, imported steel and cement products. This transformation has been institutionalized and imposed by the syndicate of engineers. There is an ambiguity in the construction permitting procedure and the relevant laws when it comes to construction with local materials, which is another major challenge.

Alternative housing models

Alternative modes of housing are thus emerging to meet local needs, depending on the resources within communities themselves and facing some of the above economic, social, and environmental challenges and risks.

Tackling Speculation on Land Values and Land Ownership: The Sakan organization is willing to initiate its work in Azaz in response to land speculation and the need for affordable housing. It aims at locking housing units out of the market, by challenging the existing ownership frames so that they cannot be sold in a speculative real estate market but only transacted through the Sakan organization among families in need. It is still seeking a legal model to register property ownership in a way that avoids speculation. It tackles liquidity by seeding money from the middle class to generate revenues to build new affordable housing that is accessible to new families. It requires capacity building between communities and local authorities and spreading public learning to build a culture that believes in housing as a right, and not as a speculative commodity.

The legal solution for its model is still under development, and its work needs institutionalization, which cannot happen without building partnerships and trust with local institutions for proper implementation. It also needs to build a collective and renegotiable intent among stakeholders including officials and private sector actors.

Urban Development in the Lack of Liquidity: In the past, local developers managed to sell land before construction. This is unlikely to happen now that the recent lack of liquidity is preventing them from keeping up with the demands of the market. Nevertheless, developers have shown some resilience in different ways to tackle the lack of liquidity through alternative models of urban management.

An alternative model of financial-spatial management is based on negotiating the value and layout of land between owners and developers. An example is a development comprising 23 residential buildings, two of which are already built whereas the rest are planned. To tackle the challenge of lack of liquidity, the model negotiated land value with the landowner to split the land and decrease the value of the plots that are not on the main street by 40 percent. This way they aim at accelerating the construction process on the affordable plots to sell for cheaper prices and create liquidity for the next phases of the development. To tackle the lack of liquidity, this model decreases the land value itself, through modifying the spatial layout of the project.

In 2019, a financial management model emerged as an adaption of market-based solutions in Turkey after the collapse of its currency in 2018. This model was adapted in Azaz to tackle the lack of liquidity in the hands of urban developers and the need for employment by construction services providers.

It is based on new forms of informal deals held between the urban developer[3] and the construction services providers such as aluminum workers, carpenters, and plumbers. The deal comprises selling contracts and receipts between parties to guarantee the exchange of values based on certain criteria. The first installment of the cost of the service is paid to the service provider in cash by the urban developer. The rest of the service cost is not paid, but the urban developer gives a receipt with the service amount stating that it is a payment from the service provider to the developer as an installment for buying an apartment from the urban developer. Service providers buy from urban developers through a regular contract. The Sarai building in Azaz is an example of this model. The carpenter received only 50 percent of the cost of the wood works. A receipt of the rest of the cost of $7500 was received from the urban developer as a first installment in buying an apartment built by the same urban developer. The cycle was closed when the carpenter paid the second and final installment to the developer through implementing the wood works in another project.

This model not only partially allowed urban developers to overcome the lack of liquidity, but also encouraged a form of local credits unlike the classical exchange of values through money in the construction process.

An important detail in this example is that all the parties engage in this alternative process except for one, the traders of cement and steel since they are providers of an external commodity, and are a part of an external value chain, unlike all the rest who provide their own local resources.

Even though construction services providers are obliged to the deal by the lack of liquidity, a fair legal and institutional development of the process would allow the emergence of contracts that measure the value and increase the trust in the market, and generate more work opportunities for them, and additional housing stock. Institutionalizing this process and monitoring it might result in a framework of local value chains that is generated and operated locally, potentially in forms of urban development cooperatives, which include urban developers, owners, services providers, and customers.

Housing in the future

If these models find their way as established institutional and legal frames, they will likely spread further and get adapted by other practitioners. People with limited income and skillful services providers are local resources in the process, and more families can be trained to contribute to the process. Trained families might be able to exchange a housing unit with an alternative ownership model for their contribution in skills and time. Such an alternative ownership will become a practical solution for housing rather than a cultural challenge and, thus, succeed reducing speculation on land values. The feasibility of the process on the longer term needs to consider two additional factors: The energy performance of the house and the need to reduce external supply chains as much as possible depending on locally sourced natural construction materials. Construction with local materials reduces the initial cost, and improves energy and cost performance across the entire life cycle of the building since it uses less steel and cement, less fuel for transport, and less fuel and water in the initial production.

The governance of this process and its regulations needs to be negotiated on the levels of syndicates and local administrations. Moreover, other issues that would slightly increase construction cost but decrease energy performance and building running cost during life cycle are feasible. Solar energy and insulation materials must also be used by involving other actors including ones from the private sector. Comparative models for testing performance can be researched and then evidence-based knowledge encouraging legislations must be put in place. These requirements are resolved on a local community level and in the process, on a strategic level to involve more people.

Local economy and local value chains

Azaz witnessed high economic activity compared to other cities in the region; however, it depends on external factors such as trade roots, fuel prices, and the influx of IDPs. Even though the local heritage of production and processing is based on agriculture, imports from Turkey are directed towards the northwest region, 70 percent of which include food and grains (ACU, 2017). Other major imports include clothes and construction materials (ACU, 2017), even though the region is rich with local construction materials and workforce. Simultaneously, a very low trade movement is witnessed in the other direction or between cities in the region. Furthermore, for security reasons, entrepreneurs and traders on a local level hold a massive amount of capital and refrain from investing inside the area, except for the real estate sector. Much of the capacities have been exhausted by the dismantlement of many structures and moving them to Turkey. Renewal of factories, workshops, and often small-to-medium enterprises, including processing facilities, is complicated by the lack of produced rough materials and adequate infrastructure.

The city has strong pre-2011 socioeconomic centers made of hundreds of businesses. The increase in informal street shops negatively impacts markets that are more established, creating competition over skilled workers. Many small projects and family businesses are hit hard by external factors including sanctions, the devaluation of the Turkish and Syrian currencies, the increase in fuel prices, and the establishment of other trade routes.

Recovery requires awareness of these challenges and of the value that resources generate. IDPs constitute a reliable labor force; however, increase the demand for affordable housing. Additionally, agriculture and natural resources such as stone are available in quarries across the northwest region. Rough materials and their processing in agriculture and housing are essential for socioeconomic activities needed for recovery. Therefore, the city must generate value in agriculture and byproducts, and in housing, depending on the local resources and local production economy. All these processes, whether in agriculture or housing, need to be negotiated on a community scale based on local value chains to avoid disruptions from external factors. It is necessary that this process is evidence-based and accessible to communities. Such measures will provide a margin of predictability of market behavior to plan for local development. They require considering local traders, military and family affiliations, and their current and future roles in the local economy, because they are the main holders of wealth and power.

Agriculture

It is known that north of Aleppo has traditionally depended on agriculture as a pillar of its economy, including wheat, legumes, barely, and potatoes (ACU, 2017). Azaz is known to produce cherries and olives as well as byproducts such as olive oil with many factories scattered in its northern neighborhoods. However, agriculture has drastically deteriorated after 2011, and again in 2015 due to scarce rainfall and lack of pesticides, fertilizers, and water, as well as the rising cost of fuel for transport and irrigation (ACU, 2017). Nevertheless, agricultural lands and living resources such as cattle and poultry farms in the north Aleppo region still form a large asset to the economy, keeping agriculture as one of the main sectors generating value and employment (URC, 2018a).

Based on the nature of imports, food security is a pressing issue in the region, including Azaz. Main imports from Turkey include manufactured food and agricultural products (Syria Independent Monitoring (SIM), 2018). Until today, Azaz imports tomatoes from Turkey in the winter and is still in need of grain, wheat, and olive products, and subsidized bread. At the same time, agricultural lands are spread around the region. Large areas are not cultivated because of the lack of management, while crops in small and mid-size farms are supported by iNGOs. Furthermore, a massive number of families within the camps in Azaz that host around 85000 IDPs that depend on donated food baskets while engaging in arbitrary, seasonal work in agricultural fields. Accordingly, there is an urgent need to integrate both, the original residents and IDPs, in agricultural production in underutilized land and its byproducts instead of solely depending on the donation of food baskets. The impact of such a process prevents large communities from being deprived of meaningful participation in socioeconomic life. It mitigates segregation in the long run. Furthermore, it allows the original residents and IDPs to produce value in exchange for housing that they urgently need and contribute to the economic vitality of the city.

The management of such a process must consider the number of people dependent on food baskets and their working age. It can be a process of gradual transformation to being active members in economic life. Such a process needs to tackle the current agricultural land investment models to allow more accessibility to small farmers, who are not owners. It also needs to be aware of the production of byproducts and allow chains of traders to emerge from the local community. It substantially challenges the modern culture of practicing agriculture based on the overuse of harmful pesticides and fertilizers, and some irrigation systems among others that harmed the entire ecosystem including soil and underground water. Like the housing issue, agricultural engineers are promoting and practicing such non-sustainable practices with pinpoints structural shortcomings in the governance system. Therefore, it is important to establish technical knowledge among practitioners and more importantly, communities of farmers, to improve the long-term environmental and economic impacts of such practices on the city.

Being aware of the estimated amounts of consumption of energy, water, and agricultural inputs, and of the value of the production outputs is important to optimize the process within the city itself and to harmonize the exchanged commodities with other cities in the region and beyond. Furthermore, such processes much consider the risks of potential drought which caused around 1.5 million people from the rural farming communities in Syria and Iraq to move to urban areas between 1998 and 2009 (The World Bank, 2017).

Economy of housing

There is no doubt that the main challenge facing housing in Azaz is the high land value, which is still increasing. As a result, investors tend to increase the revenue by building the maximum allowed number of floors. Therefore, construction depends on concrete as opposed to locally sourced stone since it does not structurally support more than three to four floors. Stone construction requires less imported materials and allows the alternative housing models to deal with the lack of liquidity. Carpenters as construction service providers charge an equivalent of up to 30 percent of the cost of the wood works as net profit since there is value of their handwork, whereas steel and cement traders make less than 5 percent profit[4] as they provide no service, but rather a pure commodity. This makes steel and cement traders less amenable to engage in alternative housing models. Furthermore, the sole dependence on steel and cement imports leaves the whole process of construction again subject to the volatility of fuel prices and security, as well as political instability.

Thus, constraining construction height to only three floors would push developers and investors to use local materials to bring down the prices of the land and of the housing units because of two major reasons. Firstly, it reduces the value of land as a commodity for speculators by reducing the expected revenues. Secondly, it encourages the use of stone construction from Aleppo, again reducing the total prices of houses and replacing cement and steel imports with the involvement of stone producers from local quarries. The economic impact will be seen in reducing energy consumption and increasing employment in local stone production and construction sites.

Networks of investors and engineers often adapt to and benefit from the status quo, while other community actors lack an understanding of it. Therefore, such practices need to be evidence-based and well researched and accessible by communities, so they can negotiate such construction practices.

HLP registry

Space is one of the most contested commodities in Syria. Many obstacles prevent communities from accessing affordable housing such as speculation and inflation in land prices due to bureaucratic and institutional aspects. For example, the price of a property increases many times when a signature in the city administration is given for changing the land use of a property, or when a property is included in the masterplan zone. However, basic concerns as to who authorizes such a decision and what authorities are involved remain unresolved for communities facing archaic and complex systems and legal and institutional frames. This is manipulated by local traders, military forces, and family affiliations. Therefore, the CDF analyzed the complexity of the HLP system and aimed at making it more understandable to communities so that they challenge the way in which it is practiced on the local level.

While these challenges are faced all over Syria, and the land ownership is a direct indicator on holding power, the HLP Registry was given a high value as a trusted document. In July 2012, the court of Azaz nominated a group of lawyers to seal the hard copy of the HLP Registry[5] due to the absence of the central government. The HLP Registry Documents were scanned and copied on flash drives archived in the seat of the court while another soft copy was used in the HLP Office[6] in the Azaz local council. This meant that legal notices were not added to the hard copy of the HLP Registry sheets when ownership transfers happen; this is typically a legal requirement. Lawyers process property transfers through non-registered deeds (external contracts) made by lawyers, without registering the new ownership status in the HLP Registry. This resulted in problems such as duplications of market transactions on the same property while other pending ownership transfer operations such as settlement of succession accumulated.

Thus, the value of the HLP Registry as the trusted protocol on land ownership is clear, and it is understandable how its absence along the years contributed to lack of trust and therefore recession in land dealings. In February 2019, a Temporary HLP Registry was created and was managed by a Turkish administrator ensuring its direct linkage to the HLP Registry in the Turkish city of Kilis. As a result, the city faced other challenges such as registering and legalizing external contracts, including the validation of documents and the resolution of accumulating disputes, in addition to registering new land ownership transfers. This overwhelmed the city's institutions.

Technical Analysis: The visual analysis allows us to track different processes and how they are conducted and often interrelated. The following example shows an appropriation of a part of a private property to build a school before 2011, including a reimbursement to private owners as follows: At the beginning, the city council meets to propose the appropriation and then sends it to the branch of Albaath party in the city and the Ministry of Local Administration, which proposes it to the ministry council. The ministry sends a statement to the president of the city council with the Appropriation Confirmation. Then the city council forms an Estimation Committee to estimate the price of the square meter to reimburse. If the owners later object to the price, it can be raised once in accordance with the land price limits and other indicators. Generally, the government does not reimburse more than a third of the actual land price. The CAD files are sent from the technical office to the survey administration in Aleppo to identify the plots that will be appropriated as well as the plot outline of the new project. At this point, the Survey Committee is formed to demarcate the location of the school, adjust the CAD file if necessary, and give the dimensions

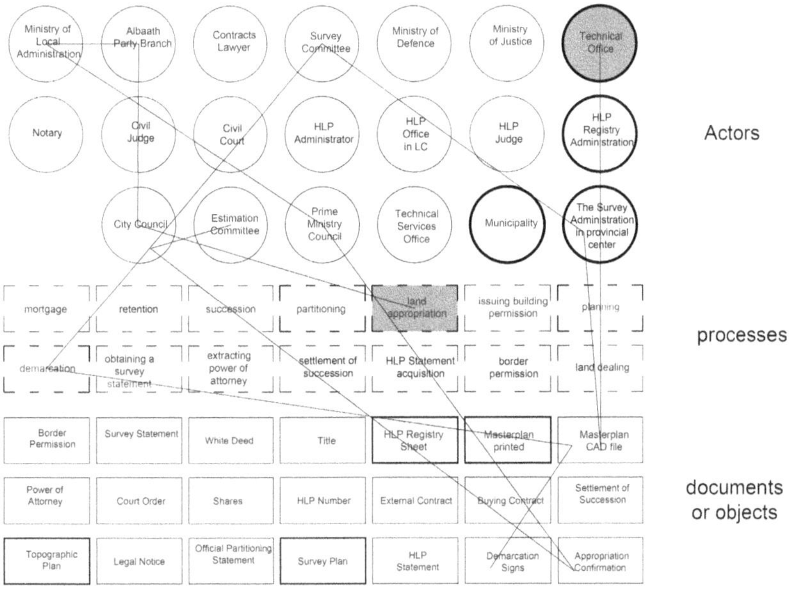

Figure 6.5 HLP complexity analysis

of the school and demarcate its location on the site. However, this process is lengthy and cannot be processed without written documents (Figure 6.5).

The analysis shows that urban development processes after 2011 were also complex and many were legally impossible due institutional segregation from Aleppo. The registries and local maps do not correspond to each other with an approximate 8 meters error in the spatial layout between the topographic plan, the survey plan, and the masterplan. However, the 2019 re-launch of the Temporary HLP Registry in Azaz did not address institutional or legal reforms, nor did it cease the opportunity to develop its information management systems (registries and maps). Therefore, the city again faces the same complexities.

Based on this, there is a need to simplify and automate these procedures to become more accessible to the community. Land plots need to be spatially documented and linked to the HLP Registry sheet[7] instead of having separate paper systems. Achieving this and linking it with the Civil Registry will optimize the flow of information between citizens and local authorities, making complex processes such as land appropriation, partitioning, and succession faster and more efficient and more inclusive, transparent, and just project.[8]

Land Values: Due to the consistent increase in the numbers of incoming IDPs from other regions in Syria and the lack of housing, land values in Azaz were on a steady increase and peaked in early 2017. Land values started to decrease again after the expulsion of the Islamic State from Al-Bab in

January 2017, increasing the availability of housing in the region. At the beginning of 2018, the Turkish military operation on Afrin expelled residents out of the region, further increasing the availability of empty houses,[9] which further decrease land values in Azaz. This was accompanied by the US economic sanctions on Turkey influencing most of the sectors in Azaz, including land dealings and values. In 2019, incoming IDPs due to the military operation of the Syrian government against the Idlib region drastically increased land values again. These external issues have changed land prices whereas many internal issues contributed drastically to the increase of land value. These are the issues that the CDF addresses.

Since July 2019, local speculators from Azaz and IDPs from Daraa and Ghuota bought affordable pieces of land 1 km to 1.5 kilometer away from Azaz, mainly towards Marea and Souran in the east. They believed that safe zones will be established, and international reconstruction funds will come soon, contributing to the further increase of prices. However, speculators who have more accessibility to valid information about main projects generated profit, causing further inflation in land values. The exclusivity of such information reduced access to housing to the wider community.

Between mid-2017 and mid-2019, pieces of lost around 50 percent of their values whereas their prices skyrocketed since mid-2019. Yet, there is no dynamic or tool that allows us to track, observe, and track land prices. The tool for observing land values aims to create a comparison between land values at different dates.[10]

It is currently not possible to conduct any form of control because of the links between local authorities and local investors. But, controlling land values should be managed on a city level. Land prices in Azaz within the masterplan zone range between US$500 per square meter in areas close to the center and US$20 per square meter near Neyara Way and South East (Figure 6.6).

On the one hand, the location on a main street and the availability of infrastructural services such as water and electricity networks often increase the value of land and property. On the other hand, land value decrease in areas that witness a poor socioeconomic condition such as Alasianeh and Tabbaleh in Azaz. Land prices in these areas typically range between US$50 and US$80 per square meter due to the narrow streets and limited infrastructural and waste management services causing more segregation. Land prices and socioeconomic segregation are interlinked.

Civil registry

The population of north Aleppo region was estimated at 400,000 before 2011, and reached 700,000 by the end of 2017, with around 200,000 IDPs scattered throughout the region and near Azaz (ACU, 2017). IDP camps near Azaz host around 90000, while the city's population itself currently amounts to more than 150,000. In addition, around half of the population

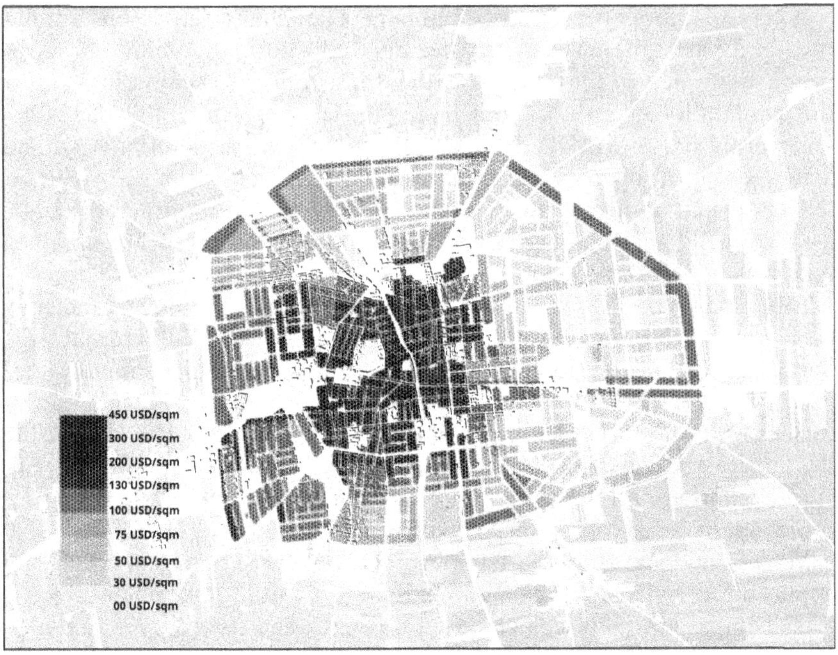

Figure 6.6 Land values in Azaz, July 2019

of the city fled to Kilis, Turkey, and Europe. Thus, around 50 percent of the population of Azaz was displaced since 2011. The 50 percent of its current population arrived from other regions in Syria seeking immediate housing.

In October 2018, the civil registry in Azaz launched new identification documents within a new civil registry system that is internally connected to the new administrative areas in the region. The aim was to avoid the overlap of registers.[11] The whole process, however, did not go beyond the gathering of data and listing the names of current residents. Cities need an efficient information system, which intersects with many other dimensions including socioeconomic data. Population growth, working age, expertise, and employment rates would explain incentives to enhance agriculture, businesses, and tax collection. Population analysis can also be connected to spatial aspects by integrating it into the HLP system and the social aspect to get an in-depth understanding of how urban population segregation occurs or develops.

Local communities and social aspect

While the project first focused on spatial planning for urban development, community dynamics also became important. These dynamics include local culture and heritage, inclusion of IDPs and other groups, as well as the civil society and its role are major factors in the development process.

After 2011, more opportunities became available for community initiatives and participation in Azaz. Formal and informal community movements attempted to fill the political and institutional vacuum in terms of the availability of services and humanitarian aid, while still depending on external funds[12] (IMPACT, 2019). Since 2017, the scene of international actors and local civil society initiatives has changed as the permission of iNGOs became issued by Turkish cities. Effective community actors and the local council formed an active network of influence. This includes traders, military, and prominent families.

Even though there is an international role in local development, the negotiation of wider communities for their rights is primarily an internal issue. Sustainable future development requires the involvement of communities as its main driver. Community dynamics play a primary role in balancing other dimensions: The manipulation and market dynamics in socioeconomic life, the understanding and applying law, and the practices of agriculture and construction are all are based on established community values that are rooted in the behavior of everyday life.

Values and local culture are important assets since they generate locally applicable practices. An example is how IDPs were able to integrate in Azaz compared to other cities since locals are accustomed to communication with people from other areas in Syria. This is due to Azaz's role as a major border city. Another example is how family and tribal structures play a major role in power balances. After the electoral cycle of the former local council ended in late 2017, civil actors failed to instigate change through local protests while families managed to put pressure on the local council and succeeded to restructure it according to their demands.

However, these local cultures in Syria were often appropriated or seen irrelevant before 2011, with a focus on national identity, strong state, and often modernism. Local cultures were often seen as dated, preventing communities from reinventing their local culture and heritage and benefiting from it in urban development.

Another major issue is the social and cultural integration of IDPs. At first, most of the IDPs were concentrated in Azaz and its neighboring camps. Other camps opened in Jarablus, Al-Bab, and Al Raai when Azaz reached its maximum hosting capacity and has not declined yet. IDPs came from the Aleppo region and other areas in Syria such as Deir ez Zor, Homs, Daraa, and Ghouta. This questions the future balance and integration of the community of IDPs in Azaz. Communities in Azaz are likely to face an extension of the current socioeconomic segregation if not remedied in its early stages through fair access to housing and employment opportunities to all people. It is evident that community divisions increase, and tension escalate every time many IDPs arrive.

Marginalization in Azaz is manifested in the dominance of men aged between 40 and 60. Women are generally excluded from most aspects of active life, while some initiatives represent how women managed to be proactive despite resistance from local dynamics. Apart from projects by iNGOs or local NGOs dedicated to women, opportunities for their participation are limited. There is an active discouragement of young people and women to play meaningful roles in Azaz due to the local system of values, which takes its legitimacy from external and local power imbalances. Women, young people, and IDPs are prime human capital and agents of change; therefore, no real recovery can take place if they are not actively included in decision-making.

Local governance

After 2011, the main challenge was providing adequate local governance after a strong grip from the central government. With the cut in administration and subsidies from the central government, local councils depended on external governmental or non-governmental actors for basic services provision.

The provision of services gave the local council legitimacy to acquire new roles and gradually replace the role of the central government. For example, when Azaz was a district center in the Aleppo Governorate before 2011, some developments required communication with the Directorate of Survey in Aleppo. Faced with institutional and legal barriers, the local council created its own the Office of Technical Services. The changes that took place were sometimes planned and officially taken, such as establishing the cultural office and the public relations office, or unofficially through the downgrade of specific offices into sub-offices. Thus, the local council demonstrated a great deal of agility in coping with new conditions and swiftly reforming its institutional structures. However, familial and personal interests continue to influence the work of the local council and its technical performance. As such, the performance of the local council and its power of execution is an important indicator of the extent to which the city can perform urban governance roles that are necessary for its recovery and development (Figure 6.7).

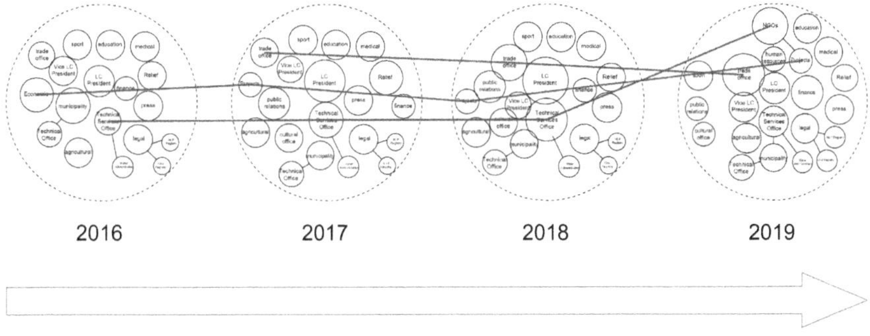

Figure 6.7 Shifting structure and power in the local council in Azaz

Governance of housing provision

Experts in Azaz believe that urban housing provision is the responsibility of the state and should be set in public legislation. While the private sector succeeded in providing a limited number of affordable housing units to people through payment installments, the local government should provide similar opportunities. Furthermore, the Legal Building Typology Framework needs to allow for the construction of smaller apartments and high-rise buildings to cope with the additional needs for housing units. However, the local council is currently not able to issue and manage planning legislation because of its limited institutional capacities. As such, the local council needs capacity building through community participation. When it can execute legislations, the adoption of project management tools and templates can organize the interaction between the involved actors to ensure that the process is efficient, affordable, and just. Additionally, unions of architects and public administrators should be equipped with evidence-based technical knowledge that reduces energy and cost to users in housing units. Architects and engineers need to question their role and produce professional protocols that engage people in the construction process through local administrations and syndicates.

Environment

Additional regulations will also optimize the environmental performance of buildings. They will encourage the use of solar energy and building insulation materials, both of which incur extra construction costs but reduce long-term operational costs that can be subsidized by external actors through funds or loans. To successfully implement such legislations, the public administration will need to reduce the fees of building and renovation permits and to provide loans and issue payment installments. Furthermore, waste management in Azaz can be done on a household level. This can also be done by the municipality through halting garbage collection, if users do not comply with waste-sorting requirements. It has been suggested that experiments can be done on some neighborhoods, and then expanding the practice if it succeeds. At the same time, the private sector, through investing in recycling materials and waste management, can contribute to the process on a household level.

The City Development Frame (CDF) in three modes

The CDF tackles the six sectors simultaneously to guarantee a multidimensional understanding of issues. Input is inserted into the two main tools that are substantially interlinked and work together as the CDF.

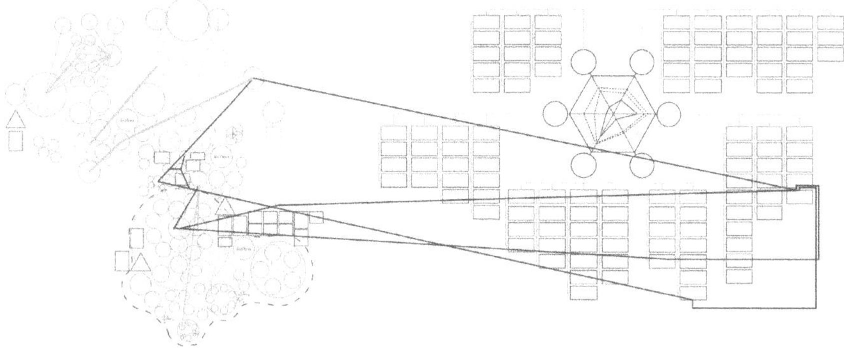

Figure 6.8 Template and elements map work together

The first one is the Template Tool, and the second one is the Elements Map tool. The content of each tool differs between cities (Figure 6.8).

The function of the Template Tool is to categorize content (problems) into sectors (pillars) and primary and secondary issues (slots), to measure them, and then highlight processes (sequences) across them. The Elements Map is the main analysis tool. It focuses on interaction between actors, protocols, and events that feed into the template tool (Figure 6.9).

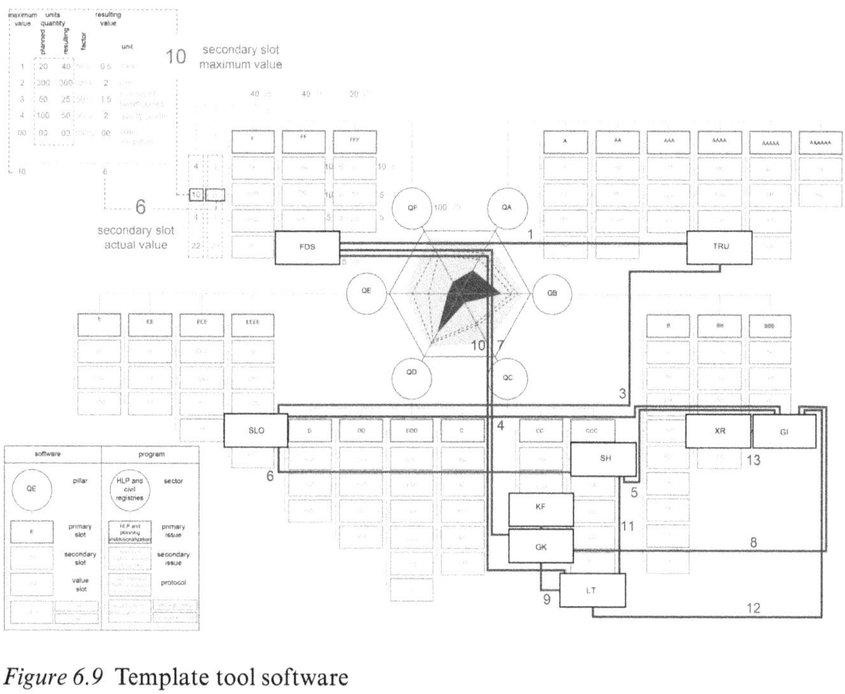

Figure 6.9 Template tool software

The protocol is the smallest unit in the template. Its content can be used for its meaning and can be quantified into numbers to set indicators for the software. Examples of protocols include:

- Values: A document of agreed sets of technical values between actors
- Laws or legislations in place
- Official agreements
- Contracts relevant to public issues
- Reports, including assessments and evaluations
- Written plans or future procedures, such as guidance on how to produce or manage a digital map for the city
- Technical tools or products such as a digital map or internal civil registry software
- Work frame tools, for example, applications for construction permits to the municipality

The CDF works in three modes; the first one is the multidisciplinary interlink between different expertise. The second mode is a governance and management tool for the city. The third mode is the community-based information management mode.

The CDF as a multi-dimensional approach to the city

While the six sectors in the Template Tool deeply interrelate, they need to evolve together keeping the corners of the hexagon balanced. The primary and secondary issues among them need to work together in parallel because processes (sequences) that are tracked within the template (plans, solutions, etc.) are cross sectorial. If there is a deficit within issues, then cross-sectorial processes that pass through them will be cut. Through its Elements Map tool, the CDF depicts actors from different sectors, protocols, and relevant events. It alters synergies between them based on the complementary roles they can play within the processes (Figures 6.10 and 6.11).

In this way, the CDF allows us to propose sequences for a holistic approach towards housing and masterplanning as examples. If we set the distant goal of initiating detailed housing project, we can follow a multi-dimensional approach and track parallel processes, among them are the following thirteen:

- (1) Resources need to be directed to groups that are directly harmed or threatened, and to groups that are interested and able to participate in the public debate on HLP issues and mobilize communities to defend their rights, and advocate on the local council level.
- (2) Tolerance to community involvement and their right to monitor in HLP and construction issues needs to be established on the local council level. Establishing a dynamic for entry points on different scales allows for a multi-stakeholder approach in the city, which involves local communities in public issues along other actors through the local council.

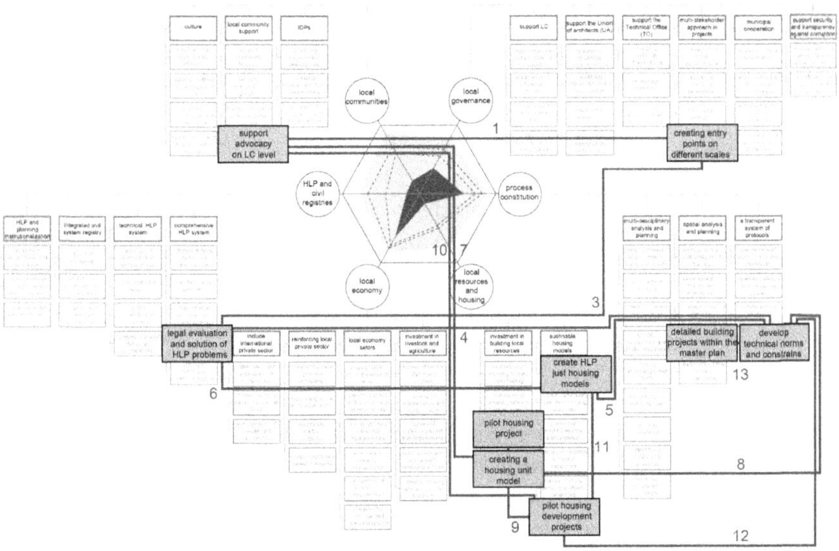

Figure 6.10 Example of the functionality of the CDF – Template Tool

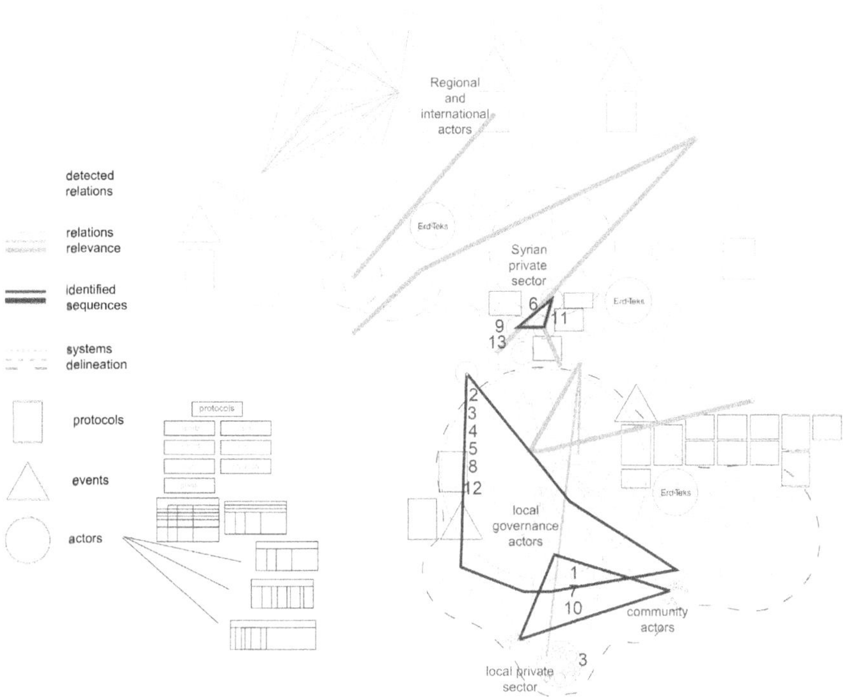

Figure 6.11 Example of the functionality of the CDF – Elements Map Tool

- (2) This approach involves communities in the debate after discussions with legal development experts, HLP judges, and the civil court in making a legal evaluation and propose solutions to the HLP problems.
- (3) Based on protocols on the legal evaluation and solutions of HLP problems, legal experts develop technical norms and constraints. (4) At the same time, legal evaluation and solutions to HLP problems feed into creating inclusive housing models; (5) they also contribute to developing technical norms and constraints when formulated jointly by legal and urban experts. The HLP just housing models should correspond to the urgent needs for housing and the current conditions of high land prices.
- (6) Other sequences feed into creating a housing unit model, which needs first an understanding of the technical properties of the local stone such as load bearing and thermal performance. This can be conducted through the union of engineers and other experts in construction technology. After then a pilot housing project is tested and building environmental and economic performance is measured.
- (7) This model is to be tested by the users from the local community. (8) Then through cooperation between legal experts and construction experts, it must feed directly into developing technical norms and constraints. (9) At the same time, it should feed into larger pilot housing development projects which happen on a scale of a cluster. These pilot housing development projects get also fed by users from local communities (11) and by the created HLP just housing models.
- (12) When the pilot housing development projects are tested their protocols feed into developing technical norms and constraints, (13) which itself feeds to the construction permissions that are issued by the Technical Office in the local council to regulate the ongoing construction projects according to the master plan.

By detecting risks, resources, problems, and needs, the CDF presents solutions and opportunities to mobilize hibernating resources; initially to self-recover later to develop and generate value. The Template Tool is dynamic, there is no one way, or a fixed plan. It helps connect issues from different sectors and prioritizes them. As it is knowledge based, it allows pre-war problems to appear.

The CDF as an alternative governance tool (informed governance programming)

For governance on a city level, the CDF needs a margin of freedom in the system to operate. It invests in the institutional and legal void by operating as a supportive tool rather than a replacement to any entity or constitution. It can be used for the coordination and management of projects, information, and knowledge between actors within an open knowledge and ownership system. Engaged actors can generate frameworks and tools that are missing because of the cuts in institutional chains, such as technical norms, regulations, and urban management systems.

Whether for a single round or as an adapted tool on a long term, the CDF can be used for research, planning, implementation, and monitoring of the process. It is substantially about knowledge-based and agile planning and decision-making. The frame can be used as an optimizer to measure a baseline and raise indicators of the current situation in a city through its six sectors to alter the self-recovery and development capacity. It can be managed by both experts and practitioners in local communities and the local administration. Accordingly, it aims at creating a consensus on sets of values and frameworks (as protocols) between actors. It must remain a dynamic and not a static tool, open to participants and not exclusive, transparent and not secret, and accessible by the local community and all those who engage in it upon their agreement on the interest of the local community and justice in the recovery process.

The CDF as a city cloud managed by the local community

The CDF in the cloud mode is an information system used and managed by local communities, through which they have capacity to monitor and participate in the governance of the city. This is not a tool for smart cities in the sense of synchronized transport networks, but the CDF is rather about equal accessibility to knowledge that facilitates mobilization for a fairer division of power and resources. Access to locally relevant knowledge raises the interest of the public on local issues. When communities find that they benefit from it, they start to use it more and it becomes more participative. Nemo who is the CDF computer scientist explains in an interview how information systems can be used on a community level:

> Data should be for everyone, so this tool should be in the hands of communities. For instance, if I am poor, living in the center, while land value is increasing, then such a tool should allow me on my phone to proof that, so that I respond, whereas if you give such a tool only to a small elite that is already in charge, then there is a higher possibility that its power will increase. There should be a contract in such a situation which concludes that if you generate power over this information, then you must redistribute this power. Especially in such a complex system, you need a community that uses the information and acts upon it. Young community initiatives for example can use such data for their projects.

Not only the communal aspect, but also the technical aspect of data management was discussed with Nemo. In the three modes, it seems appealing to translate the CDF tool into software. Particularly in this mode, where pools of data are fed and managed by communities, a cognitive machine with adequate interfaces allows the CDF to be more functional. Nemo continues:

> If we want to turn it into software, the frame already has a structure that works as a neural network. These slots are data points and problems. The start will be through initiating a linear machine learning system. We must identify problems and insert them into the program. When people use it, they will feed into it more information.

Certainly, if communities find value out of it then they will input more data to get additional value. Eventually, when communities input more data, the system will operate better and provide more accurate predictions of projected land values, for example. The machine learning algorithm will even note new correlations and provide statistical evidence. However, the CDF software cannot identify additional issues and can be used to provide consultancy to decision makers.

Conclusion

All six sectors are interrelated in the recovery process. Spatial planning and provision of housing well require the active involvement of the community. This requires an informed and knowledge-based governance that takes in consideration statistics and values in local economy, lands, and population, and gives communities accessibility to this knowledge. The CDF is an experimental project that is limited to inner city issues. It currently does not claim that it is participating in the recovery of the city of Azaz because the structures of inner and direct external actors are much more present. While it is accommodating knowledge from a city of 150,000–200,000 inhabitants and putting it into a template, its practical use might be in cities of 25,000–50,000.

The city is a living orgasm with inner metabolism and interaction with the external living system. A meaningful change in a city will remain limited unless it is equalized with other neighboring cities in the region. External actors play a role on the local level, but the CDF focuses on local-local relationships. It builds on local resources and values so that it keeps itself independent while open to partnerships with external actors to leverage resources. It is community oriented and sustained by the continuous involvement of the local community as an actor and an observer of the process. Simultaneously, counter to virtual space, the physical space is limited to local communities in terms of movement and use, therefore any practice in the city must be based on the justice of accessibility to space.

To create dynamics that mobilize communities including tens of thousands in IDPs camps, the void in institutional and legal frames should be exploited to transform running systems and established facts, values, and practices and to generate new ones. Otherwise, the cycle turns again, and root causes of the conflict will keep regenerating.

Notes

1. This data is based on an urban governance questionnaire conducted with eight key actors in the city's administration and the involved NGOs.
2. Building a legal framework that sets guidelines of the construction process and buildings specifications, i.e. heights and land uses.
3. Building developers are mostly engineers who are equipped with the legal know-how of construction processes.
4. The profit is considerably very low and estimated at only 5 percent.
5. The HLP Registry is a collection of books that include documents about the city and its administrative area. Each book has 50 sheets (70 cm × 50 cm) that are called HLP Registry Sheets. Book sheets are numbered from 1 to 50, 50–100, and so on.
6. The HLP came after the HLP Registry Administration.
7. The HLP Registry Sheet is a single, two-sided sheet given one number, the HLP Number. The HLP Registry Sheet does not contain drawings or plans but only information about the property: HLP Number, owner's name, city, village, and the known name of the property by the residents and farmers in the area. It also contains all the legal notices: retention, mortgage, appropriation, parcellation, succession, and selling and buying transactions. The sheet also includes a historical narrative about the property and the type of property transfer at each period. Names of old owners are usually crossed, and the names of the owner are added.
8. In addition to the automated process, high-resolution scans are uploaded as a historical evidence, and hard copies of the HLP Registry Documents are put in a museum or displayed at the HLP Registry hall.
9. The successful military operation of the Turks on Afrin led to a devaluation of land values caused by the internal displacement of almost 10,000 people from Homs, Talbieseh, Rastan, Ghouta, Daraa, and Afrin.
10. The assessment was conducted in consultation with individuals from the municipality of Azaz and realtors and contractors. The process included the analysis of the fluctuations in the prices of 50 pieces of land within the last two years. A comparative analysis was conducted afterwards with pieces of land in neighboring areas. The price estimation relied on several criteria that impact prices of lands in the city.
11. The records are also connected to the capital city, Ankara. These identification documents do not replace the Syrian documents but are rather auxiliary cards that validate their legitimacy as dwellers.
12. There are no clear numbers about the northwest region; however, only 3.2 percent of Syrian CSOs are self-sustaining themselves through their own income-generating projects.

References

ACU. (2017, November). *Euphrates Shield Dynamo Report*. Retrieved from Assistance Coordination Unit: https://reliefweb.int/sites/reliefweb.int/files/resources/Euphrates_Shield_DYNAMO_EN.pdf

ACU. (2019, July 24). *Statement by the Assistance Coordination Unit (ACU) on the Recent Military Escalation in Idleb, Syria*. Retrieved June 6, 2020, from Assistance Coordination Unit: https://www.acu-sy.org/en/statement-by-the-assistance-coordination-unit-acu-on-the-recent-military-escalation-in-idleb-syria-24-july-2019/

Aita, S. (2017). Economic Assessment of Urban Issues for Reconstruction in Syria. In *State of Syrian Cities 2016–2017*. Brussels: European Union.

IMPACT. (2019). *Changing Contexts and Trends in Syrian Civil Society*. Berlin: IMPACT.
Syria Independent Monitoring (SIM). (2018). *Understanding Market Drivers in Syria*. Retrieved from https://assets.publishing.service.gov.uk/media/5a65ef1fe5274a0a37ef2f15/Research_Report_-_Understanding_Market_Drivers_in_Syria__January_2018.pdf
The World Bank. (2017). *The Toll of War: The Economic and Social Consequences of the Conflict in Syria*. World Bank Group. Retrieved October 25, 2019, from https://www.worldbank.org/en/country/syria/publication/the-toll-of-war-the-economic-and-social-consequences-of-the-conflict-in-syria
UNHCR. (2019). *Syria Emergency*. Retrieved October 25, 2019, from The UN Refugee Agency: https://www.unhcr.org/syria-emergency.html
Urban Housing and Question of Property Rights in Syria. (2017). *State of Syrian Cities 2016–2017 report*.
URC. (2018a). *Azaz City Fast City Profile*.
URC. (2018b). *Population and Space Survey of Azaz City*.

Part II

Housing the displaced

7 Understanding protracted displacement through the dwelling

The temporal injustice of the not quite, not yet solutions to refugee crises

Cathrine Brun

Introduction

Protracted displacement is distinguished by experiences of stuckness, waiting and dwelling in the temporary (Brun, 2015, 2016a,b). Reflections on dis*place*ment in academic research often concern the role of place in displacement: The ways in which displacement changes the relationship between people and places (Kibreab, 1999; Stepputat, 1999). 'Place' in forced migration studies has largely been adopted as a progressive term (Massey, 2005), as changeable, consisting of relations stretching beyond a particular place and formed by the multiple interactions, practices and experiences of everyone present at that place. In academic knowledge production, the lens has then to some extent shifted from sedentarist to more mobile and transient ways of considering the world (Hannam et al., 2006; Cresswell, 2010). However, beyond academia and behind a changing language in policy and practice, not that much seems to have actually changed. Displaced populations continue to be confronted with a rigid system and essentialist notions of place. Liisa Malkki's (1992) exposure of the logic of national geographic is still relevant as policies on displacement continue to be distinguished by the need to put people back into *a* place.

What can we learn from these last 25 years of theorising forced migration? Place is still a key dimension in how displacement is conceptualised and experienced. But one oversight I can identify in my own work comes from the too narrow – or compartmentalised – understanding of place that caused an ignorance of how place must be understood as always temporal and material. In research on displacement, there have certainly been discussions about material conditions, about sustainable livelihoods and making a living as part of local integration processes. There is a rich body of work on shelter and homemaking, reflections on the past and the future, but theoretical engagements with temporality and materiality did not really take off until recently.

During the summer of 2019, as I complete this chapter, research on temporality and materiality is accelerating and gaining momentum in research on displacement (Cwerner, 2001; Brun, 2015; Pasucci, 2016; Khosravi, 2019). Temporality and materiality are, however, not yet dimensions reflected in policies relating to protracted displacement. More insights are needed to understand and make operational temporal dimensions and materiality in protracted displacement. With this chapter, I aim to contribute to this work of engaging with the temporal and material in protracted displacement. The chapter has three main aims: First, to show the significance of the temporal and the material to how we conceptualise protracted displacement; second, to show how protracted displacement and recovery in its widest sense can be understood through the dwelling; and third, to address the current policy regimes for forced migrants and how a more active engagement with temporality and materiality could help to shift policies towards more realistic protection, assistance and a better life for people living with displacement.

In contributing to these discussions, I will analyse and provide learning from one particular situation of protracted displacement, the Georgian internally displaced persons (IDPs), who were forced to move from Abkhazia during the 1990s and still live in displacement. One key takeaway from the conference that the chapter was presented at, the City Debates in April 2019 at the American University of Beirut, was the importance of understanding the urban forced migrants in the urban fabric of which they are part: We need to contextualise the experience of forced migration in new ways and this chapter is an attempt to provide a deeper understanding of temporal and material dimensions through what could be termed as a *longue durée* approach to conceptualising protracted urban displacement.

My work in Georgia is based on a number of shorter (10–20 days) periods of fieldwork from 2003 to 2018. During this time, I have revisited some of the same people but also always interviewed people I have not met before. I have worked closely with the IDP women's organisation *Consent* and had numerous discussions with Julia Kharasvili, the head of that organisation. The fieldwork started as a consultancy in 2003 to find ways in which the Norwegian Refugee Council could provide more durable solutions in the collective centres where they worked. From there, most of my interviews have focused on the experience of protracted displacement and the dwelling in the temporary in different ways. My work and interviews concentrate on the experience of displacement and what people do in the context of displacement.

In the next section of this chapter, I reflect on the limits to the current durable solutions framework and how it leads to a *solutions lite* approach for most displaced populations today, these reflections are based on ideas I developed with Anita H. Fábos (Brun and Fábos, 2015, 2017; Brun et al., 2017). The following sections consider dwelling in the temporary as a particular condition of protracted displacement, inspired by a critical geography of architecture and work I developed with Ragne Øwre Thorshaug (Brun and Thorshaug, 2020; Thorshaug and Brun, 2019). I then move to write the Georgian IDPs into the history and development of Tbilisi as a city, particularly referring to housing, and present the establishment of the so-called 'durable housing solution'. In order to understand the temporal injustice that people in protracted displacement experience, I set out to discuss the experience of durable housing solutions related to three different sets of material dwellings: The privatised collective centre, the more precarious non-privatised living spaces in collective centres and the transient dwelling in rented spaces. I reflect on the ways in which times are always out of joint in a durable solution framework before reflecting, in conclusion, on how we can start a temporal insurgency that may help to escape the solutions.

A solution is never a solution: Problem solving in the present as temporal injustice

The durable solutions were formulated to protect and end displacement for people of Europe who had been displaced during World War II and who were still displaced in the late 1950s (Milner and Loescher, 2011; Brun and Fábos, 2017). By the 1960s, the UNHCR had provided resettlement quotas for European refugees who could not return or integrate locally – a precedent for addressing the 'durable solutions' as we know them today and a way of 'putting people back into a place' (Brun and Fábos, 2017). While resettlement was the focus of the early years of the durable solutions, by the mid-1980s, the focus shifted from resettlement to repatriation (Chimni, 2004). Durable solutions are still understood as the ultimate aim of international refugee policy; however, the emphasis on the solutions has waxed and waned over the time period of their existence.

'Solution' is a puzzling word. According to the Cambridge online dictionary, it is 'a liquid into which a solid has been mixed and has dissolved' (…); 'it is an answer to a problem'.[1]

If we were to translate these definitions into the framework of the durable solutions for refugees and displaced persons, the solid to be dissolved must here refer to the experience and situation of displacement that will be dissolved into normalcy (or the normal), the liquid is the normal, it might change character, but the solid is dissolved, disappears and becomes unrecognisable. Hence, finding a solution, the answer to the displacement-problem is to dissolve the condition of displacement. A problem-solving approach, Cox (1981 in Nyers, 2019, p. 4) states, is 'the strict temporal limitations placed on how the 'problem' of refugees and their movements are defined'. It is a problem that is being solved without changing the conditions causing that particular problem. A critical approach – on the other hand – would demand an analysis of historical conditions of possibility of political problems as well as attention to change (Nyers, 2019). In this chapter, I aspire to contribute to such a critical approach by engaging with temporal and material dimensions of inhabiting the city.

There are particular meanings and expectations of solutions in the context of protracted displacement. But, these expectations are not fulfilled: Very few internally displaced and refugees can return due to protracted conflict. At the same time, only around 1% of refugees resettle each year (The Expert Council's Research Unit (SVR Research Unit), 2018). Local integration is then perhaps the only remaining alternative. But local integration does not work either.

The UN General Assembly confirmed the Global Compact on Refugees in 2018. In the Compact, local integration receives limited attention. The Compact does not pay much attention to local integration and turns instead its attention to the aim of self-reliance as an alternative solution.[2] So, with self-reliance, forced migrants can stay in the place of displacement, be self-reliant, but cannot become members of the societies where they live. This is an *integration lite* version of the solutions as Anita Fábos and I have showed. There are also other such *solutions lite* that I will discuss here. The durable solutions may then be understood as a population management strategy in the service of the state that legitimates keeping populations on hold while waiting for a solution to be realised (Brun and Fábos, 2017). Integration with its accompanying status, which would (in theory at least) ensure a certain level of legal protection and recognition of the subject's presence in a nation state, is only possible if the host state is willing to provide this route. The international community is therefore finding other ways of working in protracted situations, more removed from legal protection. It is in this context that self-reliance has come back on the agenda. Self-reliance is not a new debate (Crisp, 2003), but this version of *integration lite* becomes an alternative amid reduced opportunities for legal pathways to work, housing, poor host economies, restrictions on rights, a lack of administrative planning and limited arable land for farming (Easton-Calabria and Omata, 2018). What matters is some sort of survival mode where independence from aid is the aim, and where a final solution to displacement has not been achieved. Solutions for displacement in this case are considered place-based and devoid of a temporal dimension. They are about being put back into a place and re-establish the order of the past. However, when a solution is not possible, the alternative self-reliance is just a way of existing in the present, independent of aid but with no marks on the landscape or stakes in the future of the place in which people have lived for years. The ways in which the solutions framework is currently used with the increasing acceptance of *solutions lite*, may then be described as a 'temporal injustice' (Fontanari, 2017): An injustice that deprives individuals of control over and access to decisions that influence their everyday time and their future lives and at the same time render their biographies invisible (Thorshaug and Brun, 2019; Fontanari, 2017).

The narrative I present below shows that the temporal injustice embedded in a *lite*-version of the solutions extends to an understanding of time as reversible, or still; it does not cater for the irreversibility of displacement. Temporal injustice is then brought about by not recognising – or being aware of – how time operates. The dilemma is that the discourse on the status of displaced population as matter out of place (Malkki, 1995; Brun, 2003) has made it politically unthinkable to introduce an alternative temporal frame. An alternative temporal perspective would have to recognise displaced persons' presence in the city, their biographies and histories in that city as well as their stake in the future at that place. But time in protracted displacement is always already 'out of joint' (Mukherjee, 2019, p. 2) as long as a solution is always considered as an end-point. A definitive moment of closure is an illusion – a tenuous demarcation between past and present – which neglects temporality and instigates temporal injustice in particular ways, first and foremost by neglecting an exploration of temporality in itself. Embedded in the current policies for forced migrants there is thus a clash of temporalities.

Dwelling in the temporary: The material building

In Georgia, to which I return shortly, the experience of a process of *integration lite* can be exemplified through the processes of housing for the displaced in a state strategy termed *Durable Housing Solutions*. I will analyse the experiences of the durable housing solutions by considering different forms of dwelling and the experience of how temporal injustice materialises in a *solutions lite* approach. Ragne Øwre Thorshaug and I (Brun and Thorshaug, 2020; Thorshaug and Brun, 2019) developed an understanding of how temporal injustice operates when particular institutionalised living spaces such as asylum centres and collective centres are inhabited during displacement. In this section and in the remainder of the chapter, I build on this work to explore the micro-processes of housing for refugees and displaced in the city: The act of dwelling and inhabiting particular living spaces.

Many displaced in the city end up inhabiting living spaces that were not intended for permanent living or that were meant for far fewer residents than the number actually inhabiting them. Yet other spaces are just on the brink of collapsing after years of neglect due to the economic situation of its inhabitants, the commercialisation of living spaces and the nature of a private renting market. Common for many of the dwellings are then the substandard spaces, the crowding and the huge physical and economic restrictions involved in making these spaces more bearable spaces to live in. Displaced people dwell in the temporary and by considering the processes of dwelling, of inhabiting a material living space, we can learn more about the displacement process, about the displacement experience and about how people find their bearings: The experience of displacement, the institutional conditions and the material meet in the building.

'Dwelling' is here both understood as a noun – a residence and an abode – and as a verb – as being in the world: Being somewhere – a place where we make our world meaningful (Heidegger, [1954] 1971; Cresswell, 2009). I follow Latimer and Munro's (2009) understanding of dwelling as a relational act: It is about taking care of material and non-material relations of which we are a part. Through their dwelling people engage with their surroundings, and dwelling is the starting point for understanding how people become involved in society and can build social relations (Brun, 2016a). The buildings represent the point from which people find their bearings (Ahmed 2006, 2010). Thorshaug and I were inspired by Sara Ahmed's work, which helped us to understand the role of the dwelling in how people orient their lives: On the one hand, the dwelling represents a background to people's lives; on the other hand, the dwelling is an element of what people draw together through an everyday skilful engagement with their environment. This intimate co-dwelling of bodies and buildings help to shape people's horizons, their desires and ambitions (Brun and Thorshaug, 2020; Thorshaug and Brun 2019; Ahmed, 2006).

The material (matter), institutional arrangements, social relations and everyday practices come together in the experiences and practices of dwelling in a particular building (Thorshaug, 2018; Brun and Thorshaug, 2020; Kraftl and Adey, 2008; Lees and Baxter, 2011). By understanding the interaction between different temporal logics that come together in the building, we can also unpack the clashes of temporality that sits at the core of the experience of protracted displacement and the *solutions lite*.

War and displacement: Internally displaced Georgians from Abkhazia

To engage with the temporal and material dimensions of displacement, I will continue by presenting the displacement history of the Georgian IDPs from Abkhazia and write them into the history of urban development in independent Georgia.

Displacement

Georgia was declared an independent nation state in 1991 after the dissolution of the Soviet Union. Included in the Georgian nation state was Abkhazia, a region on the coast of the Black Sea that had occupied varying statuses during the history of the Soviet Union. The declaration of the Georgian nation state was followed by secession claims by Abkhazia from Georgia in 1992 but the Georgian authorities refused to accept. Georgian forces entered Abkhazia to regain the disputed territory but were defeated and, in the conflict that followed, about 250,000 ethnic Georgians, approximately 46% of the population in Abkhazia at the time, fled their homes and were granted the status of IDPs when they moved into the Georgian-controlled territories of the country (ICRC, 1999). Many settled in Western Georgia close to Abkhazia in and around the cities of Zugdidi and Kutaisi, but a large proportion found their way to the capital Tbilisi.

Nearly half of the Georgians who came from Abkhazia stayed in collective centres. Until 2009, the collective centres were institutionalised spaces where services such as electricity and water – where available – was provided by the state. Humanitarian organisations assisted the internally displaced to make minor improvements, such as repairing communal sanitary facilities, leaking roofs and sometimes changing windows. The collective centres were established in buildings not intended for permanent living such as student dormitories, hotels, kindergartens and old hospitals, and represented crowded, substandard and unhealthy living conditions, but nevertheless a space where people lived together. Of those internally displaced not living in collective centres, most stayed with family and friends or rented their dwellings. Living in equally squalid conditions, with less tenure security and with a very high proportion of any income going towards rent, this group was increasingly becoming part of the commercial renting market distinguished by a high level of exploitation (Brun, 2016a). A very few IDPs managed to purchase a dwelling.

At the time of writing in 2019, the people who left Abkhazia and their descendants are still considered to be internally displaced by many of the IDPs themselves and by the wider society. The conflict may be described as a frozen conflict with periodic outbursts of war and violence. In August 2008, a war between Russia and Georgia prompted many scholars to believe that Georgia has lost de facto control over Abkhazia (Kabachnik, 2012). Nevertheless, the nationalist discourse of a unified Georgia that includes Abkhazia is still strong. The internally displaced Georgians play an important role in keeping alive Georgia's hope of regaining control over Abkhazia. The popular and official attitude in Georgia is that return of the IDPs represents the only valid solution to their displacement. While the majority of the displaced may still hope for return (UNHCR, 2015), many do not think it will happen in their lifetime (Brun, 2015). Within this policy environment with a strong emphasis on return, IDPs have struggled to recreate a sense of belonging where they live. They are long-term recipients of state aid, experience lack of access to mainstream job opportunities, and face a daily encountering of public discourses that overlook their presence in Georgia and tend to misrepresent the complexities of their lives.

Urbanisation, privatisation and IDPs in Tbilisi

Displacement in Georgia is closely connected with urbanisation[3] and the post-socialist development in the independent republic. It is also closely linked to the particular ties between people and places that developed during the twentieth century. Since the early 1930s in the Soviet Union, there was an internal passport system, which largely fixed people to rural or urban areas (Manning, 2009). Categories of nationality and residence – the *propiska* or residence permit – regulated the right to reside in a specific urban district. It was through residence that people were entitled to access rights, services and privileges, including limited rights of inheritance. Until the 1950s, residence was almost as essentialised a distinction as nationality: Having a passport at all (and hence, the ability to move) was a privilege granted city dwellers and denied to villagers. The propiska system, abolished in 1996 in Georgia, but retained in Russia, was introduced for urban locations to prevent massive immigrations into urban spaces, and to keep rural populations rooted in villages (Manning, 2009). This is a discourse that has prevailed in Georgia and elsewhere: The need to keep people out of urban areas and the fear of urbanisation and population increase in urban areas.

Urbanisation cannot easily be controlled and stopped as we know, and Tbilisi experienced significant territorial growth and dramatic change of its urban space and social environment throughout most of the twentieth century (Salukvadze, 2009). The city went from a medium-size, relatively compact urban settlement into a big Soviet metropolis with a population over one million. With its complex economic profile and heterogeneous social and ethnic constitution, it was a product of Soviet city planning. Population growth was largely based on the influx of migrants from rural regions of Georgia, and, on a lesser scale, on immigrants from other Soviet republics (Salukvadze, 2009). The main driving force in the 1930s through to the 1950s was the expansion of industrial activity. Since the 1960s, however, the expansion of industrial functions slowed down and until the collapse of the Union, mass housing became the driver of territorial growth (Salukvadze, 2009).

Since independence, the regulations of the relationship between people and places became less prominent, but the strong links between people and the origin of their family's places are upheld. Notably, after independence, all cities in Georgia experienced population decline. Tbilisi's population started increasing again around the turn of the millennium and, today, it is by far the most significant city in Georgia with about 1.1 million registered inhabitants (Tbilisi City Hall, 2018).

With independence came liberalisation of the property market. The first year after independence, however, was distinguished by very poor planning and unregulated property appropriation (van der Zande, 2009). From a neo-liberal perspective, a dominant in many post-socialist countries including Georgia after the Rose Revolution in 2003,[4] spatial planning could be limited to the clear articulation of property boundaries, property rights and development rights, enabling powerful actors to continue to dominate the market. Literature indicates that these were actors that had also dominated spatial planning during the Soviets, and also representatives of industrial ministries and state enterprises (Van Assche, 2009). It was only after the Rose Revolution that an attempt was made to formulate effective legislation and urban planning. The Georgian propiska[5] system had provided tenancy, allowing occupancy on an indefinite basis and inheritance by members of the household rather than ownership (IDMC, 2009). But from 1991 to 1992, shortly after Georgian independence, the government enabled privatisation of apartments for the general population. As such, private ownership is a relatively new feature to the Georgian society. There is little official knowledge about how the current property market in Tbilisi was affected by the transformations between independence in 1991 and the Rose Revolution in 2003. What we do know is that unregulated land and real estate markets led to the collapse of social housing and affordable housing which again lead to increased vulnerability of several social groups including the internally displaced (Vardosanidze, 2009; Brun, 2016a).

Privatisation change the way of thinking about one's dwelling, it makes a dwelling unit into a commodity with a value in a market rather than an asset that only have value as long as you keep it in the family through occupancy rights (Struyk, 1996). Additionally, with the privatisation of the housing market followed an unprecedented mobility of populations both between urban neighbourhoods and between rural and urban areas (Manning, 2009). The internally displaced moving into the city was very much within this wave of migration. The privatisation processes and the in-migration created new social categories based on wealth and property.[6] A new group of wealthy property owners emerged as ownership of property in residences was transferred to the existing residents. The city dwellers became owners of real estate property, while newly arriving villagers became net renters (Manning, 2009). An exception to this pattern was the internally displaced in the collective centres who nevertheless contributed, through their inhabitation and dwelling, to transform Tbilisi space and city politics (Manning, 2009).

IDPs were inscribed into the Tbilisi landscape in a process that many established city dwellers considered a process of ruralisation and destruction of the Tbilisi culture (Manning, 2009). Their contribution in other ways to the city was, however, largely overlooked by government and citizens and the more visible section of the internally displaced – those in collective centres – represented an exceptional category as neither owners nor renters until the new government policy in 2009, to which I return below.

Solutions lite: Towards a durable housing solution for the internally displaced

For many years, it was unacceptable to talk about any other durable solution than return for the Georgian IDPs from Abkhazia. Even creating a better life for people where they lived during displacement was difficult as the policy focus was on short-term relief while waiting for return. The permanent impermanence was becoming more and more unbearable and the willingness of the government to engage with longer-term strategies increased gradually, influenced by pressure from national and international organisations. In 2007, a new state strategy for the internally displaced was formulated with a strong emphasis on housing as a way of providing a 'partly' durable solution for people while waiting for return (Government of Georgia, 2007). This was not, it was emphasised, a 'full local integration process' and as such may be understood as an *integration lite*. The government coined it a 'durable housing solution'.

The launch of the new state strategy happened just before the 2008 war with Russia. With this war, there was an increase in aid for newly displaced persons, mainly from South Ossetia, which also benefitted assistance to the so-called 'old displaced' from Abkhazia. The fresh funding enabled the beginning of implementation of the first stage of the durable housing strategy: Transfer of ownership to those IDPs residing in collective centres. The programme to transfer ownership – from now on called 'privatisation' – to the internally displaced was launched in 2009. Under the programme, the internally displaced could buy the living spaces that they occupied in the collective centres for a symbolic amount of 1 Georgian Lari. Simultaneously, buildings of commercial value, occupied by the internally displaced could be taken over by business actors and the IDPs in those buildings were given a compensation (Thorshaug, 2011; Brun, 2016a; Brun and Thorshaug, 2020). In the beginning, the privatisation process was rapid.

The second stage of the strategy would be to house the many IDPs living in rented accommodation (self-settled) and finally, the third stage of the strategy was aiming to compensate those few IDPs who had managed to buy housing after their displacement.

At the time of writing, most IDPs in collective centres where ownership can be transferred to the IDPs have been privatised. However, the process of privatising buildings of commercial value where the displaced will have to move has not been finalised. In the second stage of the durable housing solutions, some few IDPs on the private rental market have been offered new dwellings, but this process is far from done and there has been no rolling out of a compensation scheme.

From here, I will present three different co-existing and ongoing narratives of housing in context of the durable housing solutions: The privatised collective centre, the non-privatised spaces in collective centres and the privately rented dwelling.

Privatisation of collective centres[7]

Following the establishment of the State Strategy on the Internally Displaced and the introduction of a Durable Housing Policy, a Working Group on Privatisation with members from the humanitarian community and civil society established minimum standards and legal guidelines for the spaces that could be privatised. The minimum standards were recognised by the Georgian authorities but were not systematically applied on the ground and particularly not in the capital Tbilisi. In Tbilisi, the argument was that people are being transferred property on prime locations that will increase in value. Consequently, the government was not willing to increase the standard or attempt to fulfil minimum standards.[8] Hence, many of the buildings that were privatised were left in a poor condition.[9] Privatisation took place based on the family members that left Abkhazia and after 17 years of living in the same space, many families had increased their numbers but could not access larger spaces.

Formally, people had difficulties obtaining deeds for their newly owned spaces. The authorities keep the original contract whereas the IDPs retained an owner certificate that is not strictly valid (Thorshaug, 2011). Additionally, the contract only allowed one signatory on behalf of the household. However, everyone registered in the apartment are also included and there is a clause in the contract stating that the signatory renounces the right to make any further claims to the government concerning housing assistance. In effect, the second and third generation have then unintentionally given up their rights. Information concerning legal rights have become a main issue among civil society organisations in Georgia.

Despite these concerns, most people who were offered privatisation accepted. Afraid of losing this opportunity, they did not dare to decline. Some, however, were careful and reluctant: If they privatised, would they lose the right to assistance upon return and the right to compensation for the lost houses in Abkhazia? What would happen with their monthly IDP-benefits? If the spaces in the collective centres were small and sub-standard, would this mean the end to any possibilities of improving living conditions?

In many ways, privatisation was a mixed experience. With poverty and uncertainty, the privatisation came with responsibilities that were difficult to fulfil such as paying for electricity and to organise and pay for maintenance. Ownership of living spaces also left it unclear whose responsibility it was to maintain and clean stairwells, lifts, roofs and basements.

While ownership created more stability and more autonomy in some ways, people continued to feel stuck in the material dwelling. One man I interviewed had lived in a collective centre at the outskirts of Tbilisi since the initial displacement. With his family, he had managed to obtain two neighbouring rooms. The expansion of living space meant that there were two bedrooms, a living room and a kitchen in their now privately owned space (Figure 7.1). The space housed himself, his wife and his two grown-up

Figure 7.1 (a) Cooking in the corridor in a Collective Centre Tbilisi.

Figure 7.1 (b) A shared bathroom at the end of the corridor, Collective Centre Tbilisi

children with their spouses and altogether three children. He felt stuck in the building, not able to sell up and move on, not able to return to Abkhazia. He blamed his ill health on the building, stating (probably with a reason) that the air that came out of the ventilation had both Soviet pollution and bad air from the current regime making him ill. For him, the history of the material building cannot be escaped and the status of his living space as an IDP-building continues to exist.

The precarious status of the non-privatised spaces

The idea of transfer of ownership may be understood as a very convenient way for the government to handle the IDP problem. With privatisation, the Government showed substantial efforts to provide durable solutions for IDPs. However, privatisation was also understood as a first attempt to cut off assistance and see the IDPs as economically self-reliant and less dependent on government assistance (see also, Manning, 2009). The impact was particularly difficult for those IDPs living in collective centres that were defined as buildings of commercial value and where ownership of living spaces was not transferred to their residents but to original or new owners. For a country striving for economic development, privatisation of buildings with commercial value was part of an important step in opening the economy with liberal economic policies being a general trend.

In November 2018, during my last visit to Georgia so far, the process of privatising such buildings was still going on. I visited one of these spaces where some of the original residents have been offered alternative living spaces or compensation and have moved out, but some are still living in the buildings and yet others, sometimes non-displaced, have moved into the vacant spaces. Although understood as a collective centre, many of this and similar buildings were established later in the displacement process and often residents did not know who the owner was. In this non-privatised collective centre, former barracks for workers, I met a 48-year-old widow and mother of four children living in two small rooms and with a bathroom on the corridor. After their displacement from Abkhazia, they first moved to stay with some relatives near the border. Her husband stayed behind and fought in the war, but when he was killed in 1995, she decided to move to the city for a better chance of finding a livelihood. Through a relative, she learnt about her current living space and managed to occupy a space there. While raising the family, she held different jobs. The kids slept on the floor in the living room, one son has moved out, the remaining children, now grown up, still live with her, none of them in regular employment. She recently repaired the ceiling of her dwelling: She had to because rats kept coming into the flat.

The barracks are two lines, two storey wooden buildings surrounded by an increasing number of high rises. The barracks resemble ruins from the Soviet era in an increasing neo-liberal architectural landscape of the post-socialist Tbilisi. For the people still living in these old buildings, their status and security of tenure were increasingly more contested. Adding to their stress and uncertainty was the dismal conditions of the buildings: They are so run down that people are nervous about the safety of staying in them, fearing that the pillars between ground floor and second floor might collapse. But people are used to the location, children go to school nearby and livelihood is found in the city. They do not know whether they can stay put and for how long, and they have no idea who now owns the buildings.

Not knowing who the owners are, what their rights are, what will happen when, not being able to invest and to be part of the place in which they live create a particular temporal frame in their everyday lives, a temporality distinguished by uncertainty about the future. While some are still waiting to find out what will happen to their living spaces, many have been removed from privatised collective centres. The most famous example is those who lived in Hotel Iveria that later became Radisson Blu (Figure 7.2). A common pattern for this privatisation process was that the residents were asked to move and were given US$7,500 in compensation. Many found, however, that the compensation was insufficient to buy a dwelling in the city. The money they received was used for rent and other outlays, such as medical expenses in the privatised healthcare system or the deposit to rent a flat. One family I interviewed lived in a collective centre in a village close to Batumi (in

Figure 7.2 (a) Hotel Iveria Collective Centre in 2003.

Figure 7.2 (b) Hotel Iveria became Radisson Blu

Adjara, Western Georgia). When their collective centre was sold, the family was reluctant to move as they had employment and friends and relatives in the immediate vicinity. However, they were forced to do so and given the standard compensation of US$7,500. One of the sons in the family worked in the military in Tbilisi, and the whole family (husband, wife, wife's father, two sons, a daughter, a daughter-in-law and a baby) decided to relocate to the city and join the IDPs renting in the private sector (see Brun, 2016a).

Transient lives: Renting a dwelling[10]

> I do not remember how many places I have stayed for the last 10 years, maybe nine or ten? ... Sometimes I was living there for a couple of years, sometimes for a couple of months. Often when you arrive at a new place, it is not in the condition that the house owner promised, so you immediately start looking for a new place. And then you move again. Another time we rented and didn't know the house was for sale, and then soon after we had moved in, the house was sold and we had to move. It is difficult to move houses all the time.
>
> (Displaced man, aged 33, living in rented accommodation with his mother, quoted in Brun, 2016a)

Private-sector accommodation is believed to have accounted for approximately half of the dwellings of IDPs when the state strategy was established in 2007. There is little available information about the 'privately accommodated IDPs', but it is common to distinguish between three types of dwellings: Owned, borrowed and rented. According Georgia's Ministry of Refugees and Accommodation (MRA), there were 4,396 families who owned their dwellings in 2007 (Ministry of Refugees and Accommodation, 2010). The second type of dwelling in the private sector is borrowed dwellings – that is, where people stay with family and friends or in an empty house without paying rent. The latter seems to be more common in rural areas and regional capitals than in Tbilisi, where housing is scarce. The third category covers rented dwellings. In 2010, the second and third categories in the private sector are believed to encompass approximately 40,000 families (Ministry of Refugees and Accommodation, 2010).

In the rented sector, people had limited control over their biographical time: They did not know how long they could stay in the building and where they would go next. A study in 2013 showed that IDPs from Abkhazia in rented accommodation in Tbilisi had the lowest levels of housing stability and expected to change accommodation within a year (Norwegian Refugee Council, 2013). The housing instability is often caused by circumstances outside the residents' control, and renters describe considerable vulnerability and insecurity. They tend to live in fear of the owner increasing the rent, which would force them to find a cheaper dwelling. A young woman renting a flat in a suburb of Tbilisi with her parents and brothers stated that they were always looking for somewhere else to live so that they could move in a controlled fashion rather than being forced to leave at short notice. This volatility also affects individuals' social networks and social capital. Building and maintaining relations becomes difficult in the temporariness of the rented dwelling, as another young woman, Monica states: 'Everyone here knows that we are renting. They know that today or tomorrow we will be leaving. Neighbours are only considering us to be here for a short time'. (Monica, came to Tbilisi to work in a factory when she was 17, and has lived in somewhere between 10 and 15 places in Tbilisi since she arrived in 1999, quoted in Brun, 2016a.)

Living spaces are often substandard, with plumbing frequently a problem, windows draughty, walls damp, and floors stripped of wood or other covering. The flats I visited were sometimes furnished when rented out, but sometimes not. Common for all flats was that the families living there would have few of their own belongings – they were not investing in things before they knew they had a space they could control. Monica's family had bought one lamp and a baby cot for their new born baby – otherwise they were reluctant to invest money or energy in the rented dwelling. The low material standards of the dwellings form a reminder of people's status. Lack of control is felt both in relation to the immediate physical space and in controlling their future.

Times are always out of joint

Times are often felt to be out of joint because different temporal logics clash in the buildings, creating a sense of powerlessness and invisibility (Frederiksen, 2013). From these experiences, particular expressions of power may be identified as chronographies of power (Sharma, 2014, p. 15): Conceptions of time as lived experience are 'always political, produced at the intersection of a range of social differences and institutions (...)'. Embedded in these unequal power relations is then a temporal injustice that operates by depriving individuals of control over and access to decisions that influence their everyday time and their future lives and at the same time render their biographies invisible (Fontanari, 2017; Thorshaug and Brun, 2019). People's experiences in the buildings are results of deliberate politics of keeping people in displacement and an unwillingness to finding a solution where people live and have established themselves. Buildings and people's living spaces represent particular ways in which times are experienced as out of joint.

The general society, through institutions such as the Ministry of Housing and the policies formulated for assisting the internally displaced, represents and performs ideas of pasts and futures in particular ways. With the durable housing solutions formulated in the context of a strong discourse of return, the future that the IDPs from Abkhazia belong to lies in a past that can no longer be re-created. However, at the same time, the presence of the internally displaced in the city has not been recognised in its history and has led to an experience of non-recognition and production of invisibility in the different material buildings that they occupy. For those in living-spaces where their histories were lived and where bonds to their communities were nurtured, the temporal injustice embedded in the erasure of their biographies and histories is felt strongly.

In Frederiksen's (2013) understanding, Georgia is still in a post-socialist state. It is a country that is moving away from its past, but still has not reached the future she is aspiring to. The country is still 'lost in transition': People are caught in the past. While not merely living that past, we need to understand how people use the past that they are in the process of coming to terms with when developing their lives and planning their future. Being lost in transition must be seen in relation with the material realities of people's current lives. The feeling of being lost in transition dominates when displacement does not allow for coming to terms with the loss of homes that belong to a socialist past and when available solutions are based on that past: The durable housing solutions are formulated as if time stopped with displacement. It is the family members at the time of displacement that is given a *solutions lite*. With this timeless intervention comes the feeling of being stuck in crowded living spaces where the material conditions are experienced as dangerous, making people ill and being a constant reminder of their displacement. For some, these material conditions is a constant reminder of a lost past, a lost house, a lost opportunity for family meals and enough space to be hospitable.

Ownership in the former collective centres is not in itself a sufficient solution because it does not recognise the biographies of people: The ways in which families have grown and people have inhabited the city. For those not in collective centres, in addition to being rendered invisible they live with an uncertainty for the future that makes it difficult to invest for the future in the place where they live.

I have shown elsewhere that people do not really think that they can return in the near future (Brun, 2015). The hope for a return in the distant future is still alive, however. In November 2018, I spent time hanging out with the children of the displaced I have followed since 2003. Now young people in their early twenties, they were children when their families displaced, events that they often remember or relate to through their family stories: They grew up with displacement and the narrative of return. While their identities as displaced vary in strength, a common agreement between them is generally that displacement can only end when the possibility of return has been established.

The Georgian proposals for durable housing solutions contain no measures for the children of the displaced. As mentioned above, solutions are formulated on the basis of the original family that moved from Abkhazia 20 years earlier. The fact that those who were children during displacement have now grown up and started their own families has not been taken into account. Members of this 'second generation' need to find their own places in society and need to come to terms with displacement in their own ways.

Eka, the daughter of Salome whom I have known for many years (see Brun and Thorshaug, 2020; Brun, 2015), has a well-paid job. She has occupied some space on a landing of a former collective centre and built up a new flat there that she is renting out (Figure 7.3). She grew up in the building and now lives with her mother in her privatised room. However, she wants to escape the building and is saving to buy a flat in a 'normal' building – meaning a building not associated with IDPs or collective centres. She is aiming to escape the continued status and narrative of displacement, which the material dwelling becomes a particular symbol of a status that sticks with you as long as you are in that building.

However, Manana, another young woman who grew up in a rented dwelling in the city has become an activist for gender issues related to displacement and, in particular, making women's experiences in displacement more visible. She is educating to become a gender specialist. Her IDP

(a)

(b)

Figure 7.3 (a) Eka's occupied space in 2010, before building. (b) Eka's completed flat

identity was not strong when she grew up as she did not live among other IDPs and her school friends were mainly non-displaced. The IDP identity became more important to her later in life when coming to terms with her own past: The violence that she experienced as a child during displacement when she lost her father of whom she has only a vague memory. The narratives of displacement and violence became important to her in her late teens when she needed to come to terms with her loss. At the same time, she realised the losses and the struggles of her mother and understood the strength that her mother needed to mobilise to bring her up. She is not planning to return and is building a career for herself in Tbilisi. Her mother, grandmother and she herself rent a flat from relatives and she thinks she can continue to stay in this place. She firmly believes that the IDP status can only end when she has the choice to return; she might not want to return, but she insists that she will continue to fight for the right to return. Only when she has a choice of where to live can she stop identifying as a displaced.

Escaping the solutions through a temporal insurgency?

Contributing to the inability of finding a solution is the particular humanitarian and temporal logic to understanding place in dis*place*ment: The continued need to put people back into a place and the problem-solving approach that fails to take into account the temporal dynamics of displacement. The not quite, not yet continues to dominate and this inability of the solutions framework to deliver has contributed to an acceptance of self-reliance and other *lite solutions*. While these *lite solutions* are used as an argument to move away from relief and aid dependency in protracted displacement, they have also led to a normalisation and acceptance of displaced in cities as denizens or not quite full members of the society, not quite part of governance structures, planning processes and social and physical infrastructure (see Brun et al., 2017).

People in displacement are carving out spaces for themselves. It might be possible to identify a temporal strategy of resistance, what Sara Sharma (2014) terms a *temporal insurgency* that offers a political strategy based on inhabiting the world in time. In conclusion, I suggest three spheres of work for us to start a temporal insurgency.

The **first step** towards a temporal insurgency is to disturb the temporal logic in the support system for displaced and refugees. How time is lived is marked by how life is made liveable through support settings. When these settings are publicly criticised, these relationships between support and temporality become more visible (see Cover et al., 2017, p. 2). The recent discourses on the humanitarian-development nexus might be a good example: The pressure that has come from humanitarian workers and researchers to rethink how assistance is being offered in protracted displacement is promising. There is, however, a need to develop a more critical understanding of what development might mean in this context: A need to recognise the displaced as part of the history of the place of displacement, rather than the exclusionary emphasis that come with the solutions framework.

A **second** strategy for a temporal insurgency is to acknowledge and understand how hope is acted out in everyday practices during displacement, which is what Bayat (2015) terms 'the quiet encroachment in the everyday', and what Holston (2009) refers to as 'insurgent citizenship'. Because it is in the everyday, that politics take form, the potential of a temporal insurgency may lie in the most careful term of 'active waiting' (Brun, 2015). Active waiting focuses on the ways in which people simultaneously move on during displacement, feel stuck in the present, and actively relate to alternative and changing notions of the future. Although hope and waiting are not just positive, as I have shown elsewhere (Brun, 2015), the hope represents visions for the world that create possibilities for change (Zournazi, 2002) and is consequently an important perspective for identifying potential for improved lives. This active waiting takes place in a multitude of ways such as through the process of inhabiting living spaces and homemaking practices. As Anita Fábos and I have argued, homemaking practices take place for both displaced and non-displaced and can be a lens that help to escape the solutions. A recognition of these shared everyday practices that have their starting point in the dwelling connect different temporalities rather than make them clash in the building. These practices are adopted to get by, they are the changing the city and create the history of the city while also enabling carving out a space at the place of the dwelling for the future.

Third and finally, we should imagine an alternative temporality rather than concentrating on place alone. Regaining the place people lost in displacement is a major issue. I would suggest that a temporal insurgency would be a movement that mobilised around time – or rather place-time – to move away from an essentialist notion of place where people are only temporarily present in place, where people are allowed to come to terms with their past, keep the options of return to a past place open but at the same time have their biographies at the place of displacement recognised: A set of rights based on the right to a future in a place would shift the focus away from the static place and expand our vocabulary in order to capture the lived time in displacement.

Acknowledgements

City Debates in Beirut is one of the most inspiring conferences I attended, and heartfelt thanks go to Howayda Al-Harithy and her team for organising such an inspiring conference in April 2019 and for putting together this book during the challenging times. Many thanks to all the conference participants for the feedback on the chapter. A huge thanks to all the Georgian people who shared their experiences with me and helped to understand internal displacement over the years. Also, thank you to my good colleagues Ragne Øwre Thorshaug, Anita H. Fábos and Julia Kharasvili for their inspirational collaborations and finally to Mariam Naskidashvili, Lika Margania and Khatia Kardava for research assistance.

Notes

1. In Merriam-Webster: (https://www.merriam-webster.com/dictionary/solution) definition of solution:

   1a: an action or process of solving a problem.
   b: an answer to a problem: EXPLANATION *specifically*: a set of values of the variables that satisfies an equation.
   2a: an act or the process by which a solid, liquid, or gaseous substance is homogeneously mixed with a liquid or sometimes a gas or solid.
   b: a homogeneous mixture formed by this process *especially*: a single-phase liquid system.
   c: the condition of being dissolved.
   3: a bringing or coming to an end or into a state of discontinuity.

2. Following the development of the text of the Compact over its drafts, it is interesting, however, that local integration gets more attention in the text in later drafts then the first drafts.
3. In 1803, Tbilisi had 16.000 inhabitants. The population grew to 1.2 million at the time of independence, but the city saw a dramatic drop in population after independence. And despite having grown since around the millennium, the city has not caught up with the population it had in 1990 according to official statistics (http://geostat.ge/cms/site_images/_files/english/population/Census_release_ENG_2016.pdf., accessed 071017, see also Manning 2009). (Georgia is a country of decreasing population.)
4. The Rose Revolution represents the peaceful transfer of power, following popular demonstrations, from Edouard Shevardnadze representing the Soviet regime to US-educated Mikhail Saakashvili.
5. The Soviet system of 'propiska' rights was a distinctively different system where, instead of private ownership people had the right to rent from the socially owned housing provided by the state.
6. For more on privatisation in post-socialist countries, see Humphrey (2007) and Alexander (2004).
7. This section derives from Brun and Thorshaug (2020).
8. This information is based on interviews with stakeholders (government officials, international humanitarian organisations and national civil society organisations).
9. Rehabilitation of Collective Centres took place before or during privatisation in the western regions of Georgia. The concentration of money and efforts towards the western regions of Georgia must also be understood because the politics of settlement of IDPs are geared towards the western regions, which are closer to Abkhazia but also in order to avoid further urbanisation in Tbilisi.
10. This section relies on Brun (2016a).

References

Ahmed, S. (2006). Orientations. Towards a Queer Phenomenology. *GLQ: A Journal of Lesbian and Gay Studies*, *12*(4), 543–574.

Ahmed, S. (2010). Orientations Matter. In Bennett, J., Cheah, P., Orlie, M. A., & Grosz, E. (eds.), *New Materialisms: Ontology, Agency, and Politics*, Duke University Press, Durham, NC, 234–257.

Alexander, C. (2004). Critiquing Property Value, Relations, and Changing Bodies: Privatization and Property Rights in Kazakhstan. In C. Humphrey & K. Verderey (eds.), *Property in Question. Value Transformation in the Global Economy*, Berg Publishers, Oxford.

Bayat, A. (2015). Plebeians of the Arab Spring. *Current Anthropology*, *56*(Supplement 11), S33–S43.

Brun, C. (2003). Local Citizens or Internally Displaced Persons? Dilemmas of Long Term Displacement in Sri Lanka. *Journal of Refugee Studies*, *16*(4), 376–397.

Brun, C. (2015). Active Waiting and Changing Hopes. Toward a Time Perspective on Protracted Displacement. *Social Analysis*, *59*(1), 19–37. Special issue on Conflict, Mobility and Uncertainty edited by C. Horst and K. Grabska.

Brun, C. (2016a). Dwelling in the Temporary: The Involuntary Mobility of Displaced Georgians in Rented Accommodation. *Cultural Studies*, *30*(3), 421–440. Special issue on (Im)mobilities of Dwelling, edited by L. Meier and S. Frank.

Brun. C. (2016b). There is no Future in Humanitarianism: Emergency, Temporality and Protracted Displacement. *History and Anthropology*, *27*(4), 393–410.

Brun, C., and A.H. Fábos. (2015). Homemaking in Limbo? A Conceptual Framework. *Refuge*, *31*(1), 5–18.

Brun, C., and A.H. Fábos. (2017). Mobilising Home for Long Term Displacement: A Reflection on the Durable Solutions. *Journal of Human Rights Practice*, *9*(2), 177–183.

Brun, C., and R.Ø. Thorshaug. (2020). Privatized Housing and Never Ending Displacement: The Temporality of Dwelling for Displaced Georgians. In S. Pasquetti & R. Sanyal (eds.), *Displacement: Building Global Conversations on Refuge*, Manchester University Press, Manchester, 116–130.

Brun, C., A.H. Fábos, and O. El Abed. (2017). Displaced Citizens and Abject Living. The Categorical Discomfort with Citizens out of Place. *Norwegian Journal of Geography*, *71*(4), 220–232.

Chimni, B.S. (2004). From Resettlement to Involuntary Repatriation: Towards a Critical History of Durable Solutions to Refugee Problems. *Refugee Survey Quarterly*, *23*(3), 55–73.

Cover, R., M.L. Rasmussen, P. Aggleton, and D. Marshall. (2017). Progress in Question: The Temporalities of Politics, Support and Belonging in Gender- and Sexually-Diverse Pedagogies, *Continuum*, *31*(6), 767–779.

Cresswell, T. (2009). Place. In N. Thrift & R. Kitchin (eds.), *International Encyclopedia of Human Geography, Vol 8*. Oxford: Elsevier, 169–177.

Cresswell, T. (2010). Towards a Politics of Mobility. *Environment and Planning D: Society and Space*, *28*, 17–31.

Crisp, J. (2003). *UNHCR, Refugee Livelihoods and Self-Reliance: A Brief History*. UNHCR, UK, https://www.unhcr.org/cgibin/texis/vtx/search?page=search&docid=3f978 a894& query=jeff%20crisp, accessed 29.09.2019.

Cwerner, S. B. (2001). The Times of Migration. *Journal of Ethnic and Migration Studies*, *27*(1) 7–36.

Easton-Calabria, E., and N. Omata. (2018). Panacea for the Refugee Crisis? Rethinking the Promotion of 'Self-Reliance' for Refugees. *Third World Quarterly*, *39*(8), 1458–1474.

Fontanari, E. (2017). It's My Life. The Temporalities of Refugees and Asylum-Seekers within the European Border Regime. *Etnografia e Ricerca Qualitativa*, *10*, 25–54. https://www.rivisteweb.it/doi/10.3240/86886.

Frederiksen, M.D. (2013). *Young Men, Time and Boredom in the Republic of Georgia*. Temple University Press, Philadelphia.

Government of Georgia. (2007). *State Strategy for Internally Displaced Persons Persecuted*. Government of Georgia, Tbilisi.

Hannam, K., Sheller, M., and Urry, J. (2006). Editorial: Mobilities, Immobilities and Moorings. *Mobilities*, *1*(1), 1–22.

Heidegger, M. ([1954] 1971). Building Dwelling Thinking. In *Poetry, Language, Thought*. Translation and introduction by Albert Hofstadter. New York, Harper & Row, 145–161.

Holston, J. (2009). *Insurgent Citizenship: Disjunctions of Democracy and Modernity in Brazil*. Princeton University Press, Princeton, NJ.

Humphrey, C. (2007). New Subjects and Situated Interdependence: After Privatisation in Ulan-Ude. In C. Alexander, V. Buchli, & C. Humphrey (eds.), *Urban Life in Post-Soviet Asia*, University College London Press, Abingdon, 175–207.

IDMC. (2009). *Georgia: IDPs in Georgia Still Need Attention*. Retrieved February 22, 2010 from IDMC: Internally displaced persons (IDPs) in Georgia: http://www.internal-displacement.org/8025708F004CE90B/%28httpCountries %29/F62BE07C33DE4D19802570A7004C84A3?OpenDocument, accessed 15.06.2011.

ICRC. (1999). *Country Report Georgia/Abkhazia*. ICRC Worldwide Consultation on the Rules of War, ICRC, Geneva.

Kabachnik, P. (2012). Wounds that Won't Heal: Cartographic Anxieties and the Quest for Territorial Integrity in Georgia. *Central Asian Survey*, *31*(1), 45–60.

Khosravi, S. (2019). What Do we See if we Look at the Border from the Other Side? *Social Anthropology*, *27*(3), 409–424.

Kibreab, G. (1999): Revisiting the Debate on People, Place, Identity and Displacement. *Journal of Refugee Studies*, *12*, 384–410.

Kraftl, P., and Adey, P. (2008). Architecture/Affect/Inhabitation: Geographies of Being-In Buildings. *Annals of the Association of American Geographers*, *98*, 213–231.

Latimer, J., and Munro, R. (2009). Keeping and Dwelling: Relational Extension, the Idea of Home, and Otherness. *Space and Culture*, *12*(3), 317–331.

Lees, L., and Baxter, R. (2011). A 'Building Event' of Fear: Thinking Through the Geography of Architecture. *Social & Cultural Geography*, *12*, 107–122.

Malkki, L.H. (1992). National Geographic: The Rooting of Peoples and the Territorialization of National Identity among Scholars and Refugees. *Cultural Anthropology*, *7*(1), 24–44.

Malkki, L. (1995). *Purity and Exile. Violence, Memory, and National Cosmology among Hutu Refugees in Tanzania*, The University of Chicago Press, Chicago.

Manning, P. (2009). The Hotel/Refugee Camp Iveria: Symptom, Monster, Fetish, Home. In K. Van Assche, J. Salukvadze, & N. Shavishvili (eds.), *City Culture and City Planning in Tbilisi: Where Europe and Asia Meet*, Mellen Press, Lewistone, 319.

Massey, D. (2005). *For Space*, Sage, London.

Milner, J., and G. Loescher. (2011). *Responding to Protracted Refugee Situations. Lessons from a Decade of Discussion. Forced Migration Policy Briefing 6, Refugee Studies Centre*, University of Oxford, Oxford.

Ministry of Refugees and Accommodation. (2010). *IDP Housing Strategy and Working Plan*, Government of Georgia, Tbilisi.

Mukherjee, A. (2019). After the Empire: Narratives of Haunting in the Postcolonial Spectropolis. *South Asian Review*, https://doi.org/10.1080/02759527.2019. 1604061

Norwegian Refugee Council. (2013). *Privately Accommodated IDPs in Georgia. Needs Assessment*, NRC, Tbilisi.

Nyers, P. (2019). Humanitarian Hubris in the Global Compacts on Refugees and Migration. *Global Affairs*, https://doi.org/10.1080/23340460.2019.1644188.

Pasucci, E. (2016). Transnational Disruptions: Materialities and Temporalities of Transnational Citizenship among Somali Refugees in Cairo. *Global Networks*, *16*(3), 326–343.

Sharma, S. (2014). Because the Night Belongs to Lovers: Occupying the Time of Precarity. *Communication and Critical/Cultural Studies*, *11*(1), 5–14.

Stepputat, F. (1999). Dead Horses? Response to Kibreab. *Journal of Refugee Studies*, *12*, 416–419.

Salukvadze, J. (2009). Market Versus Planning? Mechanisms of Spatial Change in Post-Soviet Tbilisi. In K. Van Assche, J. Salukvadze, & N. Shavishvili (eds.), *City Culture and City Planning in Tbilisi: Where Europe and Asia Meet*, Mellen Press, Lewistone, 159–188.

Struyk, R. J. (1996). Housing Privatization in the Former Soviet Bloc to 1995. In G. Andrusz, M. Harloe, & I. Szelenyi (Eds.), *Cities After Socialism. Urban and Regional Change and Conflict in Post-Socialist Societies*, Blackwell Publishers, Oxford, 192–213.

Tbilisi City Hall. (2018). *Tbilisi in Figures*, Tbilisi City Hall, Tbilisi.

The Expert Council's Research Unit (SVR Research Unit). (2018). *What Next for Global Refugee Policy? Opportunities and Limits of Resettlement at Global, European and National Levels*, SVR Research Unit, Berlin.

Thorshaug, R.Ø. (2011). *Interminably Displaced? Home-Making Practices among Georgian IDPs in Collective Centres in the Context of Privatization*. MPhil thesis, Department of Geography, Norwegian University of Science and Technology, Trondheim.

Thorshaug, R. Ø. (2018). Arrival in-between: Analyzing the Lived Experiences of Different Forms of Accommodation for Asylum Seekers in Norway. In Meeus, B., Arnaut, K., & Van Heur, B. (eds.) *Arrival Infrastructures*, Springer International Publishing, London, 207–227.

Thorshaug, R.Ø., and C. Brun. (2019). Temporal Injustice and Reorientations in Asylum Reception Centres in Norway: Towards a Critical Geographies of Architecture in the Institution. *Fennia*, *197*(2), 232–248.

UNHCR. (2015). After 23 Years, Still Hoping for a Better Future. UNHCR, http://www.unhcr.org/uk/news/latest/2015/6/559166596/23-years-still-hoping-better-future.html, accessed 25.09.2017.

Van Assche, K. (2009). Introduction. City Culture and City Planning in Georgia. Identities and their Remaking. In K. Van Assche, J. Salukvadze, & N. Shavishvili (eds.), *City Culture and City Planning in Tbilisi: Where Europe and Asia Meet*, Mellen Press, Lewistone.

van der Zande (2009). Foreword. In K. Van Assche, J. Salukvadze, & N. Shavishvili (eds.), *City Culture and City Planning in Tbilisi: Where Europe and Asia Meet*, Mellen Press, Lewistone.
Vardosanidze, V. (2009). Social Dimensions of Urban Development in Post-Soviet Georgia. The Quest for Participatory Planning in Shattered Social Landscapes. In K. Van Assche, J. Salukvadze, & N. Shavishvili (eds.), *City Culture and City Planning in Tbilisi: Where Europe and Asia Meet*. Mellen Press, Lewistone, pp. 189–208.
Zournazi, M. (2002). *Hope: New Philosophies for Change*, Pluto, Annandale.

8 Learning to be a city

Emerging practices for housing the displaced in Bar Elias (Lebanon)

Joana Dabaj, Camillo Boano and Howayda Al-Harithy

Urbanization and displacement

This chapter focuses on Bar Elias, a Lebanese town located close to the Syrian border in the main agriculture plains of the Zahle district, the Beqaa Valley. It asks how Bar Elias is coping and transforming into an urban environment after the recent increase in incoming refugee and displaced populations. It defines the urban beyond policy interventions related to refugeehood and displacement. In doing so, it adopts the concept of 'refugeeness as urbanism' (Fawaz et al., 2018) that provides a more fluid reading of the urban. It contributes to literature that explore the 'urbanization of refuge' (Fawaz et al., 2018; Landau et al., 2016; Woodrow, 2017), the 'humanitarian city' (Sanyal, 2011, 2012, 2017), and 'refugee-refugee relations' (Berg and Fiddian-Qasmiyeh, 2018; Fiddian-Qasmiyeh and Qasmiyeh, 2018, Fiddian-Qasmiyeh, 2016; Gill, 2018). This chapter, thus, takes informal transactions in housing practices, and their plasticity and vitality, as indicators of Bar Elias 'learning to be a city'. The purpose is to understand housing as a process of inhabiting (Boano, 2019; Boano and Astolfo, 2020a) and producing an urban environment in Bar Elias, while reflecting on the challenges that Syrian refugees face while living under protracted displacement.

Bar Elias has a long-standing history of economic and social ties with Syria. The Litani River that passes along Bar Elias and the Beqaa Valley's fertile soil enhances agricultural production and commerce. During the Lebanese Civil War (1975–1990), the town of Bar Elias became an important trade hub strategically located halfway between Beirut and Damascus, at around 55 kilometres from each. It is described by locals as *Emm Al-Gharib* (the mother of strangers), currently hosting around 7,000 Palestinian refugees who arrived to Lebanon in the 1950s and around 45,000 Syrians displaced since 2011 (UNDP, 2018).

Therefore, Bar Elias is 'learning to be a city' where rural and agricultural forms of inhabiting adapt to a new urban future. The urban is crafted through modalities of housing provision and socio-spatial practices that emerged in response to the arrival of Syrian refugees. Multiple micro-practices of doing, undoing, renting, partitioning, subdividing, roofing and occupying become the dominant modes of producing the city.

The chapter investigates these housing and socio-spatial practices in Bar Elias where protracted displacement is the central feature of urbanization. It is based on research findings using mixed methods – ethnographic, historical and spatial – by the authors, part of the Vital City[1] research strand at the RELIEF Centre.[2] The chapter is structured in two main parts. It introduces the historic development of Bar Elias and elaborates on the different modalities of inhabitation, dwelling and socio-spatial practices in two areas in Bar Elias: The expansion area of the city centre and the peripheral agriculture fields. It then presents the multiple systems that Syrian refugees use to negotiate, secure and access housing. Sanyal (2012) describes these systems as the manifold and multiple practices of consolidation, transgressions and expansions that allow displaced populations to seek refuge in Bar Elias.

The different practices that are discussed in this chapter are not only modes of 'emergence' from everyday activities and politics of actions. They respond to a wider spectrum of institutional and normative 'adaptations' to a new urban territorial reality. Mr. Mayta, a board member of the Bar Elias Municipality, states that the municipality is struggling to meet the needs of local people. The Beqaa Valley has long been neglected by the Lebanese government that prioritizes investments in big cities. Therefore, humanitarian aid plays a crucial role in channelling money to Bar Elias.[3] This series of humanitarian led investments in different areas of urban infrastructures, among other geographic, historic and political factors, have accelerated the process of Bar Elias 'learning to be a city'.

Urban materiality in Bar Elias

Bar Elias is the second largest town in the Beqaa Valley after Zahle and the largest local Sunni town. The Beirut–Damascus Highway cuts across the town's urban fabric, creating a bustling commercial spine that makes Bar Elias an important commercial hub. The town is located at a nineteen-minute drive by car from Lebanon's border with Syria, separated by the towns of Anjar and Masnaa.

The majority of people living in Bar Elias throughout its modern history have secured their livelihood through agriculture production and commerce, by growing wheat, barely, potatoes, onions, beets and garlic. On the one hand, the agricultural sector was severely affected by ongoing local and regional conflicts. On the other hand, the strategic location of Bar Elias enhanced its commercial role during the Lebanese Civil War (1975) and the Syrian war since 2011. Bar Elias is currently a mixed community: It is home to around 60,000 to 70,000 Lebanese people, 7,000 Palestinian refugees (PRL) and 45,000 displaced Syrian since 2011[4]. It also hosts Iraqi, Bedouins and Palestinian refugees from Syria (PRS). The wave of displacement of Syrian refugees to Bar Elias since 2011 resulted in a rapid population growth. Bar Elias currently hosts the highest number of registered Syrian refugees in the Beqaa Valley, recorded at 30,550 in 2019 (UNDP, 2018; UNHCR Inter-Agency Information Management Unit, 2019).

There is no documentation of the historical evolution of the urban fabric of Bar Elias. According to an excavation between 1933 and 1966, *Tell* Bar Elias, the historic centre of Bar Elias, is an archaeological site that dates to the Early Bronze Age. Satellite images show how *Tell* Bar Elias developed in the 1970s with small agricultural areas in the buffer zone. By 2001, the agricultural areas started to disappear until 2010 when they were fully developed (Savage and Rempel, n.d.).

During the Lebanese Civil War, the economy of the Beqaa Valley thrived as battles in and around Beirut encouraged economic decentralization (Harris, 1985). As a result, Bar Elias experienced high economic growth. Muslims no longer visited the primarily Christian town of Zahle amidst sectarian tensions and instead, relied on secondary retail shops in Bar Elias. During this period, the fabric of Bar Elias grew to a point where the Beirut–Damascus Highway from both sides was lined up with shops and rudimentary services. This economic growth heavily impacted land prices in Bar Elias which increased by 388 percent (Table 8.1).

The predominance of Sunni populations is one of the main factors that attracted displaced Syrians fleeing the conflict in 2011 to settle in Bar Elias. The Beqaa Valley also has enduring economic and social ties with Syria. Syrian migrants constitute the majority of seasonal labourers in agricultural fields. Syrians moved to Bar Elias in high numbers since 2011 and developed social and familial networks. This familiarity facilitated their access to work opportunities, basic services, and protection. These enduring clan and family ties offset national political divides (UNDP, 2017).

Table 8.1 Changes in land prices in the central Beqaa Valley between 1976 and 1984.

	Locality	*1976*	*1984*	*% Price Change 1976–84*	*% Price Change Adjusted for Inflation 1976–84*
(i)	*Zahle (within the Syrian encirclement)*				
	Industrial park	65	300	362	49
	Boulevard	1300	3000	131	–25
	Upper Boulevard	2000	5000	150	–19
	Hammar (hills)	30	250	733	170
	Qa'a al Rim	50	500	300	30
(ii)	*Outside Zahle*				
	Furzul	125	500	300	30
	Furzul hills	3	16	433	67
	Ksara	400	600	50	–51
	Ksara Highway	450	1800	300	30
	Bar Elias	50	750	1400	388

Source: Adapted from Harris, 1985.

This is evident in an interview with Sabah, a senior Syrian woman from the Darayya suburb of Damascus. She is currently renting a flat with her husband, two sons and grandson in an unfinished building in the Makkawi neighbourhood in Bar Elias. Makkawi is part of the expansion area of the city centre. Sabah stresses the importance of families and social networks in providing Syrian refugees with tenure security:

> We haven't paid rent for 2 months, but the landlady understands our situation. She visits us and speaks to us, so she is aware of our circumstances. I told her that when the summer arrives, there will be more work opportunities and we will be able to pay the rent. She also stores her belongings in one of our rooms, so she is nice. She knows our family very well because my eldest son came to Bar Elias before us and stayed at her mom's house for 4 years. He was renting a room in her house.[5]

At the same time, the Bar Elias Municipality is mostly positive and welcoming to Syrians. During the recent waves of resettlement in Bar Elias, the municipality welcomed more than 3,000 refugees who were evicted from sites such as the Riyak airbase in 2017 (UNDP, 2018). The mayor of Bar Elias, Mawas Araji,[6] describes positively the impact that Syrian refugees had on Bar Elias. They contributed to trade and commercial activities. International humanitarian agencies also provided large amounts of funding. According to the UNDP (2017), economic activities were revived and roads were repaired. Médecins Sans Frontières (MSF) founded a new hospital in Bar Elias that cost around 3 million dollars. Funds were used to build a new solid waste sorting and treatment plant that cost around 4 million dollars, including repairs to the water canals. In the process, the role of the municipality becomes increasingly important in managing the livelihoods of the refugees, including their security, mobility, shelter, economic activities and access to humanitarian aid (UNDP, 2017). Even though the municipality received funds from external donors for development projects, it still struggles to manage an increased reliance on urban public services due to the high population increase (UNDP, 2017).

This population increase densified the built fabric of the Bar Elias old residential centre. Lebanese employers benefited from the increase in low-paid labour (UNDP, 2017), which also decreased local construction costs. The following text illustrates the transformations in Bar Elias due to the densification of neighbourhoods, including the proliferation of informal tented settlements (ITSs), investments in infrastructural developments, and incremental growth and transformations in building practices. The text explores these transformations through what Ortega (2016) calls 'hometactics' and the multiple types of inhabitations that give voice to objects, their materiality and spatial assemblages. These 'hometactics' mediate actions and politics that Caldeira (2017) calls 'auto-construction' and 'transversal engagements' with the official logic of land property, state regulations and formal labour markets.

Architectural and ethnographic research methods were used to gather the needed data, supplemented by the analysis of satellite images and field transect walks and observations to document the evolution of Bar Elias. Qualitative interviews as part of the ethnographic research provided an in-depth account of housing modalities and socio-spatial narratives in Bar Elias.

The territorial complexity of Bar Elias: 'Hometactics' and 'auto-constructions'

The juxtaposition of different satellite images narrowed down the research to two main areas where spatial transformations became visible: The expansion area of the city centre and the peripheral agriculture fields. The aim is to highlight key patterns of transformations and the emerging modalities of housing provision in response to the arrival of Syrian refugees. These patterns are explored through socio-spatial narratives that highlight the multiple micro-practices of doing, undoing, renting, partitioning, subdividing, roofing and occupying Bar Elias.

Bar Elias expanded from Tell Bar Elias and the old residential zone towards the Litani River to the north and on both sides of the Beirut–Damascus Highway to the south, on both sides of the Beirut–Damascus highway, and along the Bar Elias–Riyak Road. It further expanded between the Pan–Arab Highway and the Beirut–Damascus Highway (Figure 8.1). Detached urban pockets developed on the outskirts of the agricultural fields.

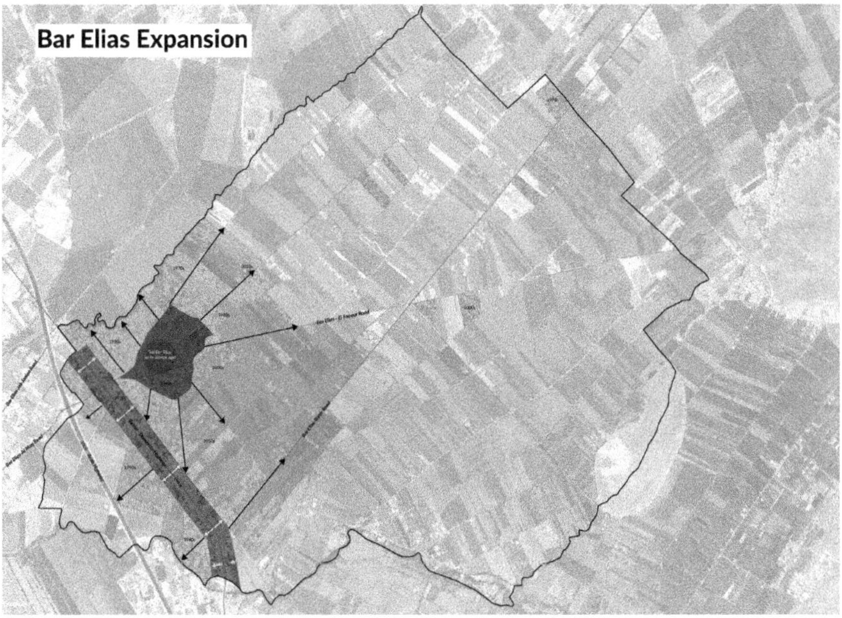

Figure 8.1 The expansion of Bar Elias

Bar Elias stretches over 35 square kilometres of predominantly agricultural lands, with 7 square kilometres of residential areas that are studied in this chapter (Figure 8.2). The city centre expanded into a larger residential zone, while the peripheral agricultural fields were densified with housing structures.

Satellite images were used to map permanent and temporary transformations in 2010, 1 year before the Syrian war, and again in 2014 and 2017. Concrete is the most commonly used construction material in permanent structures, while temporary ITSs are made of wood and tarpaulin.

Construction activity was mostly concentrated in the expansion area of the city centre due to a high surface investment ratio of 35 percent, while the maximum surface investment ratio in the agricultural fields is 5 percent. The majority of construction work took place in 2014. A key informant at the municipality attributed this spike in construction activity in 2014 to quicker and cheaper municipal building permits. In 2014, the municipality unofficially approved many building permits that typically require a lengthy and costly permission from the Lebanese Directorate General of Urban Planning. Furthermore, public officials allowed land subdivision of agricultural fields during the Lebanese Civil War. These pieces of lands are currently granted municipal building permits that allow higher surface investment rates without the proper infrastructural networks.[7] Even though

Figure 8.2 The 'city centre expansion' and the 'peripheral agriculture fields' in Bar Elias

land subdivision on agricultural fields became harder and permits became more controlled, the impact of permitting practices during the Lebanese Civil War is still evident until today. For example, satellite images depict a growing residential neighbourhood in the peripheral agricultural fields on the northern border of Bar Elias. The neighbourhood is witnessing an increase in construction activity. The owner of the first house that was built in this neighbourhood in 1983 explained that locals are primarily Bedouins from *Ashira*[8] *Al-Hrouk*. They originated from a nearby town called *Faour* but had to relocate after a conflict with their *Ashira* members. Their *Ashira* was able to buy and build on the owner's land that was subdivided during the Lebanese Civil War. The neighbourhood continues to grow with a population of around 3,000 people, without a proper waste water network, drinking water, roads or sidewalks. *Ashira* communities that live in the neighbourhood can vote in other towns in Lebanon as opposed to Bar Elias, a condition for them to gain the Lebanese nationality. According to the house owner, the Bar Elias Municipality does not provide neighbourhood upgrades because *Ashira* communities are not allowed to vote in Bar Elias.

The following text illustrates the socio-spatial narratives as emerging housing practices in two territories.

Emerging housing practices in Bar Elias

According to Bahn (2019), 'housing is not about a house'. As Boano and Astolfo (2020b) suggest, a house is a form of 'inhabitation'. Mapping housing practices in Bar Elias requires an understanding of the multiple forces that impact the built environment, including complex daily strategies of learning, navigating and governing. When discussed in the context of displacement, such an understanding allows us to explore the current and past experiences of those who 'have to integrate', by recognizing the right of displaced populations and refugees to determine their housing futures. These experiences are evident in the housing provision and socio-spatial practices in Bar Elias that is 'learning to be a city'. People shape, negotiate and secure dwelling space. Different socio-spatial practices emerge from everyday activities and micro-acts of doing, undoing, renting, partitioning, subdividing, roofing and occupying. These socio-spatial practices are explored in this chapter through ethnographic recordings of narratives in two territories.

Territory 1: City centre expansion

The city centre expanded to house Syrian refugees through long-term family investments. This is the case for buildings that were constructed after 2011. The rent is free for the first year and a half under the Occupancy Free of Charge[9] programme by humanitarian agencies such as the Norwegian Refugee Council (NRC) and the Danish Refugee Council (DRC). Through a private and direct relationship with property owners, the rent of an apartment ranges between 180,000 LBP and 400,000 LBP per month. Fahim, a Lebanese property owner, explains in an interview:

> I built this house in 2017 for my children and their families when they get married. It was a good time because we had some money saved and building construction costs were cheaper. Last year I rented out the first floor through the NRC that provided money for me to finalize some construction works. I am now looking for Syrian refugees to rent the apartment for 400,000 LBP per month[10] (Figure 8.3).

For example, the NRC supports the hosting process of displaced people from Syria in Lebanon by offering a package of upgrades to unfinished buildings and other structures in exchange for the rent-free hosting of displaced households. The rent-free period is usually up to a year and a half during which the Syrian household is assumed to become established and financially stable and therefore, able to pay rent (Norwegian Refugee Council, 2014). This programme provides occupancy free of charge or rent reduction along with tenure security. It establishes a market for unfinished buildings or apartments. Property owners of unfinished buildings or new constructions benefit from the upgrades that NGOs are offering. Syrian families benefit from the contribution to their rent, while tenure security that depends on the property owners. In this case, tenure security is not always possible.

Figure 8.3 Fahim's new building in Bar Elias

Syrians refugees who were interviewed in this research were attracted to unfinished buildings. A senior Syrian woman from Darayya expressed her interest in the programme:

> I saw a new floor being built. It was unfinished so I asked if the owners can allow the NRC to house us there. The landowner looked into it but did not approve our request. It would have been good to live one year in a house without having to pay rent.[11]

It is important to highlight that new adaptations and expansions in the built environment and new housing units emerge to meet existing demands that are fuelled by humanitarian practices as well as interventionist policies that expand existing real estate market trends.

As such, organic additions to already existing structures are another housing modality in the expansion area of the city centre. Full or partial floors are added to apartment buildings with permanent or temporary additions to the upper floors. These additions include shared and single, finished and unfinished housing units with internal partitions. The rent is free for up to a year and a half under the Occupancy Free of Charge programme implemented by humanitarian agencies. Through a private and direct relationship with the property owner, the rent ranges from 150,000 LBP to 450,000 LBP per month.

Mostafa, a Palestinian man in his 50s, has been living in a house that his father built in the Berbayta neighbourhood in Bar Elias in 1974. The house expanded into a complex three-storey building. His family lives on the first floor. His sister lives on the ground floor and his brother and his family on the second floor. Mostafa and his brother built the last part of the second floor in 2017 to include two apartments for their children. Mostafa describes his house in an interview:

> We did not build the apartment for the Syrians refugees. They are only here for one year. We built the apartment for our son. The NRC gave me 6,000,000 LBP for a year to complete these apartments and house 4 Syrian families. The money covers the cost of door installations, plastering, tiling, electrical wiring, and water supply among other basic necessities.[12]

At the time of the interview, each apartment was divided into two housing units that share a kitchen and bathroom. Four Syrian households were renting the housing units through the NRC. Both apartments are 90 square meters in size – the first apartment accommodated eight people and the second one four people. They accessed their housing units through the same entrance as the property owners. The relationship between host and displaced in this case was problematic. These four households were under a 6-month lease. The property owners mistreated the previous families. Surveillance cameras are set up over the staircase as a form of control and families are not allowed to use the shared courtyard on the ground floor.

Other housing practices and modes of inhabitation emerged in the interiors of buildings through temporary internal subdivisions and adaptive reuse strategies. The aim of these strategies is to provide short-term housing to Syrian refugees. They include the adaptive reuse of garages, businesses and farms. Syrian refugees communicate directly with property owners. The rent ranges from 150,000 LBP to 250,000 per month. The partitioning is either temporary using wood or textile dividers or permanent using concrete hollow blocks. The subdivision of a big apartment or villa, as well as the conversion from a farm or garage, includes the provision of shared or single bathrooms and water points.

A villa in the Maderes neighbourhood was subdivided into seven housing units with four shared bathrooms. The Syrian refugee households rent from the brother of the property owner who lives in Germany. Each household pays a rent of 200,000 LBP per month excluding electricity and water bills. Housing units on the ground floor have windows installed, while on the first floor the window openings are closed off with plastic bags, tarpaulin or wood. 'We added a wooden partition to create a separate kitchen area. We also added a private shower', described the tenant.[13]

Fahim in the Makkawi neighbourhood transformed his horse stable to housing units that he rents to three Syrian refugees' households. He mentioned: 'I added a small bathroom, a kitchen, I painted it, installed tiles, and fixed the ceiling, I even decorated it'.[14] Each household is composed of three to five people and pays a monthly rent of 150,000 LBP excluding electricity and water bills. The maximum area of each housing unit is 24-square meters and includes a living room/bedroom, a small kitchen and a bathroom. The households furnished their apartments depending on their financial circumstances.

As discussed earlier, one of the main characteristics of Bar Elias is the agricultural land. Empty pieces of agricultural land between buildings and neighbourhoods are rented to include ITSs. In most cases, access to ITSs is secured through renting what is almost exclusively private land, directly leased by a property owner (UNHCR and UN-Habitat, 2014). The rent of each ITS ranges from 100 000 LBP to 150 000 LBP per month. Humanitarian agency in some cases cover the rent for a specific period of time. The agreement is directly with the owner of the land on which the ITS is set up, or rented through the *shaweesh*. The organization of ITSs usually happens through the *shaweesh* whose role varies from one settlement to another; however, he remains the central point of contact for the resident Syrian refugees. The *shaweesh* is responsible for collecting the rents and paying the landowner. He is usually the first person to arrive to a location and negotiate the rent agreement with the property owner. According to a UN-Habitat and UNHCR study on ITSs in Bar Elias in 2014, the *shaweesh* maintained the agricultural land prior to the start of the Syrian war in 2011 and before his family and extended family moved in.

According to UNHCR and UN-Habitat (2014), ITSs should be recognized as parts of the refugee housing market despite their temporary structures. The most common type of housing production in ITSs is self-help where people set up their own tents using materials typically provided by NGOs: wooden poles, tarpaulin sheeting and, in some cases, a basic construction kit. Rental agreements are verbal and unwritten contracts, in most cases, expire every month. This creates a sense of insecurity with common threats of eviction. In a few cases, the rent fee is covered by an organization that selects which refugees can move in the ITS (UNHCR and UN-Habitat, 2014). With the increase in population and the government halting any decision to establish formal camps for displaced Syrians since the start of the war in Syria in 2011, ITSs were set up in neighbourhoods and along the vast agricultural fields in Bar Elias. A statement appears in the Lebanon Crisis Response Plan (LCRP) 2015–2016, 'The Government of Lebanon stresses on all occasions its longstanding position reaffirming that Lebanon is neither a country of asylum, nor a final destination for refugees, let alone a country of resettlement' (Government of Lebanon and the United Nations, 2014). Even though this statement does not appear in subsequent LCRPs (Overseas Development Institute, 2018), it strengthens the government's position against the permanent settlement of foreigners on its territory.

Within neighbourhoods close to the city centre, ITSs are set up on agricultural or empty pieces of land. As the city expands, agricultural lands are disappearing. The shift from agriculture to housing of ITSs represents a greater economic benefit to landowners. As one interviewee mentioned:

> The agriculture died. The government does not recuperate the losses for farmers. It does not encourage the farmers anymore. It does not care about the agricultural sector, hence, the opportunity came to make money, and we took it. We do have a social responsibility but also a larger economic benefit.[15]

Territory 2: Peripheral agriculture fields

New constructions and organic additions to already existing structures constitute an emerging housing practice within the agricultural fields in Bar Elias (Figure 8.4). The aim is to house Syrian refugees in the present and imagined as long-term family investment. With the increase in housing demand and economic movement, landowners aiming to benefit economically invested in their properties. Landowners are selling their agricultural pieces of land or constructing new housing units.

Renting abandoned and empty houses and wealthy ruins in the agricultural fields in Bar Elias recalls, at least aesthetically, the town's thriving past. An extended family of 24 members lives in a dilapidated villa with a unique architectural style (Figure 8.5). The house was occupied by the Syrian army before 2005. The army burnt the structure before leaving. The Syrian family

Figure 8.4 New construction within the agriculture fields in Bar Elias

saw remnants of the fire that they cleaned when they moved in. They have been living there for 4 years now. They rehabilitated the villa to fit their personal needs, connected the electricity wires themselves and tapped into the water well located in the front yard. They pay a rent of 750,000 LBP per month to the Palestinian concierge who maintained the land for 23 years.

Figure 8.5 An abandoned villa that is rented to a Syrian refugee family within the agriculture fields in Bar Elias

The owner is from Kuwait but manages the land through a local lawyer to benefit from the harvesting of wheat and potatoes. The Syrian family bought a car to commute to the city centre and seek job opportunities.

ITSs are a part of the humanitarian mode of production of an urban landscape and are likely the most immediately visible production of local 'inhabitations'. The highest number of recent tents is located on the agricultural pieces of land at the outskirts of Bar Elias. It costs between 70,000 LBP and 150,000 LBP per month to rent a tent. Tell Serhoun (Figure 8.6) is an area located on the North-Eastern border of Bar Elias that now hosts six ITSs. A total of around 450 tents were set up on lands that belong to big families from Bar Elias living away from Tell Serhoun. Tell Serhoun used to be a military base for the Syrian army. In 2005, the Syrian army withdrew from Lebanon and the military base was removed. The military base transformed into a temporary settlement for Syrian agricultural labourers who set up tents seasonally. With the start of the Syrian war in 2011, displaced Syrians started to move to Tell Serhoun and ITSs started to form incrementally. Locals from Bar Elias and neighbouring towns started building in Tell Serhoun because the prices were convenient. The number of permanent units doubled after 2011, with houses rented to Syrians, a mosque and various shops to serve the needs of people living there.

Despite the government's position around setting up camps for Syrian refugees in Lebanon, *Waqf* land in the agricultural fields in Bar Elias was transformed into a camp-like village with unique housing practices. This typology of 'makeshift' settlement is set up by the Union of Relief

Figure 8.6 ITSs in Tell Serhoun in Bar Elias

Figure 8.7 Al-Awda URDA camp in Bar Elias

and Development Agencies (URDA) that is primarily funded by the Gulf Cooperation Council. The URDA provides pre-fabricated houses as well as built-up rooms in gated sites. The built-up rooms are made of concrete blocks and galvanized roofs and are equipped with a kitchen and a bathroom. The ready-made mobile units are characterized by thermal and voice insulation capabilities (Figure 8.7) (URDA, 2015). For example, the URDA refers to one of their settlements in Bar Elias as a 'productive village' and the 'Al-Awda model of a productive village'. According to URDA representatives, Al-Awda is the largest camp in the country. The permission to have *Al-Awda* set up as a camp was described as 'a blind eye from the government'.[16] The camp is fenced off and includes 400 residential units for 1,500 to 2,000 displaced Syrians. The housing units include pre-fabricated structures and built-up rooms. They include a clinic, a school, a mosque, a kitchen, a sewing atelier and a women centre. They are serviced by two water wells, one for potable water and another for service water. A market is also set up, including a butcher's shop, a grocery store, a bakery, two eateries, and clothing, vegetables, and sweets stores that can serve the whole community. The administration of Al-Awda provides internal guards and security around-the-clock (URDA, 2015). According to the URDA, Al-Awda camp is connected to the public sewage system and has an independent water filtration system. Syrian refugees are not required to pay rent. They have been required since 2016 to pay monthly electricity and heating bills from US$20 to US$35 (Sanyal, 2017). As such, despite the government's position on banning the set-up of camps for Syrian refugees, local conditions and modes of governance in Bar Elias have produced an alternative camp typology that serves as another form of inhabitation.

These examples of 'hometactics' (Ortega, 2016) give voice to built objects, their materiality and complex spatial assemblages in the different territories of Bar Elias. People and families, whether displaced or not, 'auto-construct' their presence and their inhabitation while engaging with the official logic of property markets, land and house owners, formal labour, and informal and sectarian brokers and local powers. Such practices exemplify the manifold and multiple processes of consolidation, transgression and expansion that allow displaced populations to secure a place to live in Bar Elias. However, when considered as an assemblage, they confirm their agency in shaping, forging and producing a new urban territory. This territory is in the process of 'learning to be a city' while exerting power over the urban expansion process. It emerges from what Lancione (2019) defines as 'uncanny and uninhabitable places' and what Deleuze describes as an 'immanent and perpetual becoming urban'. As such, housing Syrian refugees in Bar Elias reflects the multiple modes of producing an urban environment that is in constant struggle to adapt to protracted displacement.

Diagramming spatial practices and housing transactions

In order to visualize the complexity of economic opportunities and housing practices, Figure 8.8 depicts the physical transformations of different types of housing practices within the expansion area of the city centre. The following text presents some of the complex assemblages explored during the research, presented in the form of spatial narratives.

Saleem, a Lebanese landowner in his 70s, explained:

> We initially thought that the Syrians will be here for one or two years– we did not expect this to last until today. There was an increase in the number of Syrian refugees in Lebanon. People who own two houses are renting one of them to Syrians. People were doing patch work on their houses to rent them out. Syrians needed to work and secure a living. They wanted to manage shops: restaurants and tailor, plumbing, or mechanics shops. Therefore, the demand for housing and commercial facilities increased. A shop on the Beirut-Damascus Highway was initially rented for US$100 a month. When the demand increased, some of the Syrian refugees were able to pay the rent of the same shop for US$300. This demand inspired locals to build and rent so they can secure additional income.[17]

Saleem was born and raised in Bar Elias. He owns two plots of lands in Al-Nahreya neighbourhood; a two-floor house with a garden that he shares with his son's family; a two floors building that was added in 2017, the ground

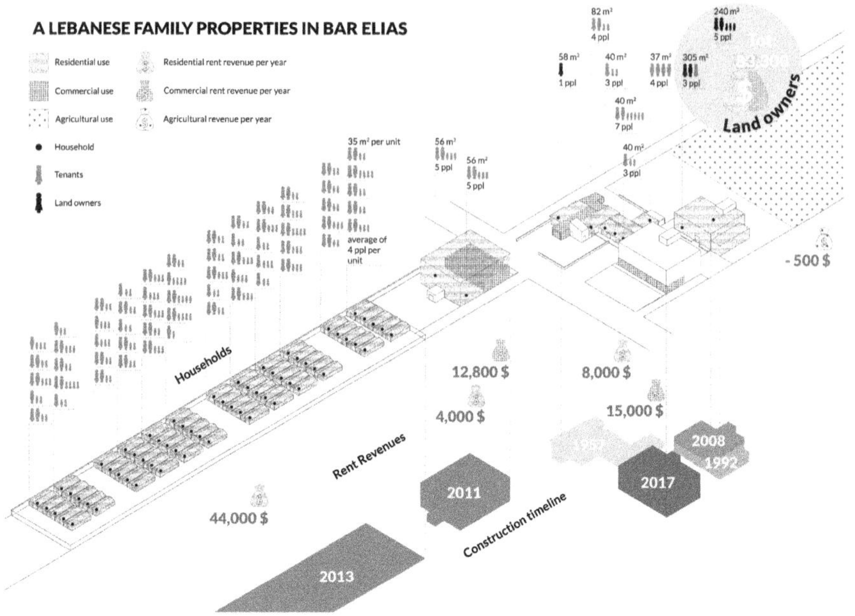

Figure 8.8 Diagram of spatial practices and housing transactions

floor is used for commercial shops while the first floor is under construction; a carpentry operated by his son; an old farm that was subdivided to rent out to Syrian refugees; an agricultural field that grows plums; and a piece of agricultural land across the street now rented out to an ITS, and include a two-floor building that was constructed in 2011.

In 2013, Saleem subdivided in his farm into seven rooms that can accommodate four Syrian families. He added wooden panels and nylon sheets to the roof structure, created a small kitchenette and a bathroom linked to the sewage system in each unit. The old house was subdivided into a carpentry for his son and an apartment rented to a Syrian family. Saleem is renting housing units to five families for 200,000 LBP per month.

The Syrian families have separate entrances to their houses in order to give Saleem and his family some privacy. Saleem mentions, 'your house is here; your entrance is here'. The Syrian families are also required to pay monthly rent. Saleem limits their access to water and electricity, or asks them to vacate the premises, when they default on their rent payments for a long period.

In 2017, Saleem constructed a new building for his eldest in front on the main road. The construction is not yet complete; however, five garage spaces are now rented to Syrians businesses. Since 2017, each business has paid a monthly rent of US$250. The Syrians have activated this type of rental market. Saleem owns another piece of land across the street. His son who lives in Saudi Arabia built a house on this piece of land in 2011 during the first year of the Syrian war. The aim was for the son to get married and start a family in the new house.

With the population increase and the government halting any decision to establish formal camps for Syrian refugees, ITSs were formed within neighbourhoods and along the vast agricultural fields in Bar Elias. An NGO rented the plot of land that used to be a wheat field and set up a refugee settlement. They agreed to pay a rental fee of US$4,000 per year in 2013. The landowner increased the rent to US$20,000 in 2015. The NGO accepted to pay the increased rent for one year only. The Syrian refugee families negotiate and pay their rent directly to the landowner since the NGO is not able to cover the costs any longer. There are 60 tents. The rent for each tent is 100,000 LBP per month. According to the landowner, 'the settlement is in front of our house, in front of our eyes, and we are in control'.[18]

This modality that recognizes informal provision of housing unfinished, just finished or ad-hoc through informal renting agreements confirms what Fawaz (2017) concluded in her research on Nab'ah in Beirut. She argued that the rental market provides access to shelter through a series of rental transactions and social negotiations that transform people's aspirations in life. Zahraa, a Syrian refugee in Bar Elias, explains the agency of refugees in such social negotiations:

> My dreams were big back in the day when I had a decent life in Syria. Now that I have lived through a tough crisis, my dream became modest: I dream that my house is clean, that I have a pillow I can rest my head on, that I am happy when I dress my children with nice clothes and provide them with decent meals, that I feel that my children are living their childhood, and see my daughters strong and equal to their peers. It is the silliest dream in the world: to wake up, drink a cup of coffee in my clean house, my belongings are not expensive but are decent, and feel like I have a decent house. I eat onion and bread and want to see my children laughing their hearts out.[19]

'Living a life' and 'learning to be a city'

McFarlane (2011) reminds us that the urban – as a space and as a terrain – is constituted by 'the actual and the possible'. He stresses the importance of the processes that shape urban trajectories, economies and identities, and habits and practices. Such processes in Bar Elias included informal exceptionality, popular construction, land subdivisions, migrant, low-paid labour markets as well as mass displacement. Humanitarian infrastructures also generated encounters with people and objects to produce new spaces of 'inhabitation' (Boano and Astolfo, 2020b).

The reality of mobility in which the becoming-displaced unfold has forced Syrian refugee families 'to endure the intolerable and struggle to repair and heal [...] themselves from the known and establish new relations (or not), [and] negotiating detours'. They 'make use of these very realities to craft new viable forms of life and project themselves into the future' (Biehl and Locke, 2017, p. 4).

The different narratives that were presented in this chapter allowed for a 'plasticity' in the production of Bar Elias. They shaped the multiple systems and forces in which refugees and hosts interacted to produce housing practices in an agrarian terrain. Syrian refugees and their hosts in Bar Elias materialized what Ingold (2000) describes as 'wayfinding'. They activated micro-practices of doing, undoing, renting, partitioning, subdividing, roofing and occupying that allowed them to negotiate their presence in Bar Elias. As such, the complex entanglements between different forms of spatial arrangements, incremental aesthetics and rent transactions contributed to Bar Elias 'learning to be a city'.

The vitality, experimentation and plasticity that emerged in Bar Elias unfold through unique socio-spatial practices and housing arrangements. They constitute creative ways, which hosts and refugees mobilize to construct a liveable life in a space that is often neither formal nor individual (McFarlane, 2011). In the process, the urban is reconstituted through encounters between individuals and their perceptions, memories and ways of inhabiting the agricultural fields of Bar Elias. Bar Elias's increasingly urbanized environment produced different infrastructures of refuge and 'learning to be a city' as it engaged with non-urban textures and rhythms.

Bar Elias is therefore, both, a double product of denser, more fragmented and infilled spaces and a process of doing and adapting to the hardships of displacement. Due to the protracted nature of displacement, the newly assembled Bar Elias through subdivisions, expansion, Informal Tented Settlements and new constructions can be explored in the *longue durée* of historical evolution. Learning to be a city is a lengthy process that deals with power imbalance and the sharing of resources that denote the 'processual, excessive, and performative [...] sociomaterial rationalities and processes' (McFarlane, 2011, p. 31). Displacement and coping mechanisms produced multiple modalities of housing provision and socio-spatial practices that emerged in response to the arrival of the Syrian refugees.

Notes

1. The R1 'The Vital City' is one of the RELIEF Centre research themes. R1 explores the complex entanglements in space and time of displacements, crisis and city futures focusing on the actual vitality of cities in Lebanon, taking into consideration the local diverse context of space and people as agents in the reproduction of space, in order to understand how to make more resilient and better-quality living environments.
2. The RELIEF Centre is a centre for research and learning focused on inclusive growth and prosperity. It is about the prosperity of Lebanon in particular, but is also part of a larger agenda for developing sustainable ways to improve

the quality of life of people throughout the world. The RELIEF Centre brings Lebanese and UK institutions and expertise together to address this challenge using cutting-edge research and innovation. www.relief-centre.org
3. R1 workshop, 2018.
4. The numbers of different populations living in Bar Elias varies according to different sources. According to a publication by the Lebanese Center for Policy Studies and Safeworld, Bar Elias hosted in 2018 from 60,000 to 70,000 inhabitants, including 30,000 to 45,000 Syrians and 5,000 Palestinian refugees.
5. Interview with Sabah, Bar Elias, 2019.
6. Interview with Bar Elias mayor Mawas Araji, Bar Elias municipality, 2018.
7. Interview with a Lebanese key informant at the Bar Elias Municipality, 2019.
8. *Ashira* is a term that refers to Bedouin tribes.
9. The OFC approach is ostensibly a rent-free shelter intervention, securing 12-months of occupancy for vulnerable refugee households, in accommodation meeting or exceeding minimum standards, in exchange for a conditional financial investment provided to Lebanese landlords to upgrade their properties. The types of accommodation under OFC are either unfinished houses or unfinished apartment buildings and should meet NRC eligibility criteria (Norwegian Refugee Council, 2018).
10. Interview with Fahim, Lebanese property owner, Bar Elias, 2019.
11. Interview with Sabah, Bar Elias, 2019.
12. Interview with Mostafa, Palestinian property owner, Bar Elias, 2019.
13. Interview with a Syrian mother, Bar Elias, 2019.
14. Interview with Fahim, Lebanese property owner, Bar Elias, 2019.
15. Interview with Saleem, Lebanese landowner, Bar Elias, 2019.
16. Meeting with URDA staff, Al-Awda camp, Bar Elias, 2018.
17. Interview with Saleem, Lebanese landowner, Bar Elias, 2019.
18. Interview with Saleem, Lebanese landowner, Bar Elias, 2019.
19. Interview with Zahraa, Syrian female tenant, Bar Elias, 2019.

References

Bahn, G. (2019). Delivering Housing for All – Indian Cities. LSE Urban Age Conference *Development Urban Futures, Addis Ababa*, Addis Ababa. https://urbanage.lsecities.net/talks/delivering-housing-for-all-indian-cities#slides.

Berg, M. and Fiddian-Qasmiyeh, E. (2018). Introduction to the Issue Encountering Hospitality and Hostility. *Migration and Society: Advances in Research*, *1*, 1–6.

Biehl J. and Locke P. (2017). *Unfinished. The Anthropology of Becoming*. Durham: Duke University Press.

Boano, C. (2019). From Exclusion to Inhabitation: Response to Gray, Benjamin. Citizenship as Barrier and Opportunity for Ancient Greek and Modern Refugees. *Humanities*, *8*(125), 1–9.

Boano, C. and Astolfo, G., (2020a). Notes around Hospitality as Inhabitation. Engaging with the Politics of Care and Refugees Dwelling Practices in the Italian Urban Context. *Migration and Society: Advances in Research*, *3*, 222–232.

Boano, C. and Astolfo, G., (2020b). Inhabitation as More-than-dwelling. Notes for a Renewed Grammar. *International Journal of Housing Policy*, *20*(4), 555–577.

Caldeira, T. (2017). Peripheral Urbanization: Autoconstruction, Transversal Logics, and Politics in Cities of the Global South. *Environment and Planning: Society and Space*, *35*(1), 3–20.

Fawaz, M. (2017). Planning and the Refugee Crisis: Informality as a Framework of Analysis and Reflection. *Planning Theory*, *16*(1) 99–115.

Fawaz, M., Gharbieh, A., Harb, M., and Salame, D. (2018). *Refugees as City-makers*. Beirut: Issam Fares Institute for Public Policy and International Affairs, American University of Beirut. http://website.aub.edu.lb/ifi/publications/Documents/research_reports/20180910_refugees_as_city_makers.pdf.

Fiddian-Qasmiyeh, E. (2016). Refugee-Refugee Relations in Contexts of Overlapping Displacement. *International Journal of Urban and Regional Research*. http://www.ijurr.org/spotlight-on/the-urban-refugee-crisis-reflections-on-cities-citizenship-and-the-displaced/refugee-refugee-relations-in-contexts-of-overlapping-displacement/.

Fiddian-Qasmiyeh, E. and Qasmiyeh, Y. M. (2018). Refugee Neighbours & Hospitality. https://refugeehosts.org/2018/03/20/refugee-neighbours-hostipitality/.

Gill, N. (2018). The Suppression of Welcome. *Fennia*, *196*(1), 88–98.

Government of Lebanon and the United Nations. (2014). *Lebanon Crisis Response Plan 2015–2016*. https://www.unocha.org/sites/dms/CAP/2015-2016_Lebanon_CRP_EN.pdf.

Harris, W. (1985). The View from Zahle: Security and Economic Conditions in the Central Bekaa 1980–1985. *Middle East Journal*, *39*(3), 270–286.

Ingold, T. (2000). *The Perception of the Environment: Essays on Livelihood, Dwelling and Skill*. Abingdon: Routledge.

Lancione, M., (2019). Radical Housing: On the Politics of Dwelling as Difference. *International Journal of Housing Policy*, *20*(2), 273–289.

Landau, L., Wanjiku-Kihato, Misago, J.P., Obot, D., and Edwards, B. (2016). *Becoming Urban Humanitarians. Engaging Local Government to Protect Displaced People*. Research Report. Washington D.C.: The Urban Institute.

Lebanon Support and UNDP. (2015). *Conflict Analysis Digest*. https://www.alnap.org/system/files/content/resource/files/main/349-ls-cad-may2014-saida.pdf.

Lebanon Support in Partnership with UNDP. (2015). *Between Local Patronage Relationships and Securitization: The Conflict Context in the Bekaa Region*, Conflict Analysis Reports. https://civilsociety-centre.org/sites/default/files/resources/conflict-anlss-rep-jan2015.pdf.

Lebanese Centre for Policy Studies and Saferworld. (2018). Building Peace into Refugee Responses, Syrian Refugees in Lebanon. https://www.saferworld.org.uk/resources/publications/1171-building-peace-into-refugee-responses-syrian-refugees-in-lebanon-briefing.

McFarlane, C. (2011). *Learning the City. Knowledge and Translocal Assemblage*. London: Wiley-Blackwell.

Norwegian Refugee Council, Lebanon and Jordan, Field Exchange. (2014). *Programming Experiences and Learning from the Nutrition Response to the Syrian Crisis*, Issue 48. https://www.ennonline.net/fex/48/nrcshelter.

Norwegian Refugee Council. (2018). *NRC Occupancy Free of Charge (OFC) Programme*. Washington DC: Norwegian Refugee Council. https://www.daleel-madani.org/sites/default/files/calls_documents/ofc_impact_evaluation_tor.pdf.

Overseas Development Institute. (2018). Dignity and Displaced Syrians in Lebanon 'There is no Karama here'. https://www.odi.org/sites/odi.org.uk/files/resource-documents/12532.pdf.

Ortega, M. (2016). *In-Between: Latina Feminist Phenomenology, Multiplicity, and the Self*. Albany: SUNY Press.

Sanyal, R. (2011). Squatting in Camps: Building and Insurgency in Spaces of Refuge. *Urban Studies*, *48*(5), 877–890.

Sanyal, R. (2012). Refugees and the City: An Urban Discussion. *Geography Compass*, *6*(11), 633–644.

Sanyal, R. (2017). A No-camp Policy: Interrogating Informal Settlements in Lebanon. *Geoforum*, *84*, 117–125.

Savage, S.H. and Rempel, S.G. (n.d.). *Climate Change and Human Impact on Ancient and Modern Settlements: Identification and Condition Assessment of Archaeological Sites in the Northern Levant.*

UNDP. (2017). *The Burden of Scarce Opportunities: The Social Stability Context in Central and West Bekaa.* Conflict Analysis Report. https://data2.unhcr.org/en/documents/download/59757.

UNDP. (2018). *Below the Surface: Results of a WhatsApp Survey of Syrian Refugees and Host Communities in Lebanon.* https://reliefweb.int/sites/reliefweb.int/files/resources/67579.pdf.

UNHCR and UN-Habitat. (2014). *Housing, Land and Property Issues in Lebanon: Implications of the Syrian refugee Crisis.* https://unhabitat.org/housing-land-and-property-issues-in-lebanon-implications-of-the-syrian-refugee-crisis-august-2014/.

URDA. (2015). *2015 Annual Report of the Achievements of URDA-Lebanon, a Comprehensive Humanitarian Vision.* http://urda.org.lb/uploaded/Items/da168194e7f7448cb391636059c44268en%20Enjaz%20BOOK%202015.pdf.

Woodrow, N. (2017). City of Welcome: Refugee Storytelling and the Politics of Place. *Continuum: Journal of Media & Cultural Studies*, *31*(6), 780–790.

9 The recovery of Baghdad's neighborhoods in the aftermath of 'Al-Taifiyah' sectarian conflict

Namariq Al-Rawi

Introduction

Between the end of 2005 and 2008, Baghdad under 'Al-Taifiyah' went through one of the most devastating conflicts in its modern history. The urban and social fabrics deteriorated because of the war following the 2003 occupation of Iraq by the United States and multinational forces (MNF) that generated power vacuum and statelessness. Ongoing socio-political turmoil by violent, antagonist militants transformed the city through the territoriality of conflictual powers.

It contributed to the sectarian identity in Baghdad and turned its population into living targets for violence that serves political agendas. In addition to the physical destruction of the built environment and the deactivation of Baghdad's infrastructural networks that hindered livability, the city witnessed large waves of external and internal displacement. A high percentage of internally displaced persons (IDPs) were relocated within the neighborhoods of Baghdad, leading to a socio-spatial shift in the city.

Under normal conditions, cities are characterized by their geographic context, institutional setups, urban functions, and socio-economic status, among other factors. These factors collectively shape the physical and social structures of cities. They play a substantial role during and after a conflict ends, not only in terms of physical rebuilding, but also in the process of long-term social recovery. They foster the right conditions for reestablishing co-existence and social cohesion that allow for physical reconstruction. However, such characteristics lead to the disruption of social norms and socio-spatial practices within the neighborhoods and the city. The neighborhoods of Baghdad, however, witnessed processes of resilience throughout and after the different phases of conflict brought on by various actors reactivating physical and social networks in the city.

This chapter focuses on the city of Baghdad that underwent socio-spatial destruction during 'Al-Taifiyah' period between 2005 and 2008. It explores the physical changes that caused social transformations, while examining processes of resilience during and after the conflict. It also examines the socio-political interventions that facilitated people's return to the city and discusses the roles of the actors that were involved in the process. A major consideration in the research and creating a concrete timeline of events and socio-spatial transformations of neighborhoods in Baghdad was the lack of up-to-date census data on the city. The complexities of such research required an iterative data collection process that relies on secondary data and various methods including in-depth interviews and consensus-group surveys. A total of 62 residents from neighborhoods in Baghdad agreed to participate and were asked open-ended questions to narrate their account of the changes that occurred in the city during and after 'Al-Taifiyah.'

Impact of conflictual powers over cities

Cities are constituted of dynamic networks of changing land-use distributions and activities (Bourne, 1982). Changes in urban infrastructures can impact the control of urban spaces and effect social interaction and human security. Whyte (1980) studied the relations between social practices and the built environment in a research project called Street Life (1958). While observing the behavior and movement of individuals within the built environment, he concluded that people interact with the space in relation to several urban variables. Accordingly, changes to the urban variables influence social practices in urban spaces and movement patterns within the city (Lehnerer, 2013). Urban environments, however, are constantly under threat from violent acts against the city, including physical destruction caused by conflicts, political interventions to create state sovereignty, and spatial systematic militarization that lead to physical and social disruptions of urban connectivity. Such practices work against urban unity, by fragmenting the urban environment into territories, in a manner that serves conflicts. Socio-spatial destruction, caused by physical interventions on the urban infrastructures of cities, is unquestionably an advance of war. Powers competing over space, as well as the activities of militants against the built, natural, and social environments, are part of the brutality of war, compromising livability in the city and obliging people to relocate (Graham, 2003; Hanafi, 2004).

Geography, state territoriality, and power have always been linked to different levels of political sovereignty. They contribute to the displacement of people and force defensive measurements on public life in the form of surveillance systems, checkpoints, and intensified security measures. Despite their militarized nature, these measurements can also reflect on the inseparability of politics and conflictual powers (Amend, 2010; Galula and Nagl, 2006, 1964; Sparke, 2007). Furthermore, conflicts promote processes of chronic disruption of urban practices. The destruction of socially oriented environments due to war and political influence resembles the effect of displacement of individuals. The elements of such environments arc thc footprints of each individual and community, thus affecting socio-urban values such as unity, integrity, and identity. Such political disruptive attacks are not only a side effect of wars and conflicts, but rather a premeditated assault through militant violence, impacting spaces and their qualities such as density, cosmopolitanism, and heterogeneity (Coward, 2010; Graham, 2003). Meanwhile, political and military controls over the city require the recognition of the social and physical obstacles, reshaped to guarantee army control, and meet the objectives of conflict. Curfews, raids, checkpoints, walls, sieges, roadblocks, and surveillance systems are measures of social strangulation and urban deterioration (Graham, 2003). They promote securitization and the need to alleviate violence, however, obstruct the potential for a cohesive and inclusive society.

Nonetheless, social cohesion and the strength of people are measures of social resilience during and after a conflict ends. Such approaches to resiliency vary across dwelling, neighborhood, and metropolitan scales. Thus, the process of urban recovery differs based on the quality and strength of the socio-characteristics of the inhabiting community. Baghdad, as a case study, was repeatedly the subject of conflicts and power struggles reflected through social and spatial practices inside the city during its recovery process.

Socio-economic understanding of the urban development of Baghdad

Being one of the densest Iraqi cities with a population count estimated at 8.4 million in 2019, Baghdad developed around the Tigris River. Its strategic location of environmental and economic importance reflects political visions since the Abbasid Caliph Abu Jaafar Al-Mansur moved the Capital of the Islamic Empire to Baghdad in the year 755. The most evident traces of the historical parts of the city and its neighborhoods are in Risafa on the eastern bank of Tigris defining the urban fabric of the city. Neighborhoods were characterized by 50 to 150 square meters of uneven plots connected by narrow alleys and occupied by houses with deep courtyards and high roof walls (Al-Ashab, 1974). Nevertheless, following the foundation of the Kingdom of Iraq in 1921, planning process of Baghdad followed political ideologies starting with the modern visions of the Iraqi Monarchy[1], the communist reflections of Abdulkarim Kasim[2], to the socialist visions of the Baathist regime[3].

Moreover, the social composition of Baghdad is heterogeneous, encompassing a religious mix of Muslims, Christians, Mandaens, Jews, as well as Arab ethno-communities, Turkmen, Kurds, and other minorities (Soane, 2007). This heterogeneity was renowned at the time of Baghdad's modernization and afterward through policies of residential land management and allotments, and the physical distribution of infrastructural services in the neighborhoods of Baghdad to sustain socio-economic diversity of the different districts. This was mostly evident in the mid-1950s when policies no longer focused on addressing the housing crisis due to population growth resulting from migration from rural areas. Each district in Baghdad included low- to high-income neighborhoods, adopting socialist principles of solidarity. Residential lands were allocated and named based on profession and homogeneity regardless of ethno-religious backgrounds. Between 1956 and 2000, the state granted 88 percent of the allocated residential lands to state employees from different professions and academic specializations, while private stakeholders invested the remaining 12 percent in residential projects. At the same time, affordable housing was allocated to low-income residents (Figure 9.1) (Al-Sa'di, 2005).

A new legal framework for the comprehensive planning of Baghdad was also essential not only in maintaining the power of the state, but also in regulating development projects in relation to the urban structure of the city. Before 1971, all efforts to administer the land tenure system in Baghdad had failed, causing a rapid population growth in scattered settlements, and instigating unregulated organic urban growth. A heavy burden fell on the state and the local administration to provide services and public utilities to

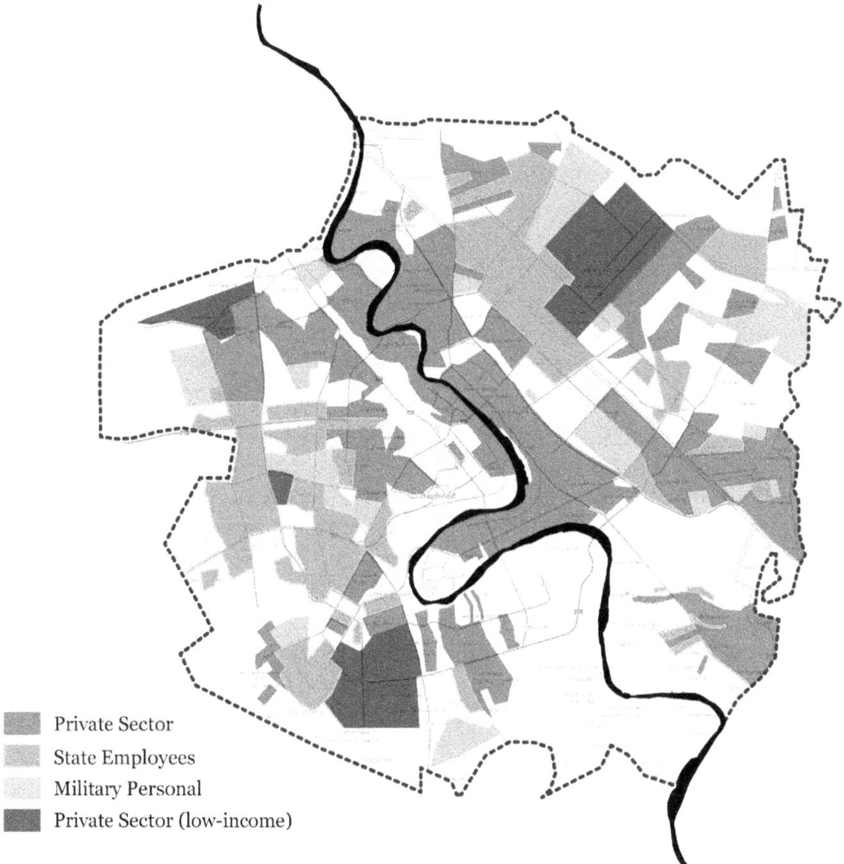

Figure 9.1 Residential land management, allotment, and management after the mid-1950s

the city until Law No. 156[4] was set in place in 1971. The law regulated the growth and development of the city, arranging transportation and infrastructural networks and land-use according to the needs of the city. The law required city planning, administrative systemization, and foundations and general rules for reconstruction. According to Law No. 156/1971, land-use should comply with the basic design guidelines and should be executed under the control of the relevant municipality (Mahmoud, 2014; Relief International, 2016).

Law No. 156/1971 also affirmed the categorization of neighborhoods according to basic design guidelines. It also delineated land-uses in the neighborhoods as follows: 20 to 30 percent of the land should be designated for public usage and circulation, including roads, pathways, and passages; 15 to 20 percent should be allocated to semi-public uses, including schools, parks, and service; and the rest to semi-private and private uses, including residential units and cul-de-sacs. The law also sets a hierarchical distribution of open green spaces and specified volumetric standards for each residential unit, including built-up areas, setbacks, green spaces per capita, and population densities (Table 9.1) (Jawad, 2013).

Another regulation related to residential units that had an impact on urban density was land allotment. Law No. 850[5] in 1979 allowed for the allotment of lands zoned as residential for the purpose of building more than one house on the same parcel as long as the area designated as a residential unit is not less than a minimum of 120 square meters. Amendments to law No. 940[6] in 1987 increased the minimum to 200 square meters with a width of 8 meters. Law No. 01[7] in 1989 allowed for the building of more than one residential unit on parcels with an area of more than 800 square meters, as long as each unit is built on a parcel that is at least 200 square meters without violating the rules of setbacks set for each residential class. Law No. 51[8] in 1989 attempted to reduce the population growth through developments in nearby cities on the outskirts of Baghdad, directing land allotment to outside the city and allocating residential lands only to people from Baghdad. Between 1991 and 2000, an estimate of 120 thousand residential parcels in the different districts of Baghdad was distributed to people who were born in Baghdad and have completed higher education.

To avoid legal violations, Law No. 154[9] in 2001 defined violations against urban planning and design guidelines as acts of malice, including

Table 9.1 Classification of residential areas according to Law No. 156/1971

Residential Area	*Unit Area sqm*	*Green Area/Capita*	*Built-up Area percent*	*Setbacks*	*Density/ha*
Class A	100–200	6	75–100	None	150–200
Class B	201–400	4	65	2.5	80–150
Class C	401–800	3	55	5	50–80

encroachment and misuse of lands and construction that do not comply with the requirements of each zone. Law No. 13[10] in 2001 stated that illegal construction, land encroachment, and easement violation of residential and commercial land-uses are subject to penalty fines.

As a result, Baghdad included 89 neighborhoods distributed across nine districts. The Tigris River provided a natural boundary between the districts of Karkh, Al Mansour, Al Rashid, and Kadhimiyah on the west from Risafa, Karadah, Adhamiyah, Nisan, and Thawra on the eastern bank. These districts were connected by 14 bridges. Physical boundaries, excluding highways like the Airport Highway, were diffused, but the distinction of neighborhoods through urban and architectural characteristics was clear, providing a mix of housing opportunities and experiences within each district. There were no requirements for school district borders, resulting in a mixed pattern of interaction among the inhabitants of the different neighborhoods. The urban fabric of Baghdad incorporated areas of distinctive characteristics combining: The vernacular qualities found in the historic districts of Karkh and Risafa, Adhamiyah, and Kadhimiyah, the 1930s modern architecture houses in Al-Bataween, post-1930s architecture garden cities, contemporary garden cities such as Al Mansour, modern commercial streets, few modern high-rise residential buildings in Haifa Street and Salhiyah, and river islands and waterfronts. These characteristics created the morphology of Baghdad and played a role in shaping the socio-spatial mosaic of the city (Figure 9.2).

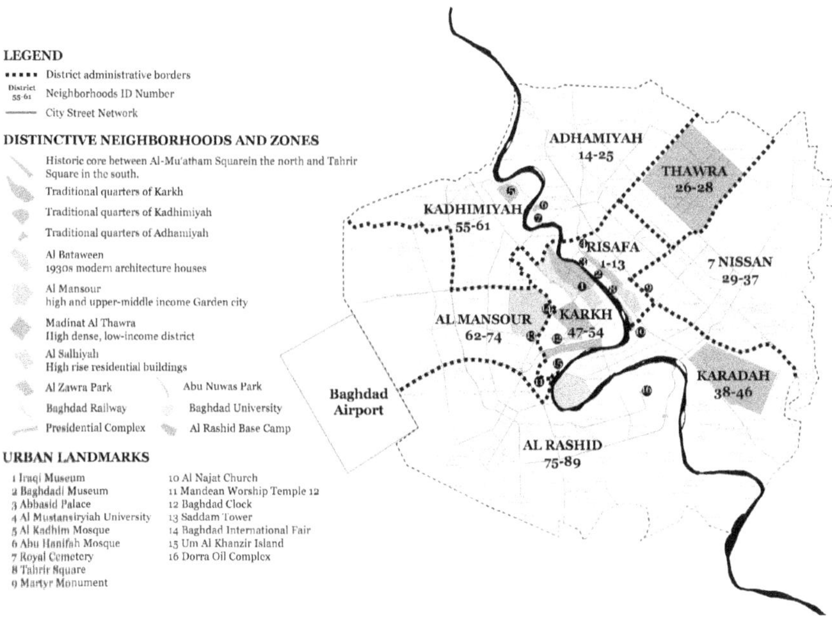

Figure 9.2 Districts and neighborhood distribution in Baghdad before 2003

The process of new sectarian political structure establishment and its impact on Baghdad's neighborhoods

The war by the MNF, and later the occupation of Iraq in 2003 by the United States, established a new power order that influenced the city after the collapse of Saddam Hussein's regime on 9 April 2003. The lack of post-war plans before the invasion proved problematic for Iraq. The ineffective Office of Reconstruction and Humanitarian Assistance (ORHA) established by the United States was replaced in May 2003 with the Coalition Provisional Authority (CPA) run by Paul Bremer. The CPA produced, with its lack of knowledge about the country, *'one of the most isolated occupations in history.'* It located its residences in a security zone in the center of Baghdad, separating them from the rest of Iraq under the premise of the *'Decapitation thesis'*[11]. The thesis was proven to be false and unsupported by efforts of state building. The Iraqi state immediately collapsed since the events relied on a group of Iraqis in exile to provide consultation to the American administration on rebuilding the country. The group fled the country during Hussein's regime and was disconnected from the social reality in Iraq since the 1970s. Nonetheless, their dogmatic view of the Iraqi society was dominant. They argued that society is divided along sectarian lines driven by deep mutual antagonism (Dodge, 2005).

Shortly after the occupation, two meetings were held with local representatives of Iraqis recently returning from exile, with very few representations of the current Iraqi society. On the one hand, given the outcomes of these meetings, Bremer decided to postpone handing over the power to Iraqis until he can find a political group that is compatible with the managerial policies of the occupation. He contradicted the principles of transition from military forces to a local civilian authority, *'an essential step in normalizing the environment and ending insurgency'* (United States Government, 2009, p. 27). On the other hand, Sergio Vieira de Mello, the UN special envoy for Iraq, articulated clear concerns around the lack of a representative Iraqi sovereign body. In July 2003, negotiations between the CPA and Vieira de Mello emanated the foundation of the Iraqi Governing Council (IGC) from seven dominants, formerly exiled, parties chosen by Bremer. This political formation was promoted as *'the most representative body in Iraq's history,'* as the CPA focus was only on sectarian and ethnic social structure of Iraq through the selection process rather than the technical skills of the members, allowing sectarianism to control the political discourse in Iraq. Despite the opposition of the UN represented by Lakhdar Brahimi and their proposal of a technocratic government, the Bush administration set the end of June 2004 as a deadline for the handover of sovereignty to the Interim Government of Iraq, which was also full of formally exiled sectarian and ethnic party representatives who politicized ethnicity and religious affiliation after the generated power vacuum. Their objective was to increase opportunities to acquire territorial expansion and ensure a base of supporters during the elections. Antagonist

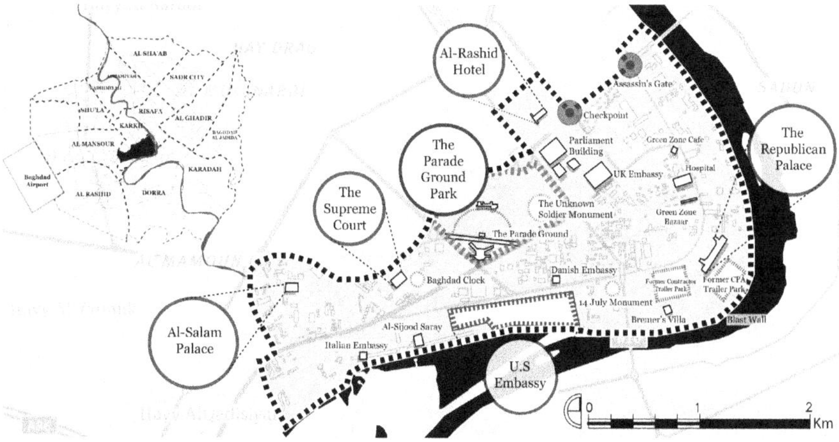

Figure 9.3 Districts distribution and the map of the Green Zone

militias emerged and political assassinations were executed during this process, reflecting political conflicts over the city and its neighborhoods, thus, creating a long-standing insurgency and forcing restrictions over human rights that were promoted before the occupation.

Although the number of neighborhoods and their characteristics remained unchanged, the nine districts were sub-divided into 14 after 2003. Nevertheless, one of the most drastic changes came with the formation of the Green Zone district in the heart of the city, carved out of the preexisting urban pattern. It encircled the former presidential palaces, public open spaces, semi-public buildings, and several neighborhoods on the riverfront. The parameters of the zone were fortified with five-meter blast concrete walls topped with razor wire. The expansion of this district included the expropriation of neighborhoods in proximity to the former presidential compound. This district was isolated and self-sufficient, causing a major change in the patterns of mobility and interaction. It required the closure of one main bridge that connected Karkh to Risafa and the strangulation of movement through other security measures. This limited parcel brought on retrograde urban and social changes in Baghdad on the different scales of the city, even at the level of neighborhoods (Figure 9.3).

'Al-Taifiyah': The eventfulness in Baghdad, 2006–2008

Conflicting powers due to statelessness emerged in Baghdad, ranked before the war and occupation in 2003 as a safe place for everyday urban activities. These powers used sectarianism as an instrument to create territorial invasion based on sectarian grounds. They expanded their territorial claims in the neighborhoods of Baghdad when the sectarian-based violence amplified drastically after the bombing of Al-Askari mosque on 22 February 2006, one

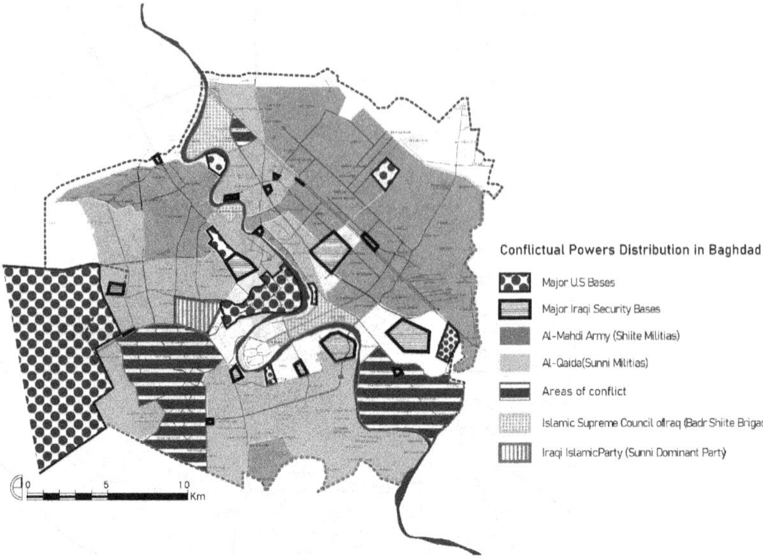

Figure 9.4 Conflictual power distribution in Baghdad after the war and occupation in 2003

of the holiest shrines for the Shiites of Iraq in the city of Samarra (Worth, 2006). This gave rise to a period, referred to by the Iraqis as 'Al-Taifiyah' (Sectarianism in Arabic), of chronic physical and socio-political disruption that lasted violently until 2008, however, had created an ongoing socio-spatial disturbance in the city (Figure 9.4).

As such, the residents of Baghdad lived under the power of subjectification. Mixed and sub-communities residing together in the neighborhoods became associated with a sectarian identity through imposed categories of individuality on the residents (Figure 9.5). Members of Baghdad's society were turned into living targets for violence that reinforced the status of statelessness. Coercion was practiced by the militias against the residents at the discretion of the ruling political agenda. Rates of violence increased 58 times higher than the period before the war of 2003 (Ismael and Ismael, 2005). Simultaneously, there was a systematic assassination and exodus of targeted scientists, intellectuals, and academics. According to the Spanish Campaign against the Occupation and for the Sovereignty of Iraq, it is estimated that around 315 scientists were assassinated in Baghdad among the total of one thousand in Iraq, while 50 percent of the 324 assassinations of academics occurred in Baghdad[12]. At the same time, direct threats spread by the armed militias against the antithetic civilian groups. A total of 80 percent of the interviewed participants for this research shared how their neighbors were forced to leave under direct threats from the antagonist militias. Those threats were filled with hatred and religious justification;

Figure 9.5 Sample area in Jehad/Baiya District showing population subjectification inside neighborhoods

the following quote is taken from a published threat note that was delivered to a Shiite resident:

> In the name of Allah the Merciful [...] Because of your shameful positions and your cooperation with the dirty Crusader occupier [...] Today you are an unwanted epidemic in our regions you rejectionist must leave from here within 48 hours Otherwise the sword is between us.
>
> On the envelope it was written: "House rent 100,000 Dinar and leave the house keys by the Sunni Neighbor, otherwise the house will be destroyed."

On the contrary, dominating militias only encouraged analogous residents to stay in contested areas. They promoted the protection and safety for those fleeing other neighborhoods and inhabitants from informal settlements or rural outskirts of Baghdad only to increase the sectarian population in one neighborhood. In one of their notes to the residents, Brigades of Pure Blood stated:

> To the followers of Aal Al-Bayt, Rest assured, followers of Ali and Hussein peace be upon them, we do not kill on suspicion. We are from your brothers, who are killing only 'Al-Nawasib' and the Baathists. Do not leave your houses nor your shops. The sinner knows himself and the good one knows himself.

This created unbalanced physical and social distribution of housings in the city as residents were evicted and families were expelled from conflict areas, whose residents mostly fled to safer neighborhoods in Baghdad supposedly until the end of conflict as a temporal stage transitioning into informally shared housing arrangements, where two to three displaced families would share the already limited-in-space house of a host family in the relatively safe neighborhood. By early 2007, around 700 thousand Iraqis were internally displaced, while 450 thousand individuals relocated within Baghdad. Moreover, all districts of Baghdad were under such power of exclusion and segregation; from the district of Kadhimiyah, for instance, 7,324 families of versified ethno-religious background were expelled, while equally 7,087 displaced families were welcomed in the district to increase the local Shiite population. Meanwhile, in the district of Adhamiyah, across the river Tigris, 2,458 families were forcibly displaced, and 1,947 displaced families moved to this district, increasing the Sunni sub-society. Some of the districts were too hazardous to receive internally displaced families. For example, 12,828 families in the Doura district were displaced due to the reasons that are mentioned, yet received only 2,244 new residents. Jihad/Baiya district, on the other hand, witnessed a relentless contesting battle that resulted in the displacement of 13,347 families. However, 3,348 displaced families preferred to have the district as their residing destination (IAU, IOM, and UN-Habitat, 2010). As such, the neighborhoods of Baghdad were forcibly transformed from a vicinity for mixed ethno-religious communities into segregated sub-groups during the state of lawlessness, imposing the superiority of the two main religious sects, Sunni, and Shiite. These two sects enforced their territoriality over the new marginalized minorities including Sunni-Shiite mixed families (Figure 9.6).

Furthermore, instead of imposing the dominance of law over the militias, the state succeeded in imposing structural power over Baghdad's population. Not only the population suffered from the fear of sectarian raiding and arrest campaigns, but also a 'gated community' strategy was considered as a suitable security measure against sectarian violence. The MNF started to construct a network of four-meter-high concrete security walls penetrated by armed checkpoints, which accentuated spatial and social divisions in the city (Damluji, 2010). Checkpoints mainly restricted movements in and out of each segregated area and exacerbated traffic congestion on these bottle-necked points. This increased the potential of car bombings at these critical entry points. Some arterial streets were labeled as 'highways of death' due to widespread random death trap checkpoints and booby-traps. This hindered social and economic interaction and limited it in the designated enclosed spaces for each sect. By April 2007, initially, five neighborhoods were selected, including Adhamiyah, Amiriyah, and Khadra to be encircled with blast walls, and later were increased to 10. It is estimated that in Baghdad alone 44 walls and 50 fixed checkpoints, and many other random security measurements had distorted the urban fabric and caused social

Figure 9.6 The sectarian landscape and shift in the social fabric in Baghdad

confusion of spatial practices in the city (Gregory, 2010). Ordinary activities like visiting family and friends became less frequent and turned into a perilous journey that can take hours to reach a destination at a short distance. Only 11 percent of Baghdad's inhabitants were able to access most of the city's neighborhoods freely. 'Al-Taifiyah' imprinted an atrocity of on-going violence that endeavored to impose homogeneity to rupture Baghdad's community solidarity, compromise its heterogeneity, and distort the emotions and meanings attached to it. Such eventfulness has gained significance to the extent of highlighting momentous milestones in any Baghdadi's life timeline. The period between 2006 and 2008 became a collective reference date in the lives and memories of people in Baghdad, referred to as 'Al-Taifiyah.'

As a result, livability was hindered in the city through restrictions through tangible and intangible measurements that obstructed the flow of people's everyday life. Schools, places of worship, hotels, commerce, and even entire neighborhoods were blocked. Shops were progressively closed since their owners were either threatened or killed, while only random stores remained open for a few hours until the early afternoon (Chandrasekaran, 2007). Municipal services were limited to the extent that garbage collectors were afraid of risking their lives every day (Luo, 2006). Furthermore, the scene of lying corpses on the streets had become part of the norm, while performing daily-life activities.

Resilience and adaptability became critical survival mechanisms. Neighborhoods became self-reliant, exclusive sub-communities with one physical access point known only to the resident. These measures facilitated the recognition of any stranger as a source of suspicion. Decentralized systems of privately owned generators providing electricity to each block appeared as a reasonable solution as the government, until recently, had failed to provide proper infrastructural services, consequently, increasing pollution levels within the urban settings of Baghdad. A limited commerce network was generated to serve only the sub-community within the enclosed neighborhood. Local markets opened after squatting and allotting parts of residential buildings, and low-tech vegetable stands invaded vacant and open green spaces, since means of punishments for violations were ineffective. Nevertheless, such resilient systems also provoked organizing self-defense guard duties by the residents of each neighborhood and setting their own checkpoints to hinder any terrorist attack.

The societal shifts also caused spatial distortions in the city, neglecting planning codes and legal requirements at a time of socio-spatial confusion, mainly through different means of encroachment. A comparative study of land-use, conducted by the Baghdad Mayoralty between 2007 and 2011, showed changes in the area in most of the land-use categories except inalienable ones including rivers and road network. There was a noticeable decrease and encroachment of around 2.44 percent of agricultural lands, 1.12 percent of open green spaces, and 1.52 percent of vacant lands.

Figure 9.7 Palm groves in Kadhimiyah District under constant threat of Allotments due to slums invasion

This violation in land-uses was done through illegal allotment and construction of mostly residential units on the new parcels. Simultaneously, an increase of 4.66 percent in residential areas was a consequence of displacement waves since the displaced sought shelter in relatively safe neighborhoods. The uncontrolled population growth and uneven distribution, exacerbated by an insufficient housing stock, resulted in the occupation of public buildings. The squatting and allotment of houses became a suitable solution at that time, transforming some of Baghdad's neighborhoods into a non-anticipated state of salinification. Lands and houses were being allotted into new residential units of less than 50 square meters, which was considered unusual in the building culture of Baghdad. Lands, especially agricultural ones, were encroached and developed into informal settlements. Unsuitable and unsustainable materials, such as wooden boxes and metal cans, were used to build residential units in the encroached lands without building permits or safety standards, generating neighborhoods suffering from infrastructure insufficiency and lack of services (Figure 9.7). The UNHCR estimated that in 2008, more than 250,000 IDPs occupied public properties, while the statistics of the year 2011 showed that 750,000 resided in informal settlements. This number constitutes 12 percent of Baghdad's population.

Post-'Al-Taifiyah': The returning process of the internally displaced

By the year 2008, the atrocious practices of 'Al-Taifiyah' started to gradually diminish inside Baghdad, and the militias were retreating from neighborhoods, thus minimizing their local power. Such statuses provoked a phase of pre-contemplation of the process of returning to neighborhoods. Conflictual violence rates decreased, security measurements were slightly relaxed, municipalities began to redeliver their services, and security forces encouraged residents to open a dialogue of trust among each other; therefore, commerce traffic increased. As part of the contemplation phase, the displaced, who originally saw displacement as a temporal spatial phase, started to track the news of their old neighborhoods, whether there was a possibility for them to return. Afterward, preparation started through initiating contact with former neighbors asking them whether the actors of sectarian practices were still in control. Eventually, a process of investigations in the neighborhood and damage assessment of the house started through regular visits of one member of the family to make sure there will be no apprehensive behaviors toward the family, if the situation allowed them to return. Sooner after, negotiations with landlords were initiated to assure the possibility of returning to the same residential unit. This process was expected to last for a few months before the decision of resettlement was made. Such decisions were also encouraged by personal values shared among the neighbors before 'Al-Taifiyah,' creating a chain of returnees and accelerating the process of social network reactivation in neighborhoods.

Post-'Al-Taifiyah:' Livability in the city and its infrastructure

Conspicuous physical traces of 'Al-Taifiyah' are still invading Baghdad's neighborhoods. Most of them, until this moment, are enclosed with high concrete walls denying physical and visual approachability. Likewise, neighborhoods can still be accessed through defined checkpoints. Such obstructing measurements stands as symbols and reminders of conflictual powers superiority and as a warning for the population of their reappearance (Figure 9.8). Urban practices and livability standards changed hugely

Figure 9.8 Checkpoint on a bridge before the main entry to Al-Qadisiya neighborhood in Baghdad

in the city. Prior to the 2003 occupation and the events of 'Al-Taifiyah,' urban spaces were auspicious with various social activities, advocating social and cultural gatherings and communal practices. The events of violence and sectarian conflict – with its outcomes of assassinations, children kidnapping, random shooting, and curfews – had only discouraged using these spaces, as they became spaces of fear. Nowadays, people feel reflexively safer in indoor public spaces and consider outdoor ones as unpleasant destinations. As reported by the Central Organization of Statistics in Iraq, only 36 percent of the youth aged from 10 to 30 were participating in outdoor activities such as playing sports, and in the year 2012, recreational open spaces were exempted from surveys on essential services in the country (COSIT, 2009, 2012).

The quality of socio-spatial life in neighborhoods and dwellings were as well substituted with another; before the war in 2003, collective social activities set good neighborliness among the inhabitants. Planning code and building regulations with certain setbacks in each residential block, created a generational sense of attachment and celebration of open green spaces in housing units; social practices in private gardens part of the daily routine including social gathering at day and night, 'teatime' in the afternoon, grilling, and some activities for children. It also incorporated individual activities like reading, raising pets, fruit harvesting, and stars gazing at night, hence flaring up competition among neighbors based on the quality and beautification of such spaces. After the acceleration of violence during 'Al-Taifiyah,' socio-spatial activities decreased, including practices inside private gardens, and were limited to the daytime, because of the transition of open spaces into ones of fear. Residents preferred to stay inside their house to avoid being exposed by the regular air patrols and random shooting by the US army at that time, and assassination threats by the militias. The sectarian violence, later, affirmed this fear. Even the garden was considered unsafe, and many had to flee, leaving behind their houses. Practices of allotment, that were required at the time of sudden uneven population growth, had led to neglect in the importance of such spaces nowadays, and to consider them as areas of potential built-up extensions and of economic benefit in the form of allotted small housing units. Baghdad has been chosen as the worst city in livability since 2008 in Mercer Quality of Living Survey, which evaluates cities based on factors such as political stability, crime, pollution, and so on, as Baghdad's residents are still suffering from many hurdles including electricity shortages and crippled infrastructure (MERCER, 2019). Still, during the interviews of Baghdadis, who lived through the eventfulness of 'Al-Taifiyah,' they were asked about the living condition in Baghdad and were acknowledged by the Mercer results, the majorly answer, accompanied with remonstrance, was: 'what is wrong with Baghdad? Baghdad is totally fine!' Such a reaction indicates a status of reconstructing local conditions of reasonableness as a mechanism of adaptability to live through such atrocity, which might not however fit the global standards.

Roles of actors in Baghdad's neighborhood recovery

In unstable urban environments undergoing post-conflict development, many actors collaborate or compete against each other over important positions in decision-making. Baghdad, with its long history of conflict primarily after the war and occupation in 2003, became an arena for different actors operating according to their interests in restructuring the urban environment for the purpose of control. Other actors endorsed development processes that promote the recovery of affected urban spaces, neighborhoods, and communities.

The involved actors can be categorized into four main groups: The state, inter-governmental organizations, civil society, and residents. The state consists of the Presidency Council, the parliament, and the administration. According to the higher committee of Basic Planning Design based on Law No. 156/1971, members of the administration must have expertise in city planning, economy, and sociology (Mahmoud, 2014). The Presidency Council is currently the highest level of authority in Iraq and Baghdad, which is indirectly controlled by the Iraqi Parliament. Most of the members of the parliament were the initiators of armed militias during 'Al-Taifiyah.' In one of the interviews, M.A. stated: 'all the different sectors in Iraq agreed on seeing the Prime Minister as the legitimate authority that would make concrete changes and reforms. The ugly truth was that he was powerless against the decisions of the corrupt parliament.' In addition to the state, according to Law No. 12[13] in 2010, the role of NGOs in Iraq was recognized and encouraged by international and national actors (NGO Directorate, 2012). Accordingly, many local NGOs were and are still directly and indirectly involved in activities that encourage community resilience and enhance the environment to accelerate the gradual process of urban recovery in Baghdad. International organizations primarily provide humanitarian emergency needs and few development activities (Relief International, 2016). Residents of Baghdad equally play an active role in the process of urban recovery.

Political interventions in decision-making and exceptional legislations have always been a core issue in urban planning to enforce state interventions with the claim that they represent the public interest, while in many cases they work for the private interest of a certain class in the community (Lehnerer, 2013). State responses in Baghdad can be broken down into authority and administration, whereby each plays a major role in enacting or repealing urban recovery. In Baghdad, the Presidency Council and the parliament exerted their authority in land expropriation, allocation, distribution, concessions, allotments, regularization, and privatization. Under normal conditions, governments aim to distribute lands through social land concession to create social equality and as a solution to the housing and property crisis. The distribution of lands in Baghdad, especially green spaces and agricultural lands that are served political purposes as opposed to proper city planning. In an interview in 2016, Dr. Al-Janabi, a Professor of Political Science at the University of Technology in Baghdad, commented on the parliamentary elections in 2005 and 2010. He explained how agricultural land was given to those who are willing to vote for a certain electoral candidate, especially in the district of Risafa. He added that the former Prime Minister Nouri Al-Maliki was among the candidates between 2006 and 2014 to get the most votes through land concessions. Many of the voters came from rural areas where negative repercussions of land concessions on residents were harder to endure.

State policies also increased unplanned expansion and urban sprawl in Baghdad. Many agricultural lands were allotted under the premise of solving the housing crisis. Between 2004 and 2007, an approximate 90.5 percent of lands, which were allotted and under easement, were agricultural and open green spaces. A total of 66.3 percent of these allotments were in 2005, mainly in the green belt on the western outskirts of Baghdad. All this happened under the control of the parliament with the consent of the Presidential Council that approved the easement of around 17 lands previously specified for recreational and green production services into different land-use. Eleven pieces of these lands were allocated to residential use (Table 9.2) (Baghdad Mayoralty, 2014; Mahmoud, 2014).

Lands were also regularized for squatters in agricultural areas and were supported by Law No. 254 in 2013, which stated that lands that were under easement and were squatted for residential use will be sold to low-income people at the price of 100 Iraqi Dinar/square meters, approximately 0.08 US Dollar. These areas were marked in eight locations on the perimeter of Baghdad with a total area of 1,125 hectares (Figure 9.9). Additionally, the parliament issued a new law[14] for the allotment of agricultural lands and orchards in 2014 under the condition that they are owned by the state, the municipality, or cooperative associations and they do not impact the Basic Planning Design of Baghdad. Still, governmental and investment projects were excluded from this law. In contradiction to the previous law, in the 38th session of the parliament on 17 November 2015, most of the members approved a new law[15], which was not consistent with public opinion. According to Ali Nsaif, a member of the Iraqi parliament in the third parliamentary session between 2014 and 2018, the law proposed the regularization and land tenure of agricultural lands for squatter settlements to eliminate the phenomenon of informal settlement (Al-Amiri, 2015).

It was not the intention of the administrative institutes to create socio-spatial disruptions in a city, but in the case of Baghdad political visions were transformed into administrative decisions. Despite its promotions of planning for a sustainable Baghdad, many haphazard changes by the Baghdad Mayoralty diverged from the Basic Planning Design. They included changes in land-use on a city scale from recreational to residential, transforming parks into educational buildings, administrative facilities, parking lots, pedestrian paths, and streets and privatizing parks and waterfronts. This opened the possibility for investment projects to take over open green

Table 9.2 Land-use changes to residential at the request of the Iraqi Parliament decisions between 2004 and 2009

No.	*Session Decision*	*Session Date*	*Land Ownership*	*Area*	*Land-use*
1	01/23/28 Tajjiyat	14.10.2004	Faw Company/ Ministry of Construction and Housing	250 ha	From Agricultural to Vertical Housing Project (3 stories)
2	01/23/28 Tajjiyat	12.02.2005	Al-Izz Company	500 ha	From Agricultural to Vertical Housing Project (3–4 stories)
3	26/01/15 Dahnah	27.11.2005 06.12.2005	Ministry of Finance request from General Secretariat of the Council of Ministries	7,500 ha	From Agricultural to Vertical Housing Project (1,250 ha) and Horizontal Housing Project (6,250 ha)
4	411/3/11 Hamdi Farms	27.01.2005 06.12.2005	De-Baathification Commission	128 ha	From Sport Fields to Horizontal Housing Project
5	Hay Tariq Block 573	17.10.2006			From Agricultural to Residential
6	1249/01/22 Sabbiat	03.06.2007	Sunni Endowment	205 ha	From Agricultural to Vertical Housing Project
7	Hay Tariq	03.06.2007			From Agricultural to Vertical Housing Project and Services
8	Lands in Block 772 and 774	04.07.2008			From Agricultural to Residential
9	11/30/17 Bazabiz Al-Fadhilah	03.12.2008		403 ha	From Agricultural to Residential
10	1800/03/19 Ghazaliyah	12.03.2009	Baghdad Mayoralty	1,828 ha	From Agricultural to Residential
11	16/207/4 Karrara	29.06.2009	Ministry of Finance	1,478 ha	From Agricultural to Residential

Figure 9.9 Eight Locations to be distributed to low-income class as stated in Law No. 254 in 2013

spaces with the intention of creating more indoor public spaces, solving the housing crisis, and boosting tourism (Raad, 2017). These changes resulted in the obstruction and weak control over the execution of plans as recommended in the Basic Planning Design. Still, the mayoralty exerted efforts to develop projects that revitalize open spaces and the urban fabric, many of which were previously closed due to alleged security concerns as a cover for state corruption.

One can argue, however, that the responses of the civil society and activists toward a more sustainable and livable environment for the community remain more effective. They were driven by community resilience and were organized through personal cooperative efforts among individuals, NGOs, and residents. Unfortunately, the Mercer survey failed to document and incorporate such efforts, while resisting the society's habitus of living with lower standards. Part of these efforts were put together to work against the memory of sectarian violence. Many movements emerged as a resilient act of camouflaging the blast walls that were used by politicians for election advertisement into canvases covered for art that communicates optimism in the city. Several NGOs, such as Imprint of Hope, carried out government roles by building houses for the IDPs and maintaining schools and sidewalks in the city. They, however, unintentionally violated the Basic Planning Design of Baghdad due to lack of planning expertise.

At the same time, activists in Baghdad reflected an image of Baghdad's neighborhoods and their public spaces as perceived by their inhabitants. They proved their capabilities in being effective actors against the image of brutality in the city. Public spaces like Abu Nuwas Street became a destination for collective events such as 'I am Iraqi, I read' and the Baghdad City of Peace Carnival (Raad, 2017). Zain Mohammed and Noof Assi, organizers of Baghdad City of Peace Carnival, decided to pursue their vision of not associating Baghdad only with wars and conflicts but more essentially with peace. They initiated the idea of a festive outdoor carnival that celebrates World Peace Day each year since 2011 on 21 of September at Abu Nuwas Street, defying the fear that haunted the city. Such seasonal events reclaimed spaces that were taken away from the public and encouraged positive social practices that are becoming increasingly popular among younger generations. R.D., a regular participant in the carnival, enthusiastically explained how the festival gave her the chance of experiencing outdoor recreational activities, something she could not practice for a long time due to forgoing insecurity that limited her everyday activities only in indoor spaces such as shopping centers, cafes, and restaurants.

Militias, nonetheless, continue to play a role in Baghdad. They have been legalized and, in some cases, have become part of civil organizations such as the Badr Organization for Reconstruction and Development that was until 2012 known as the Brigades of Badr formed by Hadi al-Amiri in 1983 as the armed wing of the largest Shiite political party in Iraq, the Supreme Council for the Islamic Revolution in Iraq (Cassman, n.d.).

Moreover, art appeared in the streets of Baghdad as an expressive tool and a means of consensual power. In 2013, Sajjad Abbas, an artist from Baghdad, went above a derelict high-rise building known as 'the Turkish Restaurant' located across the river facing the building parliament council. He drew a big eye and wrote: 'I can see you' as a powerful statement that all political corrupted agendas are exposed to the people. Four days after the installation of such artwork, the police took it down and it was completely sabotaged. This did not stop Abbas from awakening the society. Five years later, he drew another graffiti illustrating the leader of an apprising militia, stating: 'I want you with me,' the same style of the famous 'I want you' poster by James Montgomery Flagg during the first World War as a recruiting poster for the US Army (Al-Jazairi, 2016).

The efforts of the residents and civil society reflect the capacity to rebuild the societal structure and fight against the still-standing spatial borders separating both, the districts and neighborhoods. Neighborhoods are opening and defying physical boundaries of concrete walls and security checkpoints through the reestablishment of mix patterns of mobility and interaction networks. They provide neighborhoods with attractive destinations of socio-spatial daily patterns such as going to markets and schools in other neighborhoods regardless of social differences. Consequently, many neighborhoods have been gradually losing their imposed sectarian identity and returning to a more mixed one based on socio-economic values and accommodating the needs for housing opportunities of mixed families across sects and religions.

Conclusion

Cities are territories with interactive urban networks characterized by social identities, densities, and heterogeneity. They are defined by social relations and interactions between people. These territories are affected by conflicts that disturb socio-spatial practices and alter the characteristics that define cities and their neighborhoods. 'Al-Taifiyah,' a period of chronic ethnically religious conflict between the major sects of Sunnis and Shiites in Baghdad between late 2005 and 2008, produced a state of social instability. Nevertheless, Baghdad's mixed urban and social fabrics developed certain generational socio-economic attributes and rituals within societal historical phases. During 'Al-Taifiyah,' residential neighborhoods in Baghdad were redefined according to the sectarian identities of their inhabitants. The inhabitants became living targets of the conflict among antagonist militias affiliated with sectarian political parties, whose powers were amplified after the state power vacuum in 2003.

The population complied with conflictual powers due to coercion practiced by the militias in the form of direct violence, direct threats, systematic assassination, and exodus of scientists, intellectuals, and academics. At the same time, dominating militias encouraged their own sects to stay in contested areas to promote protection and safety for those fleeing other areas, from neighborhoods, informal settlements, to the rural outskirts of Baghdad. The aim was to increase the sectarian population in the dominated neighborhoods. This created an unanticipated population growth and an unbalanced physical and social distribution of housings in the city. Families were evicted from areas of conflict. Families mostly fled to safer neighborhoods in Baghdad. At the same time, the state imposed its power over the population as a co-insurgency measure. Blast concrete walls, curfews, and checkpoints among many security measures were considered suitable for protection against sectarian violence. However, they distorted the urban fabric and caused a disruption of socio-spatial practices in the neighborhoods of Baghdad with declining livability standards. Neighborhoods became a vicinity of self-reliant exclusive sub-communities. The resultant societal shifts following the waves of displacement caused spatial distortion in the city and uncontrolled encroachment of agricultural lands, open green spaces, and vacant lands, disregarding planning codes and legal legislations, creating infrastructure insufficiency and lack of services in the neighborhoods. Resilience and adaptability rose as survival mechanisms throughout various socio-spatial and systematic practices during and after the conflict, contributing in mobilizing temporal processes to reactivate neighborhoods. As such, returning to neighborhoods through a multi-dimensional process has been one of the most influential socio-spatial traces of a resilient population after such an atrocious event. This process included governmental and civil society efforts to amplify the reactivation of social networks in the neighborhoods.

Many actors have been involved in the urban recovery of Baghdad's neighborhoods, contributing to either accelerating or impairing the process. State responses were more in favor of empowering political interests over the population's need for proper planning, generating lack of short-term strategies and deficiency in the judgement of long-term effects on the city, as events of land expropriations were often used to increase the breadth of supporters for political placement during the elections, even though the state's responsibility is to address immediate spatial infrastructural needs, building social cohesion, and assuring social and spatial entitlements for the civil society and the residents. Meanwhile, local NGOs and individuals from Baghdad were involved in various activities that inspire community resilience through beautifying the physical traces of the conflict, revitalizing urban spaces, promoting outdoor recreational activities, and initiating provocative socio-political statements through art and cultural events. Subsequently, the urban recovery of Baghdad's neighborhoods has been moderated by active networks that have been gradually breaking the societal cycle of acceptance of low local conditions of livability that overpowered the population as a result of consequential reaction to the sovereignty of militias, political corruption, terrorism, insecurity, population manipulation, and the denial of their rights to dignified living within local standards. The existence of such networks should be considered as empowering organizational entities that allow for dialogue among the different sub-communities, regardless of their ethnic and religious backgrounds, to reestablish trust among the different groups and to allow them to overcome their differences and work together to rebuild their community.

It is not deniable that the socio-spatial events related to 'Al-Taifiyah' have transformed and drastically impacted the urban and architectural characteristics, boundaries of districts and the patterns of mobility and interaction. However, despite all the chronic socio-spatial disruptions embedded within, communities in Baghdad proved their ability to resume their original configurations even after the applied abrupt socio-spatial changes. The notion of social rebuilding of the neighborhoods overpowered the administrative slow progress of spatial and infrastructural rebuilding. Such social resilience assisted in breaking down physical boundaries, reestablishing mixed-community-friendly neighborhoods, and reintroducing a heterogeneous identity based on socio-economic factors. As such, it is very crucial in complex socio-environments, like Baghdad, to legitimize consensual empowerment of the society and its role in the process of rebuilding to reach a point of cohesive reactivation of the physical and social networks, which are in the interest of the process of long-term socio-urban recovery of the neighborhoods, thus reflecting on the broader metropolitan recovery scale of the city.

Notes

1. Tremendous efforts were employed to align with development progress in the West, with the rise of modernism in Europe and North America. The goal since the late 1920s was to erase the orientalist image of an Arab-Islamic Baghdad and create a new city through modern urban planning strategies. The urban spaces were regulated to be easily controlled with wide commercial streets penetrating the urban fabric (Pieri, 2008). The city became an attraction for international investments and foreign competitions were encouraged by Baghdad's city planning especially among architects such as Le Corbusier, Frank Lloyd Wright, Alvar Aalto, and Walter Gropius who link modernism to nationalist aspirations.
2. With the overthrow of the monarchy by the military coup and the declaration of Iraq as a republic under the rule of Abdulkarim Kasim in 1958, more communist visions of urban development were proposed. Among those visions was the proposal by Doxiadis for a new community district in Baghdad that can accommodate a population of 100,000 people. The goal was to address the fast growth of informal settlements of more than 400,000 migrants from the rural southern cities of Iraq to Baghdad. The newly developed district was called Al-Thawra City at the time, known now as Al-Sadr City with the highest population density in Baghdad reaching 325 person/ha according to the Mayoralty of Baghdad (Al-Ani, 2010; Theodosis, 2008).
3. After the Baath party took power in 1968, notions of nationalism and socialism advocated for economic growth in Iraq through the nationalization of petroleum companies. The economy was booming, reflected through the state setting new development priorities for the city. New comprehensive development plans were created for Baghdad, including the 1967 Polservice and the 1985 JCCF consultancy proposals. They provided development strategies based on assessing population growth rates and their impacts on the morphology of the city. Contrary to previous visions and proposals, they presented an updated analysis of the different characteristics of the urban fabric and proposed a mix of land-uses from residential, commercial, administrative, public, industrial, and recreational facilities (Al-Ani, 2010; JCCF, 1987; Stanek, 2012).
4. The Basic Planning Law of the City of Baghdad was issued on 11 December 1971 by the Iraqi Revolutionary Command Council, the highest authority in the country.
5. Specification of the Land Area Limits and boundaries was issued on 5 July 1979 by the Iraqi Revolutionary Command Council.
6. Land allotment for residential land-use was issued on 21 December 1987 by the Iraqi Revolutionary Command Council.
7. Identification of Lands suitable for Investments was issued on 23 January 1989 by the Iraqi Revolutionary Command Council.
8. Foundation of execution Comity for Development Projects Al-Hamza and Al-Aziziyah was issued on 28 June 1989 by the Iraqi Revolutionary Command Council.
9. Condemnation of Acts against State and Municipalities Properties was issued on 27 June 2001 by the Iraqi Revolutionary Command Council.
10. Violation of Municipalities Code was issued on 9 July 2001 by the Iraqi Revolutionary Command Council.
11. A concept of political actions against a regime to decapitate it through targeting the leader along with regime supporters. This approach argues that Hussein's regime depended on a small number of people to hold its dictatorial power and to assure organizational consistency, by which should be eliminated.

12. These numbers excluded failed assassination attempts, nevertheless, three quarters of these attempts succeeded. As reported by the non-governmental Association of University Lecturers in Baghdad, in the period between April 2003 and October 2008, 80 percent of assassinations were targeting employees at universities, while 62 percent of the assassinated scientists were holding PhDs.
13. Law of NGOs in 2010 issued by the Iraqi Parliament according to the provisions of articles 61 and 73 of the Iraqi constitution. The law is approved by the Presidency Council.
14. Lands and Orchards Allotments within the Basic Planning Design of the City of Baghdad and its Municipalities issued by the Iraqi Parliament in 2014, as part of articles 61 and 73 in the Iraqi Constitution.
15. There were 135 participant members: 109 voted yes, 13 abstained from voting, and 13 voted no.

References

Al-Amiri, O. (2015). *Ownership Law of Agricultural Lands as a Solution for the Housing and Renting Crisis* [Television broadcast]. Blade TV.

Al-Ani, M. Q. (2010). Continuity Urban Development of Baghdad Historical City Centre between Academic and Practice. *The Iraqi Architectural Journal*. https://www.researchgate.net/publication/322554030_Continuity_Urban_Development_of_Baghdad_Historical_City_Centre_between_Academic_and_Practice.

Al-Ashab, K.H. (1974). *The Urban Geography of Baghdad*. Newcastle: University of Newcastle upon Tyne. https://books.google.de/books?id=FBvHQgAACAAJ.

Al-Jazairi, S. (2016). Sajjad Abbas, Protest Maker on the Walls of Baghdad. *Al-Quds Al-Arabi*. https://www.alquds.co.uk/%ef%bb%bf%d8%b3%d8%ac%d9%91%d8%a7%d8%af-%d8%b9%d8%a8%d8%a7%d8%b3-%d8%b5%d8%a7%d9%86%d8%b9-%d8%a7%d9%84%d8%a5%d8%ad%d8%aa%d8%ac%d8%a7%d8%ac-%d8%b9%d9%84%d9%89-%d8%ac%d8%af%d8%b1%d8%a7%d9%86-%d8%a8/.

Al-Sa'di, J. (2005). Development of Residential Structure in Baghdad City 1956–2000: Analytical Study of Residential Market. *Baghdad University, Planning and Development Magazine*, (14), 29–54.

Amend, K. (2010). Counterinsurgency Principles for the Diplomat. *Orbis*, *54*. https://doi.org/10.1016/j.orbis.2010.01.002

Baghdad Mayoralty. (2014). Comprehensive Development Plan of Baghdad 2030: Phase 4. https://www.amanatbaghdad.gov.iq/index.php?lang=en.

Bourne, L.S. (ed.). (1982). *Internal Structure of the City: Readings on Urban Form, Growth, and Policy*, 2nd edition. New York, Oxford: Oxford University Press.

Cassman, D. (n.d.). *Badr Organization of Reconstruction and Development | Mapping Militant Organizations*. https://web.stanford.edu/group/mappingmilitants/cgi-bin/groups/view/435.

Chandrasekaran, R. (2007). *Imperial Life in the Emerald City: Inside Iraq's Green Zone*, 1st edition. New York: Vintage Books.

COSIT. (2009). *Iraq National Youth and Adolescents Survey*. Baghdad: Central Organization for Statistics.

COSIT. (2012). *Iraq National Youth and Adolescents Survey*. Baghdad: Central Organization for Statistics.

Coward, M. (2010). Urbicide: The Politics of Urban Destruction by Martin Coward. *Global Discourse*, *1*(2), 186–189. doi.org/10.1080/23269995.2010.10707864.
Damluji, M. (2010). "Securing Democracy in Iraq": Sectarian Politics and Segregation in Baghdad, 2003–2007. *Traditional Dwellings and Settlements Review*, 71–87.
Dodge, T. (2005). Iraqi Transitions: From Regime Change to State Collapse. *Third World Quarterly - THIRD WORLD Q*, *26*. doi.org/10.1080/01436590500127982.
Galula, D. and Nagl, J.A. (2006, 1964). *Counterinsurgency Warfare: Theory and Practice/David Galula*. Westport, CT and London: Praeger Security International.
Graham, S. (2003). Lessons in Urbicide. *New Left Review*, 19.
Gregory, D. (2010). Seeing Red: Baghdad and the City. *Political Geography*, *29*. doi.org/10.1016/j.polgeo.2010.04.003.
Hanafi, S. (2004). Targeting Space through Bio-politics: The Israeli Colonial Project. *Palestine Report*, *10*(32), 4.
IAU, IOM, and UN-Habitat. (2010). *Urban Baghdad: Impact of Conflict on Daily Life*. Geneva: IAU; IOM; UN-Habitat. https://reliefweb.int/sites/reliefweb.int/files/resources/Urban%20Baghdad-Impact%20of%20conflict%20on%20daily%20life-May%2022-Final.pdf.
Ismael, T.Y. and Ismael, J.S. (2005). Whither Iraq? Beyond Saddam, Sanctions, and Occupation. *Third World Quarterly*, *26*(4–5), 609–629. doi.org/10.1080/01436590500128303
Jawad, T. (2013). The Importance of the Application of Planning Standards for Open and Green Spaces and their Role in better Planning of Cities. *Baghdad University, Planning and Development Magazine*, *28*, 100–116.
JCCF. (1987). *Integrated Capital Development Plan of Baghdad: The Regional Framework, the Structure Plan of Greater Baghdad and the Comprehensive Development Plan of the City of Baghdad*. Baghdad: Japanese Consortium of Consulting Firms and Baghdad Mayoralty.
Lehnerer, A. (2013). *Grand Urban Rules*. Rotterdam: Distributed Art Pub.
Luo, M. (2006). Even Picking Up Trash Is a High Risk in Baghdad. *The New York Times*. https://www.nytimes.com/2006/10/13/world/middleeast/13trash.html.
Mahmoud, H. (2014). The Impact of Laws and Regulations in Urban Land Management and Urban Planning: Baghdad – A Case Study. *The Iraqi Architectural Journal*, *29*(3–4), 39–55.
MERCER. (2019). *2019 Quality of Living Survey*. https://www.mercer.com/newsroom/2019-quality-of-living-survey.html.
NGO Directorate. (2012). *Law No. 12/2010: NGOs*.
Pieri, C. (2008). Modernity and its Posts in Constructing an Arab Capital: Baghdad's Urban Space and Architecture. *Middle East Studies Association Bulletin*, *42*(1/2), 32–39.
Raad, Y. (2017). Diverse Socio-spatial Practices in a Militarized Public Space: The Case of Abu Nuwas Street in Baghdad. *International Journal of Contemporary Iraqi Studies*, *11*(3), 133–159. doi.org/10.1386/ijcis.11.3.133_1.
Relief International. (2016). *Iraq*. https://www.ri.org/reach/middle-east/iraq.
Soane, E.B. (2007). *To Mesopotamia and Kurdistan in Disguise*. New York, NY.: Cosimo Classics.
Sparke, M. (2007). Geopolitical Fears, Geoeconomic Hopes, and the Responsibilities of Geography. *Annals of the Association of American Geographers*, *97*(2), 338–349. https://doi.org/10.1111/j.1467-8306.2007.00540.x.

Stanek, Ł. (2012). Miastoprojekt Goes Abroad: The Transfer of Architectural Labour from Socialist Poland to Iraq (1958–1989). *The Journal of Architecture*, *17*(3), 361–386. doi.org/10.1080/13602365.2012.692603.

Theodosis, L. (2008). "Containing" Baghdad: Constantinos Doxiadis' Program for a Developing Nation. Ciudad Del Espejismo: Baghdad, De Wright a Venturi/City of Mirages: Baghdad, from Wright to Venturi.

United States Government. (2009). *U.S. Government Counterinsurgency Guide*. https://2009-2017.state.gov/documents/organization/119629.pdf

Worth, R. (2006). Muslim Clerics Call for an End to Iraqi Rioting. *The New York Times*. https://www.nytimes.com/2006/02/25/world/middleeast/muslim-clerics-call-for-an-end-to-iraqi-rioting.html.

Whyte, W.H. (1980). *The Social Life of Small Urban Spaces*. New York: Project for Public Spaces Inc.

Part III

Conceiving of cultural heritage in the recovery process

10 From recovery to resilience

Challenges and opportunities for post-crisis recovery of urban heritage

Rohit Jigyasu

The historic urban areas that have withstood the test of time are becoming increasingly vulnerable to disasters and armed conflicts. Besides the adverse impact of natural hazards such as earthquakes, floods, and cyclones, one also cannot underestimate risks to urban cultural heritage from human-induced hazards such as chemicals, war, and terrorism.

There are countless examples that illustrate increasing loss of urban cultural heritage due to disasters and armed conflicts. Some recent examples include damage to historic cathedrals in Mexico city due to the central Mexico earthquake in 2017; damage to historic settlements in central Italy due to earthquakes in 2016; damage to the World Heritage Monument Zones of historic towns of Kathmandu, Patan, and Bhaktapur in Kathmandu Valley due to the 2015 earthquakes in Nepal; fires in the World Heritage Town of Lijiang in China in 2013 and 2014; and fire in the Old Town of Edinburgh in the United Kingdom in 2002. Cultural heritage sites have also suffered enormous damage due to human-induced hazards like armed conflicts and vandalism, such as in the cases of the historic cities of Aleppo in Syria, Mosul in Iraq, and Sanaa in Yemen. In South Korea, arson damaged the Sunraemon Gate located in the centre of Seoul city in 2008, which is designated as 'cultural property number one'. More recently, urban fires have damaged the World Heritage Site of Notredam Cathedral in Paris and the National Museum of Rio-de-Jineoro in Brazil in 2019 and 2018, respectively. Extensive damage to urban cultural heritage has also adversely affected tourism-related revenues as well as associated livelihoods of local communities inhabiting urban areas. It is noteworthy to mention here that in the absence of any comprehensive legal framework for protection, historic urban areas are not only vulnerable to impending disasters, but also susceptible to damage during emergency and post-disaster recovery phases.

While disasters are a cause of great misery and irreplaceable loss of heritage, they also serve as opportunities for change though the introduction of bold policy and urban planning measures aimed at reducing the vulnerability of people, properties, and cultural heritage to future disasters and conflicts.

Recovery of cultural heritage: Emerging theoretical positions

Theoretical perspectives on the recovery of cultural heritage have undergone a remarkable shift over time. The 1964 International Charter on the Conservation and Restoration of Monuments and Sites, known as the Venice Charter, was predominantly concerned with the destruction as well as the decay of monuments (ICOMOS, 1964). Accordingly, the reconstruction of heritage buildings and sites was considered justifiable only in exceptional circumstances and only based on complete and detailed documentation and to no extent on conjecture. Obviously, retaining material authenticity was the key determining factor in the heritage-protection discourse, and reconstruction was therefore considered in the context of restoration. (ICOMOS, 2017). However, there were exceptions, for example, the Australia ICOMOS Charter for the Conservation of Places of Cultural Significance, known as the Burra Charter adopted in 1979, accepted reconstruction if it reflects a pattern of use or cultural practice that sustains cultural value. But even here, a 'cautious approach to changing a place' is advocated (Cameron, 2017).

Due to the increasing destruction of cultural heritage due to armed conflicts over recent decades, there has been a shift in the attitude towards reconstruction. This shift is also supported by the broadened recognition of intangible attributes with the inclusion of spirit as another dimension of authenticity in the 1994 Nara document on Authenticity. This shift is evident in the decision to reconstruct the historic bridge in the old city of Mostar in Bosnia and Herzegovina following its destruction due to the Civil War and its subsequent inclusion in the World Heritage List. The recent decision to reconstruct the World Heritage Site of the Kasubi tombs in Uganda following the devastating fire in 2010 clearly demonstrates this shift in the overall approach towards heritage recovery. However, in all these cases, reconstruction was based on the condition that the new structure was based on sound documentation, traditional forms, and techniques and continuing use.

In the context of historic cities, the change in the approach to conservation is evident from the key ICOMOS doctrinal texts. Charter on urban conservation adopted as the Washington Charter of 1987 (Charter for the Conservation of Historic Towns and Urban Areas) (ICOMOS, 1987) was followed by the Vienna Memorandum of 2005 and, most recently, the Valetta Principles of 2011 (ICOMOS, 2011). Rather than viewing historic cities as exclusive urban areas for protection, Valetta principles consider heritage as an essential resource, as part of the urban ecosystem. This concept seeks the harmonious development of historic towns and their settings to safeguard their values, as well as their integration into the social, cultural, and economic life of our times. This approach was further strengthened by the UNESCO recommendation on historic urban landscape in 2011, which also laid exclusive stress on the intangible heritage of cities (UNESCO, 2011). In the context of these developments, recovery of historic towns and cities was seen from a holistic and territorial perspective and no longer as physical reconstruction of town centres as in the case of Warsaw that was destroyed following the Second World War. However, the predominant emphasis of recovery was always on revival or bringing back the city to the previous state in as original manner as possible.

In the light of the challenges confronting the recovery of historic cities such as Mosul and Aleppo destroyed due to armed conflicts, the overall goal of the recovery of urban heritage in post-conflict and post-disaster situations has gradually shifted to the recovery of society. As a result, the Warsaw recommendations for World Heritage Recovery and Reconstruction articulated the aim of recovery to be consolidation of peace and security and at restoring or improving the economic, physical, social, cultural, and environmental assets, systems, and activities of an affected community or society, aligning with the principles of sustainable development and 'build back better'. An essential part of this process is the recovery of a place's heritage, which may include reconstruction (UNESCO, 2018). In the context of these theoretical underpinnings, the chapter will bring forward key challenges and opportunities for recovery of urban heritage following disasters and armed conflicts.

Reconciling safety and values of urban heritage

Critical choices need to be made regarding the basic philosophy governing post-disaster interventions undertaken as part of the recovery process. Of course, recovery should aim at reducing vulnerability and risks to future disasters, while also protecting the heritage values to the best possible extent. There is also the fundamental issue of defining urban cultural heritage: Is it only restricted to remains of the past to be admired for their pristine glory, or does it also include the living dimension that shows continuity while evolving and adapting to change? By considering it as a living and dynamically evolving historic urban landscapes comprising of both tangible and intangible, movable and immovable, as well as cultural and naturla components, the chapter seeks answers to the following critical questions for the post-disaster recovery of urban heritage.

How can we reconcile the need to safeguard lives as well as recovering lost heritage values, especially those that contribute to local identity and sense of place? What are the challenges and opportunities, failures, and success stories in achieving this? What are the possible approaches for the recovery of urban heritage and what process should be followed for reaching a decision? Also, it is worth pondering if urban heritage is only a victim of disasters to be protected for posterity or if it can also be a source of resilience through local knowledge and capacity. If so, how can we harness that potential supposedly embedded in urban heritage in the recovery process?

More importantly, the critical question is, which cultural heritage values to recover and to what extent and how? This also implies firstly, where and how to position heritage in the overall recovery process, especially when the priorities in the aftermath of a disaster or conflict are on providing the basic needs of food, shelter, infrastructure, and livelihoods for the affected communities. The second challenge is which approach to follow: Building it back as before to restore pre-crisis values at the cost of reinforcing the vulnerabilities or rebuilding to look as before but with new materials and techniques to enhance safety but restoring only the visual character. The third and diametrically opposite approach to the first one will be to replace the traditional fabric with contemporary design and construction to reflect the new opportunities provided by the disaster to break with the past.

Beside the above-mentioned approaches, the alternate one will be to embody the concept of building back better tailored to cultural heritage. This would involve deciding the threshold of safety vis-à-vis the heritage values and reimagining the past for the future by reducing vulnerability and building on existing capacities. This also implies that the recovery of heritage cannot be seen merely in terms of visual appearance. It is important to consider other factors such as traditional knowledge and skills, local affordability, and sustainability while incorporating measures to reduce vulnerability to future disasters. This would necessitate upgrading the skills of craftsmen to ensure continuity as well as evolution in response to the need to retain the values and reduce vulnerability and risks.

The biggest challenge will be in the trade-offs between new and old methods: To what extent should structurally inadequate historical building details be retained? Which details are so inherently weak that alternatives must be sought? Which characteristics are so key to the buildings' history or aesthetics that new ways to maintain them must be sought? What determines the choice between a safer modern – say steel structure inserted (whether visible or not) within an exterior of historical details, versus a less safe rebuilding of the historical building with less intrusive reinforcement measures? And of course, the basic question one may ask is: Do modern materials make traditional buildings safer? A straightforward approach would be to assess the performance of buildings that were strengthened with modern materials against the 2015 earthquake.

Which values to recover? Moving beyond authenticity debate

Recovery is also linked to the inherent nature of heritage that defines its values; for example, Isa Shrine in Japan is reconstructed every 12 months and this act contributes to the value. Therefore, in this case, authenticity of workmanship and cyclic renewal process takes precedence over preservation of physical fabric.

Understanding which heritage values are prioritized by the custodians is key to developing strategy for the post-crisis recovery of cultural heritage. For example, following the destruction of Bohol church due to 2013 earthquake and fire in Wangduaphodrang Zhong in Bhutan in 2012, the recovery of movable assets of religious and spiritual value was prioritized over the physical remains of the damaged building. In the latter case, values do not really lie in the surviving material fabric as craftsmen can reconstruct the building as before using traditional knowledge and skills that still exist as before. Moreover, the predominant consideration for the local custodian was resumption of religious practices rather than physical reconstruction of the material fabric. Following the 2015 earthquake in Nepal, the predominant national and international focus was on the repair and reconstruction of monuments that included palaces and large temples because of their symbolism and tourism revenues that they generate. Contrary to this, repair and reconstruction of local shrines and community structures were crucial for traumatized communities as these acted as social anchors for their psycho-social healing (Figure 10.1). In Sankhu village in Kathmandu valley, Nepal, young people whose traditional houses had been destroyed by the earthquake voluntarily worked to salvage traditional doors and windows from their collapsed houses and safe keep them in storage for eventual use in reconstruction of their houses, while they were living in temporary shelters a few days after the earthquake. Moreover, owing to the important socio-cultural and economic role that public open spaces and squares play in the lives of local people, their recovery was more critical rather than merely focused on buildings. Their recovery would necessitate a continuation of all the social, cultural, and economic activities that were carried out in these spaces and not

Figure 10.1 People were keen to reconstruct local shrines and community structures that were damaged in historic settlement of Sankhu following the 2015 Nepal earthquake

merely the reconstruction of the built fabric. Take, for instance, the case of Bungamati village in Kathmandu valley where the main temple located in the public square in the middle of town collapsed due to earthquake. While it is important to consider the reconstruction of temple, it is equally important to regenerate the open space around it so that the socio-cultural link with the community is restored. This was all the more important since many traditional open spaces were gradually taken over by temporary shelters for those who lost their houses and in the absence of a clear policy, there is risk of these turning into slums with inadequate living conditions (Figure 10.2).

Often during the recovery process, emphasis is given on the recovery of tangible elements while the intangible aspects are overlooked. However, disasters also impact intangible heritage values that are rooted in traditional rituals and practices collectively carried out by the community. Therefore, their continuation is indeed paramount not only for securing cultural heritage, but also for psycho-social healing of affected communities.

In many instances, tangibles carry many of their values due to intangible associations while overlooking the latter in the recovery process deprives the very essence of heritage that calls for its recovery. Moreover, in most cases, the tangibles are maintained due to their use and meaning for the local community and mere consideration of physical aspects during the recovery process makes heritage unsustainable in the long run. Therefore, recovery of tangible heritage will only carry a true meaning if associated intangibles are also restored. Macchenderanath Yatra is one of the most important ritual processions in Kathmandu valley that is carried out after every 12 years.

Figure 10.2 Traditional public space in Bungamati settlement used for constructing temporary shelters following the 2015 Nepal earthquake

In this procession, the idols of Rato Macchenderanath (Deity of Newari community) are carried out on specially made wooden chariot from village Bungamati to the centre of the historic town of Patan in Kathmandu valley. Ironically, this procession was underway when 2015 earthquake struck and got abruptly stopped. In spite of the struggle for basic needs following the disaster, the community was interested in reviving the procession at the earliest, and thus restore the space where the chariots are made, reconstruct the storage for the materials used for making the chariot, and restore the processional pathway. The community also made efforts to revive other rituals and festivals, such as the ones in Indra Jatra, within a few days of the earthquake (Figure 10.3).

The 'Tantrik' temple located in Swaymbhunath World Heritage is another case that demonstrates the importance of values rooted in local cultural context. The shrine suffered from massive damage due to 2015 earthquake. This is based on the first-hand experience of the author who joined the post-earthquake response for cultural heritage as part of joint team of ICOMOS and ICCROM. Due to local religious beliefs, only specially designated people could enter the core of the shrine that had wall paintings of immense artistic and historic value. Following the earthquake, through the support for the local community, the fragments of these wall paintings were painstakingly salvaged and stored in boxes padded with sand, under the guidance of international experts. However, the local community expressed their desire to redo these paintings afresh following the necessary rituals rather than putting back the salvaged pieces of wall paintings. Clearly, for them, spiritual value holds much more importance than the historic value for the wall paintings (Figure 10.4).

In heritage conservation charters, the term reconstruction means to return to the original or previous status of a heritage place or building based on reliable documentation, to avoid mistaken interpretations of history. Reconstruction means returning a place to a known earlier state and is distinguished from restoration by the introduction of new material (Australia ICOMOS, 2013). However, as time passes, new materials age and the difference that was once obvious in a reconstructed building tends to become blurred. In addition, natural hazards such as earthquakes and floods produce destruction and trigger reconstruction as a cyclical process. Heritage buildings located in disaster-prone areas have consequently been affected and intervened in many times after disasters. The extent of the interventions in the form of rebuilding, repair, or replacement of parts, among others – will, eventually affect our perception. In this context, does a building continue to be the same even when it has changed over the years? Or, in other words, can something be authentically rebuilt. This is similar to the case of Ship of Theseus, where planks were replaced progressively over time until the point when it was asked whether it should still be considered the same ship.

Moreover, there is a fundamental issue on whether heritage can ever be reconstructed in the real sense because it might not be possible to restore

Figure 10.3 Traditional festivals, such as Macchenderanath Jatra held after 12 years, were disrupted due to the 2015 Nepal earthquake

Figure 10.4 Santipur shrine and frescoes in the interior walls were damaged due to the 2015 earthquake. Community decided to repaint the fresco rather than restoring the damaged one

social values from day one, and historical and artistic values are bound to get adversely affected by a change.

The nature of interventions will therefore be influenced by the main conservation philosophy, which varies according to the nature of heritage and its socio-cultural context. Whether these interventions help in reading different stages of a building or the new is made to look old or in those cultural contexts, rebuilding and reconstruction are parts of the very nature of heritage as in the case of Isa Shrine. However, there are cases where craftsmen possessing traditional skills undertake post-disaster rebuilding using new traditional materials such as wood or reusing the parts from the old building completely at their discretion, sometimes also using this opportunity to demonstrate their creativity by introducing subtle or pronounced changes in design. However, in such cases, the level of craftsmanship may be questionable. A related challenge is which historical layers to retrieve while undertaking recovery. Should this retain all the historical layers that existed prior to disaster or to go back to a particular layer of history. What if a new historical layer is exposed after a disaster? What considerations should govern such decisions?

Considering the complex interaction between the heritage significance and values based on the nature of cultural heritage and the very cyclic nature of disasters, especially those caused by natural hazards, authenticity in its wider scope as defined in the Nara document (ICOMOS, 1994). It can only be one of the many considerations for the recovery of heritage sites, the others being integrity, sustainability, as well as the role in improving the quality of lives of people associated with heritage. On a final note, a disaster could be an exceptional incident that can reveal in a more profound way the limitations as well as the challenges of the discipline of conservation today, beyond established notions of authenticity towards alternative ones.

Integrating physical and social recovery

When recovering urban heritage, the essential consideration is the recovery of both the physical and social integrity of the urban fabric. While physical integrity would involve the recovery of the entire morphology of the built fabric, the social dimension would involve strategies to retain or bring back the original inhabitants rather than exposing the place to gentrification processes, which will essentially deprive heritage of its significant values that give identity and meaning. Moreover, often there is a myth of a monolithic community inhabiting a heritage place. Even seemingly, a single community may share the geographical location, religion, or ethnicity but may be divided into diverse groups based on gender, age, and shared needs and interests and power relations. In fact, each member may belong to multiple communities based on their shared identities, perceptions, and needs (Sen, 2006). Therefore, the recovery process should be mindful of the multiple narratives of heritage and ensure that heritage belonging to the powerful group does not take precedence over the heritage of the weaker and marginalized or minority groups. Otherwise, the recovery process may recreate or reinforce conflicts and the vulnerability of heritage in the long run. The recovery process should therefore promote cultural diversity and tolerance through the consideration of heritage belonging to all the sections of the community and serve as powerful means for building peace and reconciliation.

However, it is important to reflect on whether the symbolism attached to heritage really serves to achieve the well-stated purpose at the local or is merely meant to convey a message for the international community. Take, for example, the case of the reconstruction of the Mostar Bridge following the war.

All the above-mentioned examples illustrate that our approach towards post-disaster recovery of cultural heritage depends on the values attached to heritage in specific social, cultural, economic, and institutional contexts.

A more fundamental issue is the consideration of the official versus local narrative of heritage itself, where the latter may also have diverse perspectives among various sections of the community. This is even more challenging in conflict-affected societies, where there are always different narratives or perceptions on who is the winner and who is the victim, majority view versus minority views, nationalist view of heritage versus ethnic view, or communal versus secular view. Of course, short-term gains for securing political mileage often take precedence over long-term gains of rebuilding identity, achieving resilience, and building peace and reconciliation.

A critical challenge is how to deal with heritage when the socio-spatial structure of the community is significantly altered following a conflict situation since original inhabitants may have moved out or changed their location within the city, while new groups may have moved in, taking over the areas abandoned by the original inhabitants. In some instances, only a certain section of the community may have remained while others left the place for good out of choice or fear. In such situations, can old identities or values be reinforced or new identifies or values get created? Can existing identities or values go through planned transformation through post-crisis recovery process and if so, how can this be achieved and what enabling mechanisms are needed in the larger context in which recovery process is initiated?

Embedding cultural consideration in new reconstructions: Creating new heritage values

The recovery of heritage does not entail only reconstruction of the remains of the past. Heritage is the embodiment of social and cultural practices and these must be given due consideration even in the design and urban planning of new post-crisis reconstruction since the cultural dimension in general and heritage in particular also play an important role in the sustainable recovery and rehabilitation of communities following a disaster. There are many examples to show that successful reconstruction projects have taken into consideration local building traditions and way of life through deeper engagement with communities.

Take the case of owner-driven reconstruction undertaken in Gujarat following the 2001 earthquake, where NGOs and the local government facilitated development of innovative designs and technologies. It focused on traditional designs and technologies that seek to respect the local social, cultural, and environmental contexts through the sensitive use of local design, materials, and construction techniques, while introducing structural and non-structural measures that help in reducing vulnerability and risks to disasters (Jigyasu, 2010).

Contrary to this, for the reconstruction in Maharashtra state in India following the 1993 earthquake with the help of a soft loan from the World Bank, the policy of relocation of reconstructed settlements to 'earthquake safer places' was decided on the basis of the nature of soil. The policy not only distanced people from their agricultural land, but also from their cultural heritage represented by temples and mosques, wells, and old trees, which hold old, deeper spiritual values (Jigyasu, 2001).

Due to incompatibility of the new village layout and house designs with spaces needed for carrying out traditional social and cultural practices, many reconstructed houses were abandoned. Also, new materials and construction practices were encouraged at the cost of traditional skills in stone masonry. Rather than upgrading traditional building knowledge in stone, which is a locally available building material, it was condemned as the reason for destruction. While poor stone constructions were indeed responsible for the collapse of houses, the improper use of material was the reason rather than the material itself. In fact, even with subsidies provided by the government to promote new material and technology, local people could not afford building good constructions due to limited resources and the lack of appropriate skills and knowledge of new materials and technology.

Due to the fear of the use of stone inculcated by the engineers educated in contemporary construction skills, most people resorted to new materials and technologies employing heavy use of reinforced concrete and brick, albeit compromising heavily on basic safety standards. Concrete was also found to be climatically inappropriate compared to stone and wood, which kept cooler indoors during hot summer months. Perhaps the most disastrous consequence of this reconstruction policy was the nearly total disappearance of traditional stone craftsmen who slowly changed to contemporary materials and construction techniques or shifted to other occupations (Jigyasu, 2001).

However, post-crisis situations sometimes create opportunities for new values to be created. The disaster-prevention centre located in the centre of Minami-Sanriku-Cho town was destroyed by the tsunami that hit the shores of east Japan following the massive earthquake in 2011. While several people climbed to the roof to save their lives, the tsunami waves did not spare most of them. A brave woman working at the centre perished while shouting from atop the roof, telling people to escape to the safe area. The tsunami left bare the skeleton of the structure, while everything else was damaged or swept away (Figure 10.5). As the post-tsunami recovery

Figure 10.5 The house designs and the layout of villages reconstructed after the 1993 Marathwada earthquake in India had no link with traditional architecture and planning. As a result, many of these villages were not occupied by the beneficiaries, who preferred to reconstruct their traditional shelter

phase progressed, the crucial question was whether to keep this structure as a memorial or demolish it as it represents a painful memory. The views on this were sharply divided. Meanwhile, the story associated with this structure gained popularity through national and international media and there was a steady stream of visitors from Japan and abroad, visiting the site and paying tribute to the victims. Though the government was initially toying with the idea of demolishing this structure, it has been left intact ever since due to its rising popularity and is now decorated with flowers and incense for visitors to pay their respects. While most local inhabitants remain in favour of removing this structure, there is predominant acceptance of the new values this site has come to embody over time (BBC, 2016; Lin et al., 2018). Contrary to this, in another nearby town of Kesennuma, a huge fishing boat was swept inland due to tsunami, reminding people of the destructive power the waves. Similar discussion regarding the future of this wreck again started among the authorities and local communities. However, the final decision was to clear the site by removing the boat (ABC, 2013). Retaining and restoring the symbols of painful past would necessitate consensus on the accepted new post-crisis values among various stakeholders including various sections of the community. Of course, there is the likelihood of their greater recognition as heritage with increasing distance from the event. Time may serve as the great healer as in the case of Hiroshima memorial. The recovery of these remains may sometimes also need creative interpretation as in the case of Berlin wall, and, in that process, may also involve adding new values through powerful narratives linked to the event and the struggle for recovery afterwards.

Recognizing the contribution of urban heritage to resilience

Although urban cultural heritage is increasingly vulnerable to disasters, it should not be seen merely as a passive victim of disaster. In the face of disasters, traditional communities in historic cities often develop a vocabulary of resilient features in the urban environment that intentionally or unintentionally contribute towards prevention and mitigation, emergency response, and recovery. For example, many traditional buildings located in urban areas performed well during the earthquakes in Gujarat (2001), Kashmir (2005), Haiti (2010), and Nepal (2015), demonstrating traditional knowledge for earthquake mitigation that has been accrued over many generations through successive trials and errors.

There are several cases where the historic urban fabric, characterized by a series of interconnected courtyards, has helped in the emergency escape of residents from densely inhabited areas such as in the historic settlements in Kathmandu, which were recently struck by devastating earthquakes. Moreover, these traditional settlements have well-established networks of rest places (locally called *Paatis* and *Sattals*) and water sources, wells, stone water spouts (*hitis*), water tanks, and ponds that are strategically located at open squares and at street junctions and village entrances (Figure 10.6). These serve as places for settlers and visitors to carry out daily activities. In the event of a disaster, these resting places can also be used for sheltering the injured, while water sources used for drinking can double as a local fire hydrant. These public places typically used for community gatherings, playing traditional music, or just chatting can also help maintain a rapport among the local people facing a catastrophe. In this way, the tangible attributes carrying intangible and social values have the potential to enhance cooperation among residents during a crisis and may well serve as sites for disaster preparedness training (Jigyasu, 2020).

Figure 10.6 In the absence of proper knowledge, many traditional houses were wrongly repaired using new materials following the 1993 Marathwada earthquake

As mentioned before, traditional urban planning based on local geography and available natural resources have also served to build the resilience of traditional urban communities. For example, the traditional urban planning of Ayutthaya with networks of canals and check dams as well as houses built on stilts ensured that local communities have adapted their way of life to live with the risk of floods. Recovering traditional knowledge systems in urban planning and management, and identifying their potential role in disaster risk reduction, is an important challenge to be addressed for enhancing the safety and resilience of historic urban areas while undertaking recovery. Therefore, rather than only advocating solutions such as the widening of roads for reconstruction and building contemporary structures by using new technology, it is important to consider utilizing traditional knowledge by adapting it to current priorities and constraints.

Finally, traditional management systems also have tremendous potential in securing collective action among communities for post-disaster recovery. The rich expression of heritage is also a powerful means to help victims recover from the psychological impact of the disaster. In such situations, people search desperately for identity and self-esteem. Traditional social and religious networks that provide mutual support and access to collective assets, often represented by urban heritage, are an extremely effective coping mechanism for community members (UNISDR, 2013). This was well demonstrated following the 2015 Nepal earthquake as networks of traditional *guthi* (communal trusts) provided support to local communities in their transition from the response to the recovery phase.

The role of cultural heritage in building resilience during the post-disaster recovery process is well exemplified in the case of the Great East Japan Disaster in 2011. In the town of Shizugawa, Minami-Sanriku-Cho in Miyagi Prefecture, many buildings were destroyed and washed away by the tsunami that struck the eastern coast of Japan following a massive earthquake. However, the tsunami waves reached the precinct of Kaminoyama Hachimangu Shrine, but stopped before reaching the main hall due to its location on the higher ground. In fact, historical accounts mention that the shrine, which was previously located in the Shizugawa town, was gradually located on higher ground after experiencing tsunami in the past. In fact, research has shown that many temples and shrines survived the disaster, as they were located on higher grounds (Okubo, 2016). As a result, many of these temples and shrines acted as refuge shelters for the affected people in the immediate aftermath of the disaster. In Matsushima, the tsunami came halfway up the path to the Zuiganji Temple (NT) but did not reach the main hall. The religious training hall of the Zuiganji Temple attracted many tsunami victims as the nearby-designated evacuation centres were affected by the tsunami. The tatami mat-floored training hall had a large kitchen with numerous pots that were meant for preparing food for large religious gatherings and had enough food supplies given as donations. Under these conditions, as many as 300 evacuees could be provided enough hot meals from the first day onwards. There was an extended power outage after the earthquake; but as it was a temple, religious candles were used for lighting purposes. There have been previous records citing the previous use of this temple as a refuge shelter following tsunamis in the Edo Period (Okubo, 2016). Similarly, in one of the temples, more than one hundred children and teachers from nearby schools could take shelter following the disaster. These instances show the powerful role that cultural heritage played during post-disaster situations.

In the aftermath of the disaster, shrines such as Kaminoyama Hachimangu in Shizugawa town also became social anchors for bringing together communities that were split and displaced. The priest of the shrine played an important role by initiating various activities for the affected community members such as Kiriko (traditional designs made of paper) and pictorial storytelling for children (Figure 10.7).

The reconstruction process was initiated by the government after the initial response phase was over. The 10-year plan (2011–2021) entailed the restoration of infrastructure and services in the first three years, followed by rebuilding and development. Since towns like Shizugawa were swept away due to their proximity to the sea, the basic reconstruction approach was to relocate residents to a higher ground to protect them from future tsunamis. As a result, the town was split into three independent residential areas located on higher ground, while the lower terrain was reserved for commercial facilities. To protect the area from tsunamis, massive sea walls are being constructed and the topography significantly altered through landfilling.

Figure 10.7 The skeleton of disaster-prevention centre in Shizugawa town in Minami-Sanriku-Cho damaged by the 2011 tsunami in the east coast of Japan was turned into a memorial

Figure 10.8 Traditional water structures in the historic city of Bhaktapur in Kathmandu valley, Nepal, helped in providing water for essential services in the immediate aftermath of the 2015 earthquake

While this approach can be justified from the safety point of view, there are significant concerns regarding the negative impacts such interventions would have on the heritage of the area, which is closely linked to local geography and natural resources (Figure 10.8). In fact, the close interaction between culture and nature has ensured long-term social, economic, and environmental sustainability in the region. Many rituals are connected to the visual, physical, and spiritual relations to key landscape features such as the ocean, river channels, islands, and local flora and fauna. Besides the disruption of socio-spatial practices, the splitting of the town will also result in damage to community networks that are tied through festivals and rituals. The long-term ecological impacts of the sea wall are also questionable. The primary livelihood of the region relies on agriculture and fishing, also potentially impacted by the disruption in the local ecology.

Reconstruction as improvement: Building back better for heritage by reducing vulnerabilities

The term reconstruction in disaster studies is defined as restoration and improvement of facilities, livelihoods, and living conditions of disaster-affected communities, including efforts to reduce disaster risk factors. If our objective is the mere return to the former state, then we end up recreating those very

Figure 10.9 In the aftermath of the 2011 tsunami following the great east-Japan earthquake, historic shrines such as Kaminoyama Hachimangu in Shizugawa town became social anchors for bringing together communities that were split and displaced. The local priest of the shrine organized many activities to bring together communities displaced after the tsunami

Figure 10.10 In the reconstructed town of Shizugawa, the landfilling to raise the height of the village to protect the town from future tsunami has disturbed the traditional relationship with the river that was manifested through various festivals and social activities of the community

vulnerabilities that created conditions for previous disaster in the first place. The idea of resilience is behind intervention from this perspective in post-disaster contexts (Jha et al. 2010) (Figures 10.9 and 10.10).

Disasters cause misery but they also provide opportunity to get rid of pre-disaster vulnerabilities. For example, in the case of London after the Great Fire of 1666, Sir Christopher Wren initially proposed to sweep away the medieval street plan and substitute an ordered classical layout that is more reminiscent of absolutist France. However, the implemented compromise eventually retained the medieval layout stocked with completely new buildings (Glendinning, 2013).

To what extent have pre-disaster physical, social, and economic vulnerabilities been reduced in the ongoing rebuilding process? Are the affected communities better prepared for the next disaster? Clearly, the answer lies not only in creating new solutions, but also in augmenting the potential role of traditional knowledge and capacity through the careful analysis of vernacular architecture, morphology, and management systems and adapting them to the new constraints, needs, and opportunities.

This would also necessitate research and innovation in traditional building practices while keeping in mind the pragmatic realities on the ground such as peoples' aspirations and affordability, and marrying them with the responsibility for improving the quality of life, safety standards, as well as restoring local identity. This should lead to the meaningful application of feasible solutions on the ground through capacity building at various levels: Decision-makers, professionals, as well as local contractors, masons, and carpenters. Growing mistrust against traditional materials and construction techniques is another barrier to be crossed through extensive awareness among the local community. This can be achieved through effective communication made by simple experiments on the behaviour of structures to make them understand the difference between good and bad construction practices, rather than choosing between traditional or modern materials and techniques. Rather than taking extreme positions, there is a need to adopt 'in between solutions' that take into consideration multiple factors that would together contribute to improving the overall built environment.

Moreover, technology, whether it is traditional, modern, or alternative, will only be successful if it caters to multiple criteria. Of course, hazard safety is one of the primary concerns in disaster-prone areas, but equally (if not more) important are considerations of economic viability, cultural compatibility, and climatic suitability that govern the context. An appropriate solution would therefore involve necessary trade-offs between these factors to achieve viable alternatives, although this may necessitate the optimization and not maximization of earthquake safety (Jigyasu, 2010).

Moreover, technology should not be a rigid design package that is provided on a palette to the affected communities. Technology is essentially a process for which appropriate design and delivery mechanisms need to be created and institutionalized to ensure long-term sustainability. This means that technology introduced as a 'product' must be linked to this process right from the time of its conception. This is indeed a painfully slow process, and it requires mechanisms that support the local capacity to innovate and not merely duplicate what is provided to the beneficiaries (Jigyasu, 2010).

However, despite the use of the appropriate technology, physical vulnerabilities would never get reduced if underlying social and economic vulnerabilities are not addressed. In many situations, poor constructions are a result of the lack of affordability and access to resources and power for certain sections of the community. Such stark realities should be borne in mind while developing reconstruction policies, which should give due consideration to social, economic, and demographic characteristics of the affected households, which enable or inhibit their ability to undertake sound reconstruction. Therefore, rather than using standard formulas or recipes in the design, construction and rebuilding process, different options should be provided. After all, reconstruction should aim at achieving not just physical resilience but social and cultural resilience as well.

The recently adopted Sendai Framework on Disaster Risk Reduction recognizes culture as a key dimension of disaster risk reduction and the need to protect and draw on heritage as an asset for resilience through several important references (UNISDR, 2015). It also advocates for a 'Build Back Better' approach for recovery, which requires the careful interpretation of heritage in terms of improving vulnerability while impacting heritage values to the least possible impact. The challenge is to implement the framework, which requires considerable building of capacities at international, national, and local levels and the setting up of the necessary institutional mechanisms, complemented by data collection and monitoring (Dean and Boccardi, 2015). It is high time to put these well-articulated policies into tangible actions.

From physical reconstruction to holistic, sustainable, and resilient recovery of urban heritage

This takes away our focus from the mere reconstruction of physical components of urban heritage to holistic, sustainable, and resilient recovery of cultural heritage. Physical reconstruction is just part of the process and the protection of heritage values is not merely an end in itself but also strives to recognize the role heritage in building the resilience of communities and ensures the continuation of services and benefits for long-term sustainability.

- **Holistic Recovery**: Mainstreaming the tangible and intangible heritage belonging to various sections of the community in the overall recovery process. This would also entail ensuring the transversal consideration of culture in various urban recovery sectors such as shelter, infrastructure, environment, and health.

- **Sustainable Recovery**: Ensuring that the recovery process is social, economically, and institutionally sustainable and does not create dependencies on external actors in the long run.
- **Resilient Recovery**: Enabling heritage as means of reducing physical, social, economic, institutional, and attitudinal vulnerability, building peace and reconciliation, and enhancing coping capacity following the concept of 'Building Back Better'. This would also entail early recovery during the emergency response to cultural heritage. For physical recovery of the heritage fabric, a well-conceived workflow of debris management, documentation, salvage, and storage of heritage fragments from damaged buildings will aid in reusing them during the reconstruction process in the best possible way. For social and economic recovery, engagement of diverse actors and stakeholders in the initial response phase will be crucial.

References

ABC. (2013). Japanese City Kesennuma Scrapping Fishing Boat that Became Symbol of *2011 Tsunami Disaster*. https://www.abc.net.au/news/2013-09-09/japan-scrapping-fishing-boat-that-became-symbol-of-2011-tsunami/4946162.

Australia ICOMOS. (2013). The Burra Charter. http://australia.icomos.org/publications/charters/.

BBC. (2016). Shell of Building to Survive as Japan Tsunami Shrine. https://www.bbc.com/news/blogs-news-from-elsewhere-35688080.

Cameron, C. (2017). Reconstruction: Changing Attitudes. UNESCO. https://en.unesco.org/courier/july-september-2017/reconstruction-changing-attitudes.

Dean M. and Boccardi, G. (2015). Sendai Implications for Culture and Heritage. *Crisis Response*, *10*(4), 54.

Glendinning, M. (2013). *The Conservation Movement: A History of Architectural Preservation: Antiquity to Modernity*. Abingdon: Routledge.

ICOMOS. (1964). *The International Charter for the Conservation and Restoration of Monuments and Sites, Paris*. https://www.icomos.org/charters/venice_e.pdf.

ICOMOS. (1987). *Charter for the Conservation of Historic Towns and Urban Areas, Paris*. https://www.icomos.org/charters/towns_e.pdf.

ICOMOS. (1994). *Nara Document on Authenticity*. https://www.icomos.org/charters/nara-e.pdf.

ICOMOS. (2011). *The Valletta Principles for the Safeguarding and Management of Historic Cities, Towns and Urban Areas, Paris*. https://www.icomos.org/Paris2011/GA2011_CIVVIH_text_EN_FR_final_20120110.pdf.

ICOMOS. (2017). *ICOMOS Guidance on Post Trauma Recovery and Reconstruction for World Heritage Cultural Properties, Paris*. http://openarchive.icomos.org/1763/19/ICOMOS%20Guidance%20on%20Post%20Trauma%20Recovery%20.pdf.

Jha, A.K. Duyne Barenstein J., Phelps, P., Pittet, D., and Sena, S. (2010). *Safer Homes, Stronger Communities: A Handbook for Reconstructing after Natural Disaster*. Washington D.C.: The World Bank. https://www.preventionweb.net/publications/view/12229.

Jigyasu, R. (2001). From Natural to Cultural Disaster: Consequences of the Post-Earthquake Rehabilitation Process on the Cultural Heritage in Marathwada Region, India. *Bulletin of the New Zealand Society for Earthquake Engineering, 34*(3), 237–242.

Jigyasu, R. (2010). Appropriate Technology for Post Disaster Reconstruction, in *Rebuilding After Disasters. From Emergency to Sustainability*, edited by G. Lizarralde, C. Johnson, and C. Davidson. London and New York: Spoon Press, 49–69.

Jigyasu, R. (2020). Towards Safer and Resilient Historic Urban Environments, in *Routledge Handbook on Historic Urban Landscapes in Asia Pacific*, edited by K. Silva. Abingdon: Routledge.

Lin, Y., Kelemen, M. and Tresidder, R. (2018). Post-disaster Tourism: Building Resilience through Community-led Approaches in the Aftermath of the 2011 Disasters in Japan. *Journal of Sustainable Tourism*, *26*(10), 1766–1783.

Okubo, T. (2016). Traditional Wisdom for Disaster Mitigation in History of Japanese Architectures and Historic Cities. *Journal of Cultural Heritage*, *20*, 715–724.

Sen, A. (2006). *Identity and Violence: The Illusion of Destiny*. Delhi: Penguin India.

UNESCO. (2011). *Recommendation on the Historic Urban Landscape, Paris*. https://whc.unesco.org/uploads/activities/documents/activity-638-98.pdf.

UNESCO. (2018). *Recommendations for World Heritage Recovery and Reconstruction Developed in Warsaw*. https://whc.unesco.org/en/news/1826/.

UNISDR. (2013). *Heritage and Resilience. Issues and Opportunities for Reducing Disaster Risks*. https://www.unisdr.org/we/inform/publications/33189.

UNISDR. (2015). *Sendai Framework on Disaster Risk Reduction*. https://www.unisdr.org/we/inform/publications/43291.

11 The [framing] of heritage in the post-war reconstruction of Beirut central district (Lebanon)

Howayda Al-Harithy and Dina Mneimneh

Introduction

War-torn cities experience erasure of place and memory and the disruption of socio-economic and cultural practices. Post-war reconstruction strategies attempt to reestablish continuity with the past or take an opportunity to build anew. This is what Ashworth and Tunbridge (1999, p. 5) call "year zero: a new beginning founded upon the eradication of what went before." Post-war reconstruction projects that attempt to reestablish continuity with the past often miss the opportunity to integrate the recovery of cultural heritage into urban planning, resulting in further ruptures. Cultural heritage, including sites of shared memories and social significance, is partially or fully destroyed during times of conflict and violence. Mending emotional, spiritual, and socio-cultural connections becomes important to reconstructing the identity of a place (Ashworth and Tunbridge, 1999; O'Keefe, 2006; Smith, 2006). The post-war reconstruction of the Beirut Central District (BCD) is a unique case where the approach to reconstruction shifted multiple times between a conservation and a tabula rasa. It, thus, offers the opportunity to interrogate the processes, shifts, and manners in which heritage was framed in the reconstruction process.

In the aftermath of violent conflicts, post-war reconstruction plans rarely recognize the value of cultural heritage. Cultural heritage is, therefore, not prioritized in reconstruction strategies. Barakat (2007, p. 26) argues, "cultural heritage has been overlooked by mainstream post-war reconstruction practice and policy – perceived and treated as a luxury that cannot be afforded by societies emerging from the devastation of war." Even when tangible cultural assets, particularly historic buildings, are restored, post-war reconstruction projects neglect to revive intangible socio-spatial practices and rituals that constitute living heritage (Al-Harithy, 2005; Ascherson, 2007; Barakat, 2007; Wijesuriya, 2005). Confronted by the urgency to address post-war basic needs, the often-lengthy process of cultural recovery is compromised or, in many cases, not brought forward as an integral component of post-war reconstruction schemes. Stanley-Price (2007) argues that the understanding of both the cultural context and the cultural resilience of communities has a great impact on successfully meeting basic needs of post-war recovery such as food, shelter, and health.

Scholars have recognized the value of cultural heritage as a critical catalyst of post-conflict recovery. They put forth many questions around the role of cultural heritage, particularly in relation to memory, identity, and reconciliation. They link cultural heritage to the survival of a selected past and its memories, and link socio-spatial practices to place identity (Al-Harithy, 2010; Ashworth and Tunbridge, 1999; Benton and Cecil, 2010). Smith (2006) argues that heritage is a cultural process that is constantly negotiated through identity, memory, and place. This chapter attempts to contribute to these scholarly debates by addressing such questions in the context of the post-war reconstruction of the BCD. It interrogates how the diverse strategies adopted at various stages of the planning process, during and after the 15-year Civil War (1975–1990), impacted cultural recovery, in general, and the recovery of built heritage, in particular. It explores how the concept of memory was employed in the process of reconstruction. It demonstrates how the early tabula rasa approach that contributed to the production of collective amnesia failed, only to pave the way for the later and more contextual approach. The contextual approach, however, framed heritage as a commodity not as "a multilayered performance" that "embodies remembrance and commemoration" (Smith, 2006, p. 3) and, thus, failed to reconnect people to place. It therefore argues that cultural recovery was not integrated in a comprehensive manner within the scope of the project, although cultural heritage, both tangible and intangible, sustained a great deal of damage during and after the war. In this chapter, recovery, as conceptualized in the introduction of this book, is viewed as a holistic and participatory process and not mere physical reconstruction. Cultural recovery is understood as a process through which people construct meaningful connections with the past represented by tangible places and objects and intangible socio-spatial practices, rituals, and narratives. This position is grounded in the work of scholars of heritage studies who have argued that heritage is a process of ascribing meaning to a particular past to serve present and future agendas (Ascherson, 2007; Ashworth and Tunbridge, 1999; Hewison, 1987; O'Keefe, 2006; Smith, 2006).

The Beirut central district

Lebanon, before the Civil War (1975–1990), witnessed unprecedented economic, cultural, and touristic prosperity. The 1950s and 1960s, particularly the period under the presidency of General Fouad Chehab (1958–1964), were characterized by major infrastructural projects and the reform of public institutions (Salam, 1998).[1] The BCD grew as a major hub for commerce, culture, leisure, and transportation. Popular markets,[2] specialized business areas,[3] and financial institutions were established (Kabbani, 1991).

The waterfront also hosted important commercial activities following the growth of the Beirut Port (Saliba, 2004). Economic and cultural prosperity were evident in the BCD with the establishment of cultural facilities, including a number of publishing houses, theaters, and cinemas.[4] Tourism also fueled an unprecedented growth in the number of local cafés and international hotel chains.[5] The government at the time concentrated its efforts on cultural assets across the BCD and uncovered archeological sites such as the site of the Roman baths, the remains of a Roman street pattern in Nejmeh Square, and the Bronze Age tombs near Martyrs' Square (Sader, 1998). This period is often seen until today as the golden era of Beirut and is idolized in present nostalgia.

The Civil War caused severe disruptions in cultural life in the BCD. Internal clashes due to the Lebanese confessional political system resulted in intermittent episodes of violence, with devastating impacts on the BCD's cultural institutions and built environment. The BCD and the rest of Beirut were divided into two east and west sectarian enclaves separated by the green line, a north-south axis line running along Damascus Road. The intense concentration of battles along the green line caused major waves of displacement to other areas in Lebanon and abroad. The historic fabric of the BCD was extensively damaged – souks, houses, office buildings, hotels, and other landmarks appeared as wounded recollections of a prosperous past (Ghandour and Fawaz, 2010). Socio-economic disruption was also evident. The once thriving businesses and hospitality services were mostly closed. Dilapidated structures were occupied by squatters – around 30,000 families – and an uncontrolled landfill[6] heavily polluted and altered the BCD's interface with the sea (Gavin, 1998; Bădescu, 2011).

The Civil War ended in 1989 with the signing of the Taif Agreement. Due to the protracted and intermittent nature of the war, the government initiated rebuilding masterplans for the BCD as early as 1977. However, the early masterplans of 1977 and 1986 were never implemented due to renewed conflict. They were later replaced by the masterplan of 1991. The mid and late 1990s saw the revision of the 1991 masterplan and the implementation of the revised 1994 masterplan by the private real estate company SOLIDERE, tasked by the government with the urban planning, management, and redevelopment of the BCD. The masterplans primarily focused on the physical reconstruction of the BCD, which was framed as representative of national recovery given its status as the heart of Beirut. They reflected the need for a "strong image to be used as a symbol for national unity" (Tabet, 1993, p. 91). The urban planning efforts were accompanied by systematic demolitions of buildings under the pretext of clearing war debris or removing structures deemed unsafe. Known as tabula rasa decisions, these demolitions emerged first in 1983 and later at the launch of the post-war reconstruction plan of 1991 (Figure 11.1).

This chapter examines the intersection of heritage and post-war reconstruction in the BCD through the successive masterplans from 1977 to 1994.

Figure 11.1 Aerial view of the BCD in 2002

It critically analyzes their impact across social, economic, and cultural layers. The chapter studies the partially implemented masterplans of 1991 and 1994 in greater detail by exploring the multiple approaches used for the identification and promotion of built heritage. It argues that heritage was framed to serve the political and economic visions fostered by SOLIDERE for the BCD.

The war period masterplans of APUR 1977 and IAURIF 1986

Between 1975 and 1990, two masterplans, one by the French Atelier Parisien d'Urbanisme (APUR) for the BCD and one by the Institute of Planning and Urbanism of the Ile-de-France (IAURIF) for greater Beirut, were proposed but never implemented due to the intensification of the Civil War and severe fragmentation in public institutions. APUR's proposed development and new legislation paved the way for the privatization of the reconstruction process at the beginning of the 1990s.

When the war was expected to end in 1977, President Elias Sarkis and the Council of Ministers established the Council for Development and Reconstruction (CDR). At the time, the CDR was tasked with the management of reconstruction projects in Lebanon.[7] The CDR commissioned the first reconstruction masterplan for the BCD to APUR (Figure 11.2). APUR's 1977 masterplan[8] was first approved in 1978 and later in 1982 with minor revisions. Both iterations were geographically limited to the BCD and included the rehabilitation of individual buildings and clusters, street alignments, and re-parcelization. Salam, a prominent Lebanese architect and activist, explains that the APUR framework followed clear guidelines:

> The first was to maintain the urban tissue in its original condition whenever possible, and to maintain original property tenure. The second was to encourage legal owners and occupants of the BCD to return to their previous activities. The third was to accelerate the return of

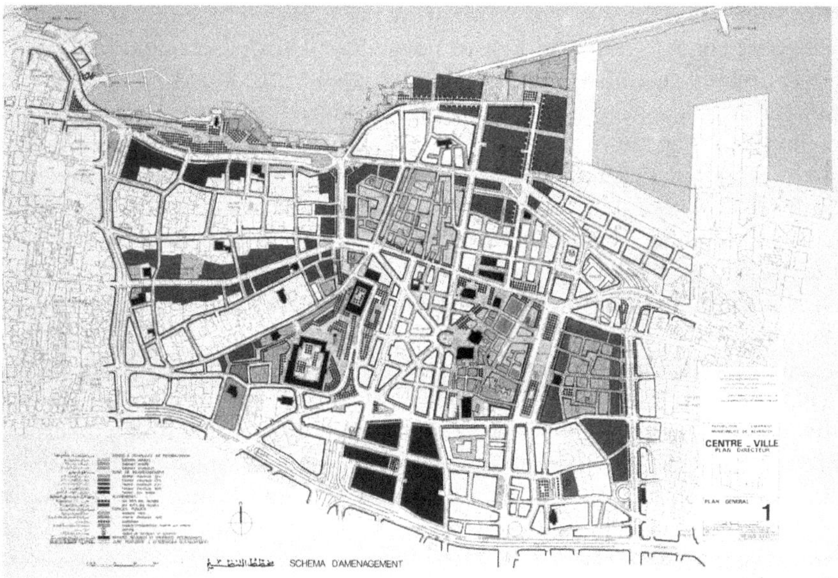

Figure 11.2 APUR masterplan 1977

> the BCD to its traditional role as a unifying ground for Lebanon's multi-confessional communal structures. The fourth was to introduce infrastructural improvements to the BCD, and finally, the fifth was to revitalize areas that had been badly damaged through the creation of real estate companies and other methods of intervention stipulated in town planning laws.
>
> (Salam, 1998, p. 130)

As a result, the APUR masterplan introduced the role of private real estate development companies in the reconstruction of heavily damaged areas in the BCD. The least damaged zones were expected to be restored by their property owners (Mango, 2004). Recognizing the impact of the war on the souks located to the west of Martyr's Square, the 1977 APUR masterplan combined restoration with reconstruction and addressed the specific conditions of the souks, stressed the need to preserve their traditional character, and gave priority to everyday activities (El-Khoury, Khoury, and Raad, 1997). The attention to continuity of socio-spatial practices and the activation of sites of shared memories was not evident in later masterplans, including the revised 1977 version in 1982. Later masterplans phased demolition of buildings and even extended demolition to cluster areas of the BCD including the souks and some residential neighborhoods.

The final APUR masterplan was set in motion in 1983 upon its approval in Decree No. 83 dated 28 December 1982.[9] The CDR commissioned the demolition of a number of buildings as early as January 1983, resulting in irreversible damage to the built heritage of the BCD. The demolition was claimed to clear the war scars, preparing the ground for reconstruction projects and encouraging a return to normality. The dense neighborhood of Saifi, directly east of Martyr's Square, as well as the souks of al-Nourieh, Sursock, and as-Sagha were largely bulldozed (Figure 11.3) (Beyhum et al., 1992). The removal of small-to-medium-size businesses caused severe socio-economic disruption. It erased an important part of Beirut's historic built fabric where daily exchanges regularly took place. Spaces of significance in the collective memory of people no longer provided continuity with the pre-war era. Economic and socio-cultural connections with the place were severed, limiting the role of cultural heritage in a potential post-war recovery.

The 1983 demolitions continued without necessarily being guided by the APUR masterplan. At the time, several private real estate companies joined the pre-reconstruction process. While the Lebanese Army isolated certain areas in the BCD to proceed with de-mining operations,[10] OGER Liban, the private contracting firm of late businessman and Prime Minister Rafiq Hariri began clearing the war debris. A special unit that was led by the Municipality of Beirut and OGER Liban assessed the condition of shops and residences. Landowners were given short-term notices to clear their property without any compensation plan in place.[11]

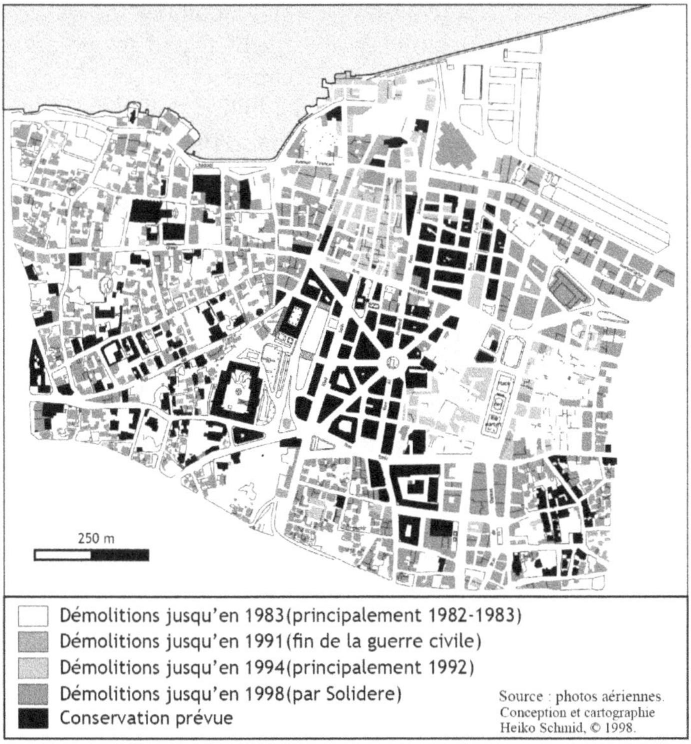

Figure 11.3 Map of demolitions in the BCD

Selective rehabilitation projects in early 1983 overshadowed the ongoing demolitions in the public eye. The CDR entrusted OGER Liban with rehabilitation projects, including the rehabilitation of Riad al-Solh and Nejmeh squares, and the implementation of a new system of infrastructure. These rehabilitation projects framed heritage as single historic monuments and proceeded to restore iconic buildings such as the Parliament Building, al-Omari Mosque, and buildings along al-Maarad Street.[12] Additionally, the Directorate General of Antiquities (DGA) resumed the archaeological excavations in the BCD.[13] These efforts, although selective and geographically limited, aimed to promote a sense of recovery by resuming normal activities in the BCD.

The scale of changes to the built heritage in the BCD was drastic and alarming. Local professionals responded by demanding urban planning revisions that integrate urban heritage in the post-war vision and ensure coordination between the different agencies involved in the project, including the Municipality of Beirut, the Governorate of Beirut, the CDR, and OGER Liban.[14] The academic community also became active in debating the reconstruction process and proposing alternatives.[15] By 1986, neither local professionals nor academics were able to counter the quick surge of demolitions that altered the built fabric of the BCD and any potential recovery of its social, economic, or cultural dynamics.

As a matter of fact, demolitions caused additional internal displacement of people. Many of the abandoned buildings in the BCD were squatted by families fleeing the Israeli invasion of the south of Lebanon in 1982. These buildings were emptied and demolished, causing further displacement and a population decrease in the BCD. This was part of a larger dynamic. Internal displacement, drastic population redistribution, and the urban expansion of Beirut became major concerns for the CDR and the Lebanese Directorate of Urban Planning (DGU). As a result, the CDR and the DGU commissioned the IAURIF and the Mission Franco-Libanaise d'Etude et d'Aménagement de la Region Métropolitaine de Beyrouth to survey and draft a masterplan for Beirut and its peripheries. In June 1986, a new masterplan, known as the Schéma Directeur, adopted the recommendations of the 1977 and 1982 APUR masterplans for the reconstruction of the BCD and its enlargement through the use of the Normandie landfill, and proposed a public park within it. The Schéma Directeur reinforced the role of the BCD as the center of public and other vital institutions. Ultimately, all efforts failed due to the severe fragmentation in public institutions and the renewed war in 1986 (Salam, 1998). None of the proposed masterplans were implemented, and the reconstruction process resumed after the end of the war in 1990.

The post-war masterplan of Dar al-Handasah 1991

By the end of the war, the start of the national rebuilding process was imbued with aspirations for nationwide reconciliation. Public officials announced that images of the war-torn BCD needed to be quickly erased. Al-Fadel Chalak, the president of the CDR in 1991, stated, "the reconstruction of Beirut City Center is of a great importance, to be considered the launching base of reconstruction across the whole country."[16] Financed by OGER Liban, the CDR commissioned Dar al-Handasah, Shair and Partners, a Beirut-based urban planning and design consultancy, to propose a masterplan for the reconstruction of the BCD. Conflicts of interest around the use of private funds from OGER Liban limited the capacity of the government to monitor the reconstruction process. In December 1991, Law No. 117 was amended to authorize the establishment of a single real estate company to undertake the reconstruction.[17] Multiple rationales were offered to justify the privatization of the reconstruction. These included the fragmented public administration unable to formulate, implement, and control development frameworks, scarce public resources to finance large-scale projects, and complex entanglement of property rights (Beyhum et al., 1992). Law No. 117 significantly reduced the role of the government in the reconstruction. It empowered a private real estate company to gain ownership of both the public and private domains in the BCD for land reassembly and to plan, finance, and implement the post-war reconstruction. Rafiq Hariri at the time was the head of OGER Liban. He became Prime Minister in January 1992 and majority shareholder in the real estate company SOLIDERE.[18]

Drafted under the leadership of architect and former Minister of Public Works and Transport, Henri Edde, the 1991 masterplan proposed drastic urban design schemes for three major axes in the BCD (Figure 11.4a). These three axes included an urban corridor replacing Martyr's Square and ending with an elliptical marina (Figure 11.4b); another passing through Place de l'Etoile and Riad al-Solh Square and ending with twin towers along the north coast; and a third clearing out the vista from the Grand Serail to a new cultural center on the reclaimed sea landfill. The total area of the BCD masterplan was around 1,200,000 square meter, in addition to 200,000 square meter of infilled sea and a 250,000 square meter structure that was meant to replace the first basin of the Beirut Port. The plan preserved only 20 percent of the preexisting urban fabric, including the western edge of Nejmeh Square, the buildings between Foch and Allenby streets, the blocks surrounding the Azariah Building, as well as several schools, hotels, and religious landmarks (Beyhum et al., 1992). The demolitions by OGER Liban between 1983 and 1991 destroyed much of the fabric of the traditional souks, the Saifi neighborhood, and around 140 buildings located between Nejmeh and Martyr's squares. These demolitions paved the way for the implementation of the 1991 masterplan (Schmid, 2006).

Local professionals, academics, and concerned citizens highly criticized the 1991 masterplan and its private financing and implementation frameworks. They denounced the legal and technical approaches to the project. Public protests and organized meetings took place on the ground. Some property right holders in the BCD, tenants, and leaseholders advocated against the project's financial framework stipulated in Law No. 117. They claimed that the exchange of buildings for shares in a real estate company "constitutes theft and that archeological treasures will be buried."[19] Against a backdrop of public rejection, authorities moved forward with the reconstruction scheme, and the masterplan was rushed into official approval despite its preliminary state. In July 1992, the CDR issued a technical review of the plan, calling for some amendments and detailed supporting documents.[20] Despite the technical review, the masterplan was approved in Decree No. 3786 issued during the meeting of the Council of Ministers on 14 October 1992. The guidelines of the masterplan were modified with a larger boundary area and an increase in Floor-to-Area Ratios leading to a dispute with Henri Edde (El-Achkar, 1998).

The public brought forward the need for heritage preservation as early as May 1991, calling for a proper study and management of archeological sites in the BCD. Architect and activist Assem Salam, one of the first scholars to exhaustively study the reconstruction of the BCD and its shortcomings, criticized the masterplan as highly insensitive to its surroundings. It was predicated on the erasure of the surviving traditional built fabric, with the

Figure 11.4 (a) The masterplan of Dar al-Handasah 1991. (b) Perspective view of BCD as per the masterplan of 1991

exception of a small area dating back to the French Mandate period and a few selected cultural and religious monuments (Salam, 1994). Lebanese archeologists demanded that public authorities organize excavations before any new construction plans to allow for the integration of archeological findings within the masterplan. Upon the request of the Lebanese government, UNESCO was brought on board to start the negotiations with urban planners. Although these efforts did not materialize within the first masterplan of 1991, a preliminary agreement was reached between UNESCO and the CDR in October 1991 to devise adequate frameworks for archaeological excavations (Abi Akl, 1999).

Proposed new-age representations of a "modernized" city with high-rise towers and ever-increasing exploitation ratios were unfamiliar and unrecognizable to the residents. They were highly dissociated from people's collective memories of a city with human-friendly scale buildings, traditional souks, and vibrant open spaces. The rush to remove the scars of the war combined with the privatization of the project, engineered collective amnesia. From a heritage point of view, the 1991 masterplan was therefore perceived as "a false start [...] with a grandiose scheme that combined *beaux-arts* 'grand planning' and a 1960's infatuation with the car and heavy road infrastructure" (Gavin, 1998, p. 218).

Salam stated that the masterplan imposed "devastation" on the BCD and ignored the archaeological heritage of the place. Salam also warned that a tabula rasa approach to reconstruction would be complicated by imminent archeological findings, diminishing or delaying the projected profit (Salam, 1994). As a result, publications, petitions, round tables, media campaigns, and international conferences were organized among professional and public circles to contest the masterplan. The Order of Engineers and Architects led these efforts, including multidisciplinary workshops with urban planners that denounced the flaws of the masterplan on urban, architectural, and socio-economic grounds. As a result, a detailed critique of the masterplan was published in local media outlets. These efforts, however, failed to have tangible impact (Beyhum et al., 1992).[21] Consultants and experts in different fields joined these resistance efforts, issuing multi-dimensional and alternative guidelines to the masterplan through a self-funded publication in 1992.[22] The guidelines prioritized the preservation of the traditional social and economic networks of the BCD as well as its built heritage, while referring to the masterplans by APUR in 1977 and IAURIF in 1986 that created connections to adjacent neighborhoods (Beyhum et al., 1992).

Despite public calls to preserve the built heritage, many buildings in the BCD were further demolished prior to the approval of the masterplan in Decree No. 15 dated 15 January 1992 by the Council of Ministers. The decree was governed by an implementation framework that was devised by a special technical team from the CDR. According to the framework, the Ministry of Interior and the Municipality of Beirut are required to assess the conditions of buildings, inform property owners of their decisions, evacuate squatters from abandoned buildings, and appoint contractors to carry out demolition works.[23] Approximately 300 buildings were demolished (Salam, 1994). From April 1992, demolitions concentrated in three areas:[24] (1) buildings along the traditional souks of Beirut[25]; (2) buildings bordering Martyr's Square, such as the Phillips Building and the Odeon Theater and between the St. George Maronite and Greek-Orthodox cathedrals; and (3) buildings along the southern side of Martyr's Square extending to Gemmeyze (Verdeil, Faour, and Velut, 2007). Despite the erasure of a significant section of the built and cultural fabrics of the BCD, the CDR claimed to only demolish buildings that were deemed unsafe, while preserving historic buildings along Maarad, Foch, and Allenby streets.[26]

Once again, the reconstruction project opted for a reductive identification of heritage that further erased significant layers of tangible and intangible heritage, which had characterized the development of the BCD for centuries. At the same time, significant archeological layers across the BCD that were uncovered due to these demolitions added to the complexity of the reconstruction project. Questions of heritage preservation and the proper management of archaeological findings emerged while public criticism mounted. For the critics, the archaeological findings fueled public demand to revisit the proposed masterplan. They favored an approach that integrates the preservation and management of heritage and archaeology in the masterplan (Perring, 2009). Moreover, political pressure and financial challenges, such as the reluctance of initial private investors to continue with the project, called for the revision of the Dar al-Handasah's masterplan in 1993 (Samara, 1996). SOLIDERE entrusted French planning consultants Sato et Associés with the evaluation of the masterplan developed and detailed by Dar al-Handasah. SOLIDERE's revised masterplan was approved in 1994 in Decree No. 4830 dated 4 March 1994 (El-Achkar, 1998).

The post-war SOLIDERE revised masterplan 1994

SOLIDERE revised the 1991 masterplan in 1994 to address the archeological findings in the BCD. The 1994 masterplan reconsidered how heritage can be integrated within a vision of a modernized city and soon introduced changes in both urban planning and design approaches (Gavin and Maluf, 1996; Nasr and Verdeil, 2008). Lead by SOLIDERE, field surveys assessed the conditions of surviving buildings. Buildings that were going to be restored were grouped under three categories: buildings of historical or architectural value; relatively large modern buildings in sufficiently good condition; and buildings of social value with legal tenants (Gavin and Maluf, 1996). Based on the surveys by SOLIDERE, the BCD comprised of 800 surviving buildings in varying states of devastation. The revised masterplan classified 291 buildings out of the 800, which "increase[ed] the number of buildings salvaged for restoration by 50 percent over those identified in the previously approved masterplan" (Gavin, 1998, p. 222).

The revised 1994 masterplan "reduced the amount of infrastructure conflicting with the historic urban fabric; maintained the original width of Martyr's square" while enhancing "some historic features like the remains of the Ottoman seashore line and jetty" (Saliba, 2004, p. 49) (Figure 11.5a). Part of this more contextual approach, additional urban planning tools were adopted such as maximizing visual permeability to the sea, limiting high-rise developments at the edge of the BCD, and maximizing mixed land use (Saliba, 2004). This shift in approach was also marked by a new type of images and representations (Figure 11.5b). SOLIDERE states specific main

Figures 11.5 (a) The SOLIDERE revised masterplan 1994. (b) Perspective view of BCD as per the revised masterplan of 1994

Figures 11.5 (Continued)

drivers to be achieved by the implementation of the 1994 masterplan. A number of these drivers conform to recommendations and proposed alternatives by local experts in 1992, such as the integration of the old with the new fabric and the design of public spaces. However, a comprehensive integration of cultural heritage or the restoration of the social fabric of the BCD was not a main priority for SOLIDERE. It adopted investor-friendly slogans: "neighborhood structures accommodating a broad mix of land uses" and "create a vibrant, 24-hour active downtown."

The discovery and maintenance of archeological findings was justified by the ambiguous goal of "unearth[ing] layers of the city center's history" and not guided by a comprehensive framework.[27] The archaeology and built heritage of the BCD were "unearthed" as constructs in the discourse of SOLIDERE around the 1994 masterplan. Concepts of "city memory" and "layers" emerged as dominant in the design. This is evident in the SOLIDERE motto "Beirut: Ancient City of the Future."[28] The contextual approach of the revised masterplan brought a shift in SOLIDERE's discourse claiming to deliver a "message of cultural continuity and pluralism, and not the imposition of foreign order" (Gavin, 1998, p. 219). This approach was exclusive to SOLIDERE's marketed definition of history and heritage, illustrated by the interventions of prominent developers (Perring, 2009). The process lacked government oversight and civic participation with regards to the re-integration of built heritage and archeology in the BCD.

SOLIDERE's identification of what constitutes heritage in the BCD was selective. Tangible and intangible heritage was framed through a market-driven approach to reconstruction. This market-driven approach manufactured a "collective memory" to attract future economic investments and tourism in the BCD, while claiming to address the public's "longing for the lost and damaged fabric of the immediate pre-war past" (Gavin and Maluf, 1996, p. 29). As a result, identifying built heritage, archaeological findings, and historic periods to be revived and integrated in the reconstruction of the "city memory narrative" remained an act of historical editing. It served a nostalgic view of the post-war era rather than the pre-war past (Al-Harithy, 2008). With the implementation of the 1994 masterplan, SOLIDERE further demolished historic buildings and neighborhoods, despite claims of a more contextual approach that valued the heritage of the BCD.

Between 1994 and 1998, SOLIDERE demolished architectural landmarks and cityscapes, which were still imprinted in the living memories of people. These included the Fattal Building, Makassed Building, Roxy Cinema, Dunia Cinema, Empire Cinema, Metropole Cinema, and Automatic Restaurant. More of Beirut's theaters were demolished, including the Rivoli Cinema, Byblos Cinema,[29] and the Phillips Building. Souk Ayyas, souk Tawileh, and souk Jamil were also destroyed, as well as a cluster of buildings adjacent to the previously razed souk Sursock and souk al-Nourieh. The demolitions revealed additional archeological sites. Unaccounted for in the masterplan of 1994, buildings were demolished in Zokak el-Blatt and Wadi Abou Jmil in 1996. The president of the Order of Engineers and Architects at the time, Assem Salam, and many heritage-protection associations such as the Association pour la Protection des Sites et Anciennes Demeures au Liban (APSAD) and Heritage without Borders strongly condemned the demolitions.[30] Furthermore, the systematic erasure of entire neighborhoods significantly altered the heritage value of the BCD. These neighborhoods included Wadi Abou Jmil and Ghalghoul to the south of Riad al-Solh Square, demolished between 1994 and 1998 while clearing the grounds for the reconstruction.

Due to the fundamental lack of a comprehensive recovery of cultural heritage, SOLIDERE's 1994 masterplan experimented with multiple and disconnected strategies that presumably aimed to preserve heritage involving different stakeholders. Two were dominant. The first strategy integrated selected archaeological findings under a complex framework that involved the official intervention of the DGA to undertake archaeological excavations as national property. The second strategy regulated the identification, assessment, and restoration of built heritage, although selective, through planning guidelines and technical briefs on a district scale.

Integrating archeology/counter narratives

Archeological excavations that uncovered ancient layers of the city were key to the shift in the masterplan. SOLIDERE mobilized to integrate archeology into their planning, thus, framing heritage in relation to an ancient past far removed from the more recent pre-war years and living memories of the people. The archeological sites were referred to as one of "the largest urban sites being excavated in the world" (Sader, 1998, p. 26). When the ancient city surfaced, the more recent history and living memory of the place was challenged. Archeological excavations were used to justify the reconstruction of the BCD, which ultimately left the heart of Beirut visibly more ancient but sterilized of its recent past. People needed to reconcile their memories in a no-longer familiar physical space. Heavy damage from the Civil War, coupled with demolitions during the early post-war reconstruction period, uncovered significant archaeological layers within the BCD. Although some sites had been studied in the pre-war period, including Nejmeh Square in the 1940s and the Roman Baths in the 1950s, the sheer scale and complexity of what emerged in the 1990s from archaeological discoveries was unprecedented. The 1994 masterplan was unable to balance the integration of archeological findings with sustaining the memory of the place.

The partial preservation of archeological findings in SOLIDERE's 1994 masterplan was marketed as a solid step towards creating an image of Beirut as an "ancient city of the future" (Gavin and Maluf, 1996). Multiple strategies and methods of integrating archeology in the reconstruction were brought forward. Responding to the outcomes of the 1994 masterplan, archeologist Perring (2009, pp. 301–302) noted:

> ... there was no attempt to assess and understand the value, potential and needs of either the material resource or the communities of interest in that resource. Instead *a priori* assumptions about the role of the historic past in framing the modern urban landscape were incorporated into the masterplan.

The management of the archeology within the BCD was inherently fraught due to the involvement of multiple local and international stakeholders in the absence of a comprehensive framework. Capitalizing upon the sheer magnitude of archeological excavations in the BCD, the Lebanese government, notably the DGA and Ministry of Culture, collaborated with UNDP and UNESCO to create project UNESCO-LIBAN/92/008: *Réhabilitation de la Direction Générale des Antiquités et Soutien à la Reconstruction du Centre-Ville de Beyrouth.*[31] The project proposed a regulatory framework with fixed objectives, budgetary modalities, and necessary resources and defined the expected results regarding the archeological findings. The CDR was responsible for the coordination of projects in the BCD, including archeological excavations, while the DGA was responsible for supervising and controlling all archaeological operations. The private sector, notably the Hariri Foundation and later SOLIDERE, was included as a financial partner and stakeholder in the process. As a result, the increased dependence on the private sector to finance the archeological excavations was problematic. It shaped how power was distributed over land, despite the legal rights of the Ministry of Culture to govern cultural property and the DGA to manage cultural property.

As a result of the systematic excavations starting September 1993, the BCD was divided into many archaeological sites, involving local and international teams from academic and professional institutions. Between 1993 and 2000, around 138 archaeological sites were excavated, uncovering archaeological layers that date back to the Iron and the Bronze ages (Al-Harithy, 2008; Saliba, 2004). Multiple approaches were set for interventions in archaeological sites. They can be categorized into three approaches: embed in a garden setting, preserve in-situ and integrate within buildings, and document and move or bury in the ground. Selecting which approach to use mostly relied on the value of the land where the archeological findings were located and the financial impact on the overall masterplan. The embed in garden settings applied in the sites of the Roman baths,[32] the Garden of Forgiveness,[33] the Ancient Tell area,[34] and the Roman Hippodrome in Wadi Abou Jamil[35] (Figure 11.6). The approach to preserve in-situ and integrate within buildings attempted to preserve and work within the original foundations and structures of the sites. It applied to the sites of the souks and the St. George Greek-Orthodox Cathedral (Figure 11.7). The document and

Figure 11.6 The Garden of Forgiveness

Figure 11.7 Archaeological remains in the new Souks

move or bury approach forced excavators to rush to collect and transfer archeological findings to museums, or face demolitions where all rubble was moved to the Normandie landfill. Examples of such archeological findings concentrated in the area to the south of Martyrs' Square,[36] the site of Banca Di Roma,[37] Mina el-Hosn,[38] and the area beneath An-Nahar Building.[39] The three approaches were ultimately hindered by delays and insufficient access to resources, despite SOLIDERE's claims to prioritize heritage.

Moreover, international cultural heritage policies ratified by the Lebanese government were not legally binding and never materialized into national regulations for the protection and management of post-war built heritage. The DGA was still unable to fund archaeological projects or enforce a regulatory framework for the compensation of property owners. Issued during the French Mandate period, Law No. 166/1933 provided a limited definition of cultural heritage and limited protection to buildings constructed before 1700.[40] Lebanon signed the Charter for the Protection and Management of the Archaeological Heritage in 1990 and the Malta Convention in 1992, and ratified the Hague Convention[41] in 1960, the World Heritage Convention[42] in 1983, and the UN 1970 Convention[43] in 1992. Despite these international conventions, conflicts of interest were fostered by a dependency on the private sector.[44] Reconstruction was an ambiguous and costly path towards an uncertain future and missed opportunities for public participation and capacity building (Perring, 2009). What SOLIDERE deemed as a BCD "charged with symbolism" was fractured, commodified, and sold as an image of a past "destroyed and rebuilt many times in its 5,000 year history."[45] The BCD was left fragmented and disassociated from living memory. Its heritage was "erased" or "museumified" without integrating ancient ruins and surviving built heritage into a multi-layered living city with a shared memory, what Naccache describes as a "memorycide" (Naccache, 1998).

Restoration of buildings/commodifying heritage

The 1994 SOLIDERE masterplan and its strategy to integrate archeology into a museum-like, ancient city of the future was not aligned with the second strategy for the restoration of the built heritage. The restorations of buildings or building ensembles revealed sub-strategies that are wide ranging in intensity from conservation to total reconstruction. To illustrate these strategies, the following sections of this chapter closely examine multiple interventions and interrogate how these unfolded and framed heritage. These are the conservation area, urban ensembles in Saifi and Wadi Abou Jmil, and the reconstructed souks.

The conservation area

The conservation area was conceived as a global destination with four areas of intervention: the Serail Hill, the Riad al-Solh Street, Nejmeh Square and Maarad Street, and the Foch-Allenby area (Figure 11.8). It encompasses

Figure 11.8 The Foch-Allenby conservation area

multiple architectural styles from the Medieval, Ottoman, French Mandate, and Modern periods. The conservation area is rich with historic buildings and is bordered by two of the largest in-situ archaeological sites. Each of the four areas of intervention represents part of Beirut's architectural history. The Serail Hill preserves two late Ottoman landmarks – the Grand Serail and the CDR building. Nejmeh Square, Maarad Street, and the Foch-Allenby area include a stock of historic buildings that belong to the French Mandate period, with older religious landmarks such as the St. George Orthodox Church and Al-Omari Mosque. Riad al-Solh Street broadens the definition of built heritage to include modern buildings that previously formed the financial district in the BCD (Saliba, 2004).

Nejmeh Square and the Foch-Allenby area were considered high-priority sites for conservation in phase one of the SOLIDERE masterplan and were maintained as urban ensembles of traditional built fabric. According to technical surveys, the conservation of Nejmeh Square and the Foch-Allenby area was possible since they were among the least damaged sites during the war.[46] Notably, Nejmeh Square and the Foch-Allenby area remain examples of SOLIDERE's response to the challenges of a market-driven conservation. These combined commercial profit with high conservation standards and mostly resulted in a hybrid building identity that integrated modern interiors within conserved historical shells. Yet, in these intervention areas, SOLIDERE successfully carried out a conservation scheme that operates at both architectural and urban scales. Clusters of buildings, rather than individual structures, were integrated into their conserved urban fabric. Therefore, the SOLIDERE masterplan positioned the conservation area as the "new" historic core of the BCD with distinct architectural landmarks infused with a "communal and national identity" (Saliba, 2004). However, SOLIDERE's conceptualization of the conservation area as the "new" historic core of the BCD was largely based on reading and conserving the tangible assets of a seemingly cohesive urban fabric. Infrastructural requirements altered the original urban fabric while the transformed land uses introduced new types of users and socio-spatial practices to the BCD. As a result, the post-war reconstruction project drastically transformed the pre-war built environment characterized by social, cultural, and economic networks and spatial practices integral to a cohesive, vibrant, and inclusive BCD.

Elsewhere within the conservation area, SOLIDERE transformed many of the surviving structures into single monuments detached from their original fabric and historic sense of place. A total of 27 buildings out of the 265 designated for restoration are civic and religious landmarks that were conserved as single monuments. Additional buildings that were not on the list by SOLIDERE also followed suit. These included government, Waqf, and church properties that were not handed over to SOLIDERE and escaped the frequent demolitions at an astounding rate before the 1994 masterplan. The Waqf, the church, as well as individual property owners and corporations, occasionally contested the transfers of ownership.[47]

At the heart of the conservation area, two major civic monuments currently retain their original purpose as symbols of national and local power. The first monument is the Parliament Building and the clock tower in Nejmeh Square built in 1934 and designed by local architect Mardiros H. Altounian (Saliba, 2004). The second monument is the Municipality of Beirut building on Weygand Street, built in 1925 and designed by the local architect Youssef Effendi Aftimos. Conceived in line with the French Mandate planning strategy for the BCD, both the Parliament and Municipality buildings symbolize a search for national identity through the built environment. The Ottoman Serail and Ottoman military hospital located near the municipality once formed the nuclei of historic political power. The Ottoman Serial was restored and repurposed into the seat of the Lebanese government and the Ottoman military hospital into offices for the CDR. Several religious buildings from different historical periods were also restored within the conservation area. They are claimed to constitute "great heritage value" that reinforces the historic identity of the BCD against what was lost (Gavin and Maluf, 1996). The restored religious structures include al-Omari Mosque, Assaf Mosque, Amir Munzer Mosque, St. Elie Greek Catholic Church, St. George Greek-Orthodox Cathedral built over a Byzantine church, St. George Maronite Cathedral, and St. Louis Capuchin Church. The buildings survive as detached from their surroundings, despite their restoration to high technical standards (Figure 11.9). They were regarded as isolated monuments that are reminiscent of a "lost society" as opposed to places of

Figure 11.9 St. George Greek-Orthodox Cathedral

"lived remembrance" (Fricke, 2005). They previously operated as part of an integrated urban ecosystem that the 1994 masterplan failed to consider. Therefore, they stand today to represent solely the power of religious institutions. Additional historic and religious buildings outside the conservation area also operate as testaments of the disintegration of urban surroundings. These buildings include al-Majidiya Mosque, Abou Bakr (Al-Dabbagha) Mosque, and the Synagogue in Wadi Abou Jmil.

SOLIDERE's approach stripped the built heritage from its essence and role of meaningfully connecting people to place and the past, which is referenced by both tangible objects and intangible narratives. It works against the conception of heritage as a process of inscribing "contemporary" meanings that connect people to places (Ascherson, 2007; Ashworth and Tunbridge, 1996; Hewison, 1987). Instead, it reduced community spaces into "iconic" monuments of a distant past with a primarily political or religious value. This framing of heritage produces frozen icons of the past and does not invest in living memory or foster reconciliation. It sharpens political and sectarian divides.

Urban ensembles of Saifi and Wadi Abou Jmil

SOLIDERE adopted a combined reconstruction and conservation strategy in Saifi and Wadi Abou Jmil. Wadi Abou Jmil constituted "clustered developments involving infill construction together with restoration,"[48] while Saifi was "conceived along Lebanese vernacular tradition; four new clusters blend in style and color with a number of existing buildings."[49] Religious and civic buildings were restored as iconic landmarks owned by the Awqaf or the government, while residential buildings in Saifi and Wadi Abou Jmil were owned by SOLIDERE and conserved, remodeled, or reconstructed as elegant clusters of houses that target the upper-class market. New buildings replicated the architectural styles and details of the restored ones to manufacture a "traditional" character for the neighborhoods. In both cases, the reconstruction project produced culturally synthetic ensembles whose original characters were lost to an easily replicable and essentialized aesthetic derived from the buildings that formerly stood in Saifi and Wadi Abou Jmil (Figure 11.10).

SOLIDERE commodified a visual representation of built heritage in sites designated for exclusive, high-end residents. Critics of the post-war reconstruction view these strategies as representations of a new political and economic era driven by neoliberal financialization and the commodification of urban space. Instead of restoring neighborhoods as social entities through public participation, SOLIDERE attracted foreign investors through tax incentives. The residential neighborhoods of Saifi and Wadi Abou Jmil rank among the most expensive in Beirut. Profit was prioritized over social inclusion and diversity in the BCD. Intended to position Beirut as a global hub for international investors, the Hariri mandate in Lebanon

Figure 11.10 Saifi village neighborhood

(1992–1998, 2000–2005) fostered radical destruction and change with policies that nearly doubled the total built-up area of the BCD and rendered it inaccessible to most city dwellers (Ghandour and Fawaz, 2010; Verdeil, Faour, and Velut, 2007).

The reconstruction of the Souks

The traditional souks of Beirut were the target of several waves of demolition due to the war and the subsequent reconstruction. In 1992, the implementation of the first masterplan by SOLIDERE required the demolition of the site of the current Beirut Souks. Prior to the war, this historic section of the BCD included souk al-Tawileh, souk Ayyass, souk al-Franj, souk Arwam, souk Sayyour, and souk al-Joukh (Makdissi, 1997). For a long period, the traditional souks were symbols of the history of Beirut and served as inclusive and active commercial nodes. Strategically positioned, the traditional souks connected other commercial areas in the BCD such as the port and the financial district to residential areas in Beirut. They fostered economic and social exchange across Beirut and its entangled cultural and historical layers.

SOLIDERE prioritized the reconstruction of the souks in phase one of the 1994 masterplan and envisioned "new" and modernized souks (Figure 11.11). SOLIDERE launched an international design competition in 1994 calling for proposals that provide a contemporary interpretation of the vibrant commercial activities in the souks. Having demolished the souks,

Figure 11.11 The new Beirut souks

SOLIDERE claimed to frame heritage in relation to memory and socio-spatial practices. The competition brief introduced that "notion of the souk, used to mean not only a commercial space, but an image, a memory, a place that honors tradition and enhances kinds of interaction between merchants and customers." However, it stressed that reconstruction does "not imply an 'honest architectural reconstruction' of the old souks, [that] the spirit and the architectural character of the souk should not be confused" (SOLIDERE, 1994, pp. 10–11). This framing failed to address the archeological findings under the souks, which were used by SOLIDERE to market the BCD as an ancient site with spatial patterns that date back to the early Phoenician period (Moneo, 1998).

The competition claimed to "open up the post-war reconstruction process to the public and to encourage professional engagement" (SOLIDERE, 1994, p. 12). The area of the site was approximately 60,000 sqm. According to SOLIDERE (1994), different functions were proposed, such as souks, offices, entertainment venues, residential units, and underground parking. The competition brief specified the structures that will be restored within and around the souks to link the souks to its surroundings, including buildings in the Foch-Allenby area. Around 300 design proposals from local and international architects were submitted, but the competition did not declare a winner. Instead, a different masterplan for the new souks was commissioned to Jad Tabet, a Lebanese architect, while the architectural design was awarded to Rafael Moneo, an international architect from Spain.

The design replaced "the traditional markets" with a modern-day mall. Although it preserved the street patterns, it primarily catered to car access and the provision of underground parking rather than pedestrian networks. Preserving old street patterns does not necessarily maintain "the spatial social practices that animate the souks, especially if elements on the site and beyond are not interdependent in their programmatic distribution, access, and circulation" (Al-Harithy, 2008, pp. 225–226). In the process, both tangible and intangible heritage was neglected and claims to maintain the traditional souks as form, memory, or social practice failed. As a result, the current Beirut Souks project is considered an internationally recognized, high-end shopping mall. It represents a constructed notion of global modernity in the BCD, and erases what was one of the most inclusive space for exchange in pre-war Beirut.

It becomes clear that SOLIDERE's strategies for the integration of the archeology and the restoration of the built heritage were fragmented. The 1994 masterplan for the BCD failed to frame restored buildings in relation to their previously active urban environments produced and reproduced over centuries by cultural practices and historical narratives. It also failed to weave the archeological findings into these urban environments. The extreme strategies of restoring single monuments as icons of a distant past and the total reconstruction of urban ensembles contributed to such ruptures. The lack of a comprehensive framework for cultural heritage prevented the activation of the BCD as a social, cultural, and economic hub. These same strategies suffered from a longstanding absence of public participation and prioritized private interests over public needs.[50] To this day, the public interest continues to be absent in the development of the BCD. The area which once formed a vital interface between the people, the land, and the sea became an urban island. Separated from its direct surroundings by major arteries, the BCD is now a mere product of the 1994 SOLIDERE masterplan. The typology, shape, and essence of large blocks that cater to the metropolitan elite stand in contrast to the low-rise, dense built fabric that once characterized the BCD and is still present in some of its surrounding neighborhoods (Saliba, 2004).

Conclusion

This chapter illustrated how the multiple masterplans and the shifting strategies leading to the one implemented in 1994 failed to integrate heritage into a comprehensive vision and, thus, was unsuccessful at cultural recovery. The 1991 masterplan and its revised version resulted in the erasure of memory instead of celebrating local diversity and cultural richness. It adopted the tabula rasa approach in reconstruction. It created urban and social rupture through the enclaving of an exclusive BCD with major arteries and commodifying heritage that fosters tourism and manufactures a false "image" of a unified post-war Lebanon. These negative consequences of the post-war reconstruction were largely due to the absence of a participatory strategy that engages directly affected stakeholders such as property owners and local experts in devising alternative visions for urban planning and heritage integration. Particularly evident is the prioritization of a profit-based neoliberal ideology that minimized the role of the government and attracted private investment in luxury housing and commerce. Ultimately, the implemented 1994 masterplan fell short of accounting for the tangible and intangible heritage in the BCD and reconnecting people to the place through living memory. By linking to the discourse on memory, both individual and collective, this chapter argued that reconstruction could have been viewed as a heritage process of "construction and negotiation of meaning through remembering" (Smith, 2006, p. 10). The act of remembering was the first casualty of the post-war reconstruction of the BCD that compromised the whole process.

It is important in the reconstruction of a post-war city to consider the pre-war socio-spatial networks that link people to a place through their practices and memories. In the BCD, the erasure of the built environment, combined with the "unearthing" of the city during the reconstruction process, contributed to the erasure of the memories of city dwellers and engineered collective amnesia that mirrored the political attitude at the time. Both the social and the urban fabrics, with all the memories, identities, practices, and historic significance that they embody, disappeared. The complete destruction of historic neighborhoods such as Wadi Abou Jmil and Saifi, as well as the old souks and the main public space of Martyr's Square, resulted in the erasure of the social memory. It uncovered an uninhabitable museum-city of "quasi-archeological presentations and staging, as well as restored buildings" (Al-Harithy, 2008, p. 224).

Heritage-management policies valued economic stimulation in relation to touristic goals but not other economies such as traditional trades and industries. Social practices that were disrupted during the war were not re-integrated in a comprehensive vision of cultural heritage management. The process of "cultural recovery" was not prioritized as an instrument for the creation of an "inclusive space of reconciliation" across the scarred cityscape that witnessed spatial and physical expressions of 15 years of Civil War.

The outcome that we witness today in the BCD, a luxurious and largely "abandoned" downtown 30 years after the end of the Civil War, is evidence of the failure of the reconstruction project to address the social and cultural dimensions of the BCD. Firstly, the adopted strategy was primarily market-driven and market-responsive, reducing the reconstruction of the whole country to the BCD entrusted to the private real estate company, SOLIDERE. The absence of the role of the government in decision-making and monitoring did not empower agencies such as the CDR, the DGA, and the DGU to play an instrumental role in the urban planning and implementation of the project. The dissolution of hierarchical public authority consolidated the overarching neoliberal paradigm brought forward by the creation of SOLIDERE as the sole planner, developer, and manager of the

BCD post-war reconstruction. This strategy was not accompanied by a plan for social and cultural recovery nor a plan that extended beyond the BCD and, thus, failed to facilitate steps towards a true post-war reconciliation. The BCD today features exclusive spaces that are not accessible to all socio-economic groups, leaving behind empty apartments and struggling businesses.

As argued in the introduction of this book, a true process of post-war recovery is holistic and participatory. It is directed by a shared vision and not an imposed one. Urban recovery is conceptualized as an open-ended social process that does not begin or end with a masterplan. It is rather open to reclaim by the people. This is exactly what we are witnessing today in the BCD through the public protests that started in October 2019. People reclaimed spaces, inscribed practices, and re-appropriated cultural landmarks through occupation, performances, and other tactics. The BCD became an opportunity where the spatial, socio-cultural, and even the imaginative dimensions of urban existence are included and translated.

Notes

1. Infrastructural projects included the development of the Beirut Port and the construction of the Beirut International Airport in 1951. The laissez-faire approach to economic planning and an important influx of Arab investments were among the main drivers of economic growth in Lebanon, both regionally and internationally (Tabet, 1998).
2. Some of the souks in the BCD included al-Franj, al-Jamil, al-Tawileh, Ayass, and al-Khodar to the north of Weygand Street, and al-Bazerkan, Sursock, al-Nourieh, as-Sagha, and Abou al-Nasr to the south of Weygand Street.
3. Office buildings such as Fattal, Philipps, Makassed, and the City Center were known well in the BCD. These buildings served as landmarks in the mental maps of people from different age groups (Saliba, 1990).
4. Newspaper publishing houses such as An-Nahar and L'Orient and Le Jour (which merged to become L'Orient-le-Jour in 1970); theatres such as the Farouk Theater and the Grand Theatre; and cinemas such as Rivoli, Byblos, Opera, Empire, Dunia, and Roxy.
5. The Holiday Inn, Hilton, Phoenicia Intercontinental, or local hotels such as the St. George, Normandie, Excelsior, and Alcazar.
6. The Normandie landfill is the product of waste and rubble accumulated during the war. It reached an area of 250,000 sqm and a height of 20 m above sea level at the end of the war.
7. The creation of the CDR in Decree No. 5, dated 31 January 1977, was among the early seeds of reconstruction efforts. The CDR was conceived as an autonomous public institution, freed from the typical bureaucratic constraints of governmental bodies (http://www.cdr.gov.lb/eng/profile.asp). It remains accountable to the Council of Ministers. However, like other government agencies, the CDR was deeply affected by political fragmentation, which rendered it largely ineffective (Salam, 1998).
8. APUR submitted the masterplan for the reconstruction of the BCD in 13 June 1977. It was officially approved in Decree No. 1163, dated 15 April 1978. It was developed by the Municipality of Beirut under the direction of Mayor Mitri Nammar in collaboration with local consultants.

9. *As-Safir*, 19 March 1983, إعمار الوسط التجاري: الفرنسيون وضعوا الخطط...واللبنانيون التفاصيل
10. *An-Nahar*, 16 January 1983, الجيش يفجر ألغاماً وأبنية وأوجيه لبنان تنظف أسواقاً
11. *An-Nahar*, 13 January 1983, تمهيداً لاعادة بناء الوسط التجاري تنظف أسواق سرسق والطويلة والجميل ودعوة أصحاب العقارات إلى المراجعة
12. *As-Safir*, 22 August 1983, دعوة أصحاب المؤسسات إلى التريث انتظارا إلى إشارة من مجلس الإعمار
13. *As-Safir*, 23 January 1983, رفع الأنقاض في الأسواق التجارية
14. *An-Nahar*, 16 January 1983, البلدية دعت الى اجتماع موسع لتنسيق أعمال الوسط التجاري
15. The symposium "Beirut of Tomorrow: Planning for Reconstruction" was organized in January 1983 by the American University of Beirut and the Goethe Institute under the auspices of H.E. Pierre el-Khoury, Minister of Public Works, Transport, and Agriculture. It was an open call to citizens and public agencies to halt the demolitions within the BCD. It included roundtables and discussions on the reconstruction process.
16. *As-Safir*, 21 July 1991, "Chalak: The city center project is the launch of the country's reconstruction"
17. Law No. 117 gave "the municipal administration the authority to create real estate companies in war-damaged areas, and to entrust them with implementation of the urban plan and promotion, marketing, and sale of properties to individuals or corporate developers" (Salam, 1994, p.11).
18. SOLIDERE stands for the Lebanese Company for the Development and Reconstruction of the Beirut Central District. A series of regulations delineated the boundaries of the BCD and the urban planning frameworks under which SOLIDERE operated: Decree No. 2236, dated 19 February 1992, delineated the boundaries of SOLIDERE's intervention. On 14 July 1992, the Council of Ministers approved SOLIDERE's Articles of Incorporation. As such, investors were invited to take part in the foundation of the company. For more details, see Beyhum et al. (1992).
19. *Financial Times*, 21 May 1992, "Peace in Beirut hit by Storm over Reconstruction".
20. The detailed technical review is published in *As-Safir*, 25 July 1992: "للوسط التجاري: "سيؤدي إلى تغيير أساسي في البنية الاجتماعية تقرير المجلس الأعلى للتنظيم المدني حول المخطط التوجيهي
21. *As-Safir*, 26 September 1991, دعوة إلى طاولة مستديرة
22. In 1992, Nabil Beyhum, Kamal Hamdan, Toufic Shanbour, George Corm, Toni Choueiri, Baha' al-Din Bsatt, Pierre el-Khoury, Assem Salam, Jad Tabet, and Ziad Akl published a self-funded report entitled اعمار بيروت والفرصة الضائعة وسط بيروت التجاري و الشركة العقارية: الوقائع - المرتكزات - البدائل. They criticized the series of events leading up to the creation of the reconstruction framework and put forward a series of alternative guidelines, stressing on the necessity to empower the state to intervene, minimizing the role of real estate companies, and establishing other public-private partnerships.
23. *An-Nahar*, 17 March 1992, بحجة أنها تشكل خطر على السلامة العامة: مجلس النماء والإعمار يطلب إزالة الأبنية المتصدعة في الوسط التجاري
24. *An-Nahar*, 28 April 1992, الوسط التجاري: اليوم المنطقة الثالثة-الانماء والإعمار ينفي التعرض للأبنية الأثرية
25. *An-Nahar*, 24 April 1992, هدم 40 بناء في المنطقة الأولى واليوم الثانية ثم خطوط التماس
26. *An-Nahar*, 28 April 1992, الوسط التجاري: اليوم المنطقة الثالثة-الانماء والإعمار ينفي التعرض للأبنية الأثرية
27. SOLIDERE Masterplan, http://www.solidere.com/, 17 March 2019.
28. According to Gavin (1998, pp. 222–223), "Historical or 'city' memory became a fundamental concept of the master plan. Beirut is an ancient, 'layered' city [...].We need to respect and preserve the continuum and allow such a pattern of layering to survive and evolve into the future."
29. *An-Nahar*, 05 July 1994, اخلاء الريفولي و5 مبان في الأسواق وسوليدير تسلمتها وباشرت تفجيرها
30. An-Nahar, 21 May 1996, نداء لحماية تراث بيروت and *An-Nahar*, 25 April 1996, الحائط المراسيم والقوانين سلام: سوليدير تهدم مباني تراثية ضاربة عرض
31. The project UNESCO-LIBAN/92/008 was primarily concerned with archeological excavations in the BCD, and was then extended to provide essential support to the protection and rehabilitation of other historic centers across the country, including a contribution to the reconstruction of the National

Museum of Beirut. An International Scientific Committee consisting of regional and international archaeology consultants was appointed to monitor and advise on the works.

32. The plan of Roman Beirut was constructed by the French architect and archaeologist Jean Lauffray for the first large-scale excavations in the BCD between 1941 and 1947. Located in and around Parliament Square, these excavations uncovered the structure of the Roman city (Sader, 1998).
33. This 20,196 sqm open archeological site is located between Martyr's Square and Maarad Street. The Garden of Forgiveness was initially partially excavated between 1950 and 1952. It was one of the main archeological projects at the time, stretching for seven years between 1993 and 2000. Excavations on this site took place across four divisions (Badre, 1992).
34. Given the lack of protection by successive governments over the years, many of the archeological layers of the site were destroyed due to the construction of modern buildings such as the Cinema Rivoli and the Byblos Building, as well as their later demolition.
35. All squares and gardens are listed in: SOLIDERE, Open Spaces, http://www.solidere.com/, 07 April 2019,
36. Excavations of this site spread over an area of 1,800 sqm covered for the creation of an open public space, later repurposed as a parking lot. Findings were reported by Mountaha Saghiye and Hans Curver (Abi Akl, 1999).
37. Located behind the Parliament Building on Nejmeh Square, this site has been the subject of heated controversy. Upon the demolition of the Banca Di Roma Building by SOLIDERE in October 1994 to clear the site for a Parliament annex, important archeological findings were uncovered, mainly a Roman arch. No proper plan for in-situ preservation was ever determined (Abi Akl, 1999).
38. Roman mosaics found on this site were unearthed, documented and transported to the National Museum of Beirut. See SOLIDERE, Archaeology, http://www.solidere.com/. 12 April 2019.
39. Two of the twelve Byzantine mosaics uncovered on site were relocated and housed within the new An-Nahar Building. The Museum of Archeological at the American University of Beirut showcases fifty years of excavations in the BCD. https://www.aub.edu.lb/museum_archeo/Pages/fieldwork.aspx, 12 April 2019.
40. "Article 1: Antiquities are products of human activity, no matter the civilization they belong to, prior to the year 1700 (1107 Hijri). Their conservation is important to the historic or artistic public interest [...]." (Author's translation) For reference see Arrêté N.166 portant règlement sur les antiquités. http://www.unesco.org/culture/natlaws/index.php and Sader (1998).
41. The 1954 Hague Convention for the Protection of Cultural Property in the Event of Armed Conflict.
42. The 1972 World Heritage Convention Concerning the Protection of the World Cultural and Natural Heritage.
43. The 1970 Convention on the Means of Prohibiting and Preventing the Illicit Import, Export and Transfer of Ownership of Cultural Property.
44. According to Massena (2004), "A special solution in decrees No. 2786/92 and No. 4830/94: For very important monuments or ruins by SOLIDERE and kept in situ, the company must compensate the developer by giving him a substitute land instead at no extra charge to him nor the Lebanese state." This gave rise to many disputes. The Landmark Beirut project is an example. The 149 million USD development was designed by Jean Nouvel and stopped in 2013, after the pictures of archeological findings dating back to the Byzantine period were leaked by activists to the media. For further details see:

Florence, M., (2014), "Beirut's Archaeological Heritage Threatened by Construction", Al-Monitor, 17 October 2014 http://www.al-monitor.com/pulse/originals/2014/10/beirut-lebanon-city-construction-historical-heritage.html#ixzz55HMQXGfw and *The Daily Star*, 21 May 2013, "Archeological ruins halt $149 million Landmark Project" http://www.dailystar.com.lb/News/Local-News/2013/May-21/217804-archaeological-ruins-halt-149m-landmark-project.ashx#axzz2TG9673Dn.

45. SOLIDERE Garden of Forgiveness, http://www.solidere.com/, 17 March 2019.
46. Technical surveys were undertaken in 1993 and 1994, not only to assess damages of the still-standing buildings, but also to locate alterations and additions to original structures. For a detailed description of these surveys see Saliba (2004).
47. "Much like it dealt with the opposition of the financiers on 'Bank Street,' SOLIDERE generally gained the favor of the public religious institutions such as Mosques by preserving their right to waqf or, with their consent, relocating them to an acceptable position." "The private waqfs however, suffered as they were neither acknowledged nor upheld by the firm." (Mango, 2004, pp. 25–26).
48. http://www.solidere.com/city-center/urban-overview/districts-main-axes/wadi-abou-jamil-zokak-el-blatt, 13 July 2020.
49. http://www.solidere.com/city-center/urban-overview/districts-main-axes/saifi, 13 July 2020.
50. Established property owners were also not given flexible development options based on comprehensive guidelines. For example, Fadi Khoury, the owner of the landmark 1929 St. George Hotel next to the sea, was denied the right to restore the property. The hotel's access to the sea was limited to make way for a boardwalk along the new Zaitunay Bay.

References

Abi Akl, M. (1999). Beirut City Center between Discoveries and Bulldozers. *An-Nahar Files*, 13–15.

Al-Harithy, H. (2005). [Reframing] "World Heritage." *Traditional Dwellings and Settlements Review*, 7–17.

Al-Harithy, H. (2008). Weaving Historical Narratives: Beirut's Last Mamluk Monument. In *Muqarnas Volume XXV: An Annual on the Visual Culture of the Islamic World. Frontiers of Islamic Art and Architecture* (Vol. 25, pp. 187–214), edited by G. Necipoglu and J. Bailey. Essays in Celebration of Oleg Grabar's Eightieth Birthday. Leiden: E.J. Brill.

Al-Harithy, H. (ed.). (2010). *Lessons in Post-war Reconstruction: Case Studies from Lebanon in the Aftermath of the 2006 War.* Abingdon: Routledge.

Ascherson, N. (2007). Cultural Destruction by War, and its Impact on Group Identities. In *Cultural Heritage in Postwar Recovery* (pp. 17–25), edited by N. Stanley-Price. Rome: ICCROM.

Ashworth, G. J. and Tunbridge. J. E. (1999). Old Cities, New Pasts: Heritage Planning in Selected Cities of Central Europe. *Geo Journal*, *49*, 105–116.

Bădescu, G. (2011). Beyond the Green Line: Sustainability and Beirut's Post-war Reconstruction. *Society for International Development*, *53*(4), 358–367. http://www.palgrave-journals.com/development/journal/v54/n3/pdf/dev201153a.pdf. 10 July 2020.

Badre, L. (1992). *Bey 003 Preliminary Report, Excavations of the American University of Beirut Museum 1993–1996, BAAL- Volume 2*. Beirut: American University of Beirut. Retrieved from http://ddc.aub.edu.lb/projects/museum/baal/1997/page6.html. 12 April 2019.

Barakat, S. (2007). Postwar Reconstruction and the Recovery of Cultural Heritage: Critical Lessons from the Past Fifteen Years. In *Papers from the ICCROM Forum held on October 4–6 2005. 6* (pp. 26–28), edited by N. Stanley-Price. Rome: ICCROM Conservation Studies.

Benton, T. and Cecil, C. (2010). Heritage and Public Memory. In *Understanding Heritage and Memory* (pp. 7–43), edited by T. Benton. Manchester: Manchester University Press.

Beyhum, N. et al. (1992). *I'amar Beirut Wa'l Fursa Al-da'i'a:* إعمار بيروت والفرصة الضائعة. Beirut.

El-Achkar, É. (1998). Chapter 4. La Reconstruction du Centre-ville de Beyrouth. In *Réglementation et Formes Urbaines: Le Cas de Beyrouth (pp.* 101–134). Beirut: Presses de l'Ifpo. Retrieved from http://books.openedition.org/ifpo/4315. 12 April 2019.

El-Khoury, P., Khoury, K., and Raad, A. S. (1997). *Plan Directeur d'Aménagement du Centre de Beyrouth 1977.* République Libanaise: Municipalité de Beyrouth.

Fricke, A. (2005). Forever Nearing the Finish Line: Heritage Policy and the Problem of Memory in Postwar Beirut. *International Journal of Cultural Property, 12*(2), 163–181. doi: https://doi.org/10.1017/S0940739105050150.

Gavin, A. (1998). Heart of Beirut: Making the Masterplan for the Renewal of the Central District. In *Projecting Beirut: Episodes in the Construction and Reconstruction of a Modern City* (pp. 217–234), edited by P. G. Rowe and H. Sarkis. Munich, New York: Prestel.

Gavin, A. and Maluf, R. (1996). *Beirut Reborn: The Restoration and Development of the Central District.* London: Academy Press.

Ghandour, M. and Fawaz, M. (2010). Spatial Erasure: Reconstruction Projects in Beirut. *Architecture Publications.* https://lib.dr.iastate.edu/cgi/viewcontent.cgi?article=1047&context=arch_pubs. 12 April 2019.

Hewison, R. (1987). *The Heritage Industry.* London: Methuen.

Kabbani, O. (1991). The Reconstruction of Beirut. In *Morphological Investigation of Downtown Beirut: Towards an Urban Design Framework* (pp. 39–41), edited by R. Saliba. Beirut: American University of Beirut.

Makdissi, S. (1997). *Reconstructing History in Central Beirut.* Middle East Research and Information Project.

Mango, T. (2004). *Solidere: The Battle for Beirut's Central District.* Thesis for the degree of Master of City Planning. Massachusetts Institute of Technology.

Moneo, R. (1998). The Souks of Beirut. In *Projecting Beirut: Episodes in the Construction and Reconstruction of a Modern City* (pp. 263–274), edited by P. G. Rowe and H. Sarkis. Munich; New York: Prestel.

Naccache, A.F.H. (1998). Beirut's Memorycide: Hear no Evil, See no Evil. In *Archaeology Under Fire: Nationalism, Politics and Heritage in the Eastern Mediterranean and Middle East* (pp. 140–158), edited by L. Meskell. Abingdon: Routledge.

Nasr, J. and Verdeil, E. (2008). The Reconstructions of Beirut. In *The City in the Islamic World* (Vol. 1, pp. 1115–1141), edited by S. K. Jayyusi, R. Holod, A. Petruccioli, and A. Raymond. Boston: Leiden, Brill.

O'Keefe, R. (2006). *The Protection of Cultural Property in Armed Conflict.* Cambridge: Cambridge University Press.

Perring, D. (2009). Archaeology and the Post-war Reconstruction of Beirut. *Conservation and Management of Archaeological Sites, 11*(3–4), 296–314. https://doi.org/10.1179/175355210X12747818485529.

Sader, H. (1998). Ancient Beirut: Urban Growth in the Light of Recent Excavations. In *Projecting Beirut: Episodes in the Construction and Reconstruction of a Modern City* (pp. 23–40), edited by P. G. Rowe and H. Sarkis. Munich and New York: Prestel.

Salam, A. (1994). The Reconstruction of Beirut: A Lost Opportunity. *Architectural Association School of Architecture, AA Files, 27*, 11–13.

Salam, A. (1998). The Role of Government in Shaping the Built-Environment. In *Projecting Beirut: Episodes in the Construction and Reconstruction of a Modern City* (pp. 122–134), edited by P. G. Rowe and H. Sarkis. Munich and New York: Prestel.

Saliba, R. (1990). *Morphological Investigation of Downtown Beirut: Towards an Urban Design Framework*. Beirut: American University of Beirut.

Saliba, R. (2004). *Beirut City Center Recovery: The Foch-Allenby and Etoile Conservation Area*. Beirut: Steidl and Solidere.

Samara, R. (1996). *Urban Reconstruction in the Twentieth-century: The Postwar Deconstruction of Beirut, Lebanon*. Thesis submitted to the Faculty of Graduate Studies and Research in partial fulfillment of the degree of Master of Architecture, School of Architecture. Montreal: McGill University.

Schmid, H. (2006). Privatized Urbanity or a Politicized Society? Reconstruction in Beirut after the Civil War. *European Planning Studies, 14*(3), 365–381, DOI: 10.1080/09654310500420859

Smith, L. (2006). *Uses of Heritage*. Abingdon: Routledge.

SOLIDERE. (1994). *The Reconstruction of the Souks of Beirut: An International Ideas Competition, La Reconstruction des souks de Beyrouth: Un Concours D'idees International*. Beirut: Board of Founders, SOLIDERE.

Stanley-Price, N. (2007). The Thread of Continuity: Cultural Heritage in Postwar Recovery. In *ICCROM Forum held on October 4–6 2005. 6*, (pp. 1–16), edited by N. Stanley-Price. Rome: ICCROM Conservation Studies.

Tabet, J. (1993). Towards a Master Plan for Post-War Lebanon. In *Recovering Beirut: Urban Design and Post-War Reconstruction (Social, Economic, and Political Studies of the Middle East)* (Vol. 47), edited by S. Khalaf and P. S. Khoury. Boston: Brill Academic Publishers.

Tabet, J. (1998). From Colonial Style to Regional Revivalism: Modern Architecture in Lebanon and the Problem of Cultural Identity. In *Projecting Beirut: Episodes in the Construction and Reconstruction of a Modern City* (pp. 94–95), edited by P. G. Rowe and H. Sarkis. Munich and New York: Prestel.

Verdeil, É., Faour, G., and Velut, S. (eds.). (2007). *Atlas du Liban: Territoires et Société*. Beirut: Presses de l'Ifpo, CNRS Liban.

Wijesuriya, G. (2005). The Past is in the Present: Perspectives in Caring for Buddhist Heritage Sites in Sri Lanka. In *Conservation of Living Religious Heritage; Papers from the ICCROM 2003 Forum on Living Religious Heritage: Conserving the Sacred* (pp. 30–43), edited by H. Stovel, N. Stanley-Price, and R. Killick. Rome, ICCROM.

12 The politics of urban recovery in a Soviet-era spa resort town

Heritage tourism and displaced communities in Tskaltubo, Georgia

Suzanne Harris-Brandts and David Sichinava

Introduction

In February 2018, the Georgian Ministry of Culture proposed placing fourteen spa resort buildings in the small town of Tskaltubo on a national heritage protection list (Department of Cultural Heritage, 2018). Spanning the early decades of the twentieth century Stalinist era (1928–1953), the buildings reflect a unique moment in the history of the town and are indicative of the spread of vacation culture around restorative health in the Soviet Union more broadly. After decades of physical deterioration and partial closure, the town's historic bathhouses and hotels (called *sanatoria*,[1], as shown in Figures 12.1–12.3) were again being positioned for tourism, recalling the town's zenith as one of the most sought-after balneological[2] destinations in the Soviet Union. Throughout the Soviet era, the landscape of the Caucasus mountain around the eastern Black Sea had been a particularly lively destination for restorative health (in Russian: *otdykh*). With the collapse of the Soviet Union, the tourist industry faltered, leaving behind thousands of unused guest rooms and resort facilities.

In the Fall of 1992—less than a year and a half after the Soviet Union's collapse—the sanatoria of Tskaltubo were again being filled with new inhabitants. This time, it was thousands of internally displaced persons (IDPs) seeking refuge from conflict in the nearby secessionist region of Abkhazia. Almost three decades after their arrival, Tskaltubo's resort facilities remain the protracted de facto homes of many who were displaced, taking on adapted uses as residential buildings and places for community interaction (Figure 12.4). Built in the mid-twentieth century, these buildings have now equally spent their lives as resort complexes and spaces of humanitarian shelter. As sites of dual memory, they embody the divergent histories of tourists and IDPs, while further reflecting the politico-economic upheavals that caused this transition in history.

Considering current government efforts toward Tskaltubo's urban recovery, this chapter discusses the complex dynamics of promoting heritage tourism in areas occupied by vulnerable populations. As has been demonstrated in other geographic contexts, a market-based approach to heritage

Figure 12.1 Sanatorium of the Georgian Institute of Resort Studies, in Tskaltubo, Georgia

that operates by selectively framing the past for tourism risks doing so at the expense of present-day inhabitants (Timothy, 2007a, 2007b, 2007c). In the specific case of Tskaltubo, marketing for heritage tourism has omitted decades of IDP inhabitation, framing their presence instead as an aberration. Tskaltubo thus mirrors other global examples of development-induced evictions and displacement brought about by heritage tourism. Despite claims by the Georgian government to be effectively managing the resettlement of IDPs, there has yet to be proper community engagement or public consultations. This stands in contrast to extensive state tourism feasibility studies

Figure 12.2 Sanatorium Imereti in Tskaltubo, Georgia

Figure 12.3 Sanatorium Iveria in Tskaltubo, Georgia

Figure 12.4 Sanatorium Tbilisi transformed by IDPs

and heritage assessments. By recounting the specifics of the government initiatives to manage urban renewal and resettlement in Tskaltubo, we demonstrate how the welfare of the displaced has been compromised in favor of other state political and economic priorities.

Using data gathered through eight focus groups, in this work we further aim to give voice to the often-overlooked local resident and IDP populations. We show how government aspirations for a revival of the town's Soviet-era prosperity are generating complacency toward IDP eviction and resettlement. For IDPs who have continued to face social stigmatization and poor living conditions, the state's promises of new purpose-built housing and employment are additionally driving eviction complacency. As the testimonies of IDPs who have already been resettled demonstrate, however, these promises have been overstated: IDPs continue to suffer from physical isolation and a lack of social integration. Likewise, Tskaltubo continues to experience poor employment prospects. The three sanatoria vacated in 2012 as part of initial urban renewal efforts have also fallen short on their promises, remaining vacant and un-renovated. The likelihood that heritage tourism will improve IDP livelihoods in Tskaltubo thus remains questionable.

This chapter begins with a brief case description, providing an overview of Tskaltubo's changing history since its establishment as a Soviet resort town in the 1920s. We look at the growth of resort tourism during the Soviet era (1953–1990), a period when many bathhouses and sanatoria were constructed and when the town reached its zenith as a destination for restorative health activities. The demise of Soviet tourism and the arrival of IDPs in the early 1990s is then contextualized relative to the production of almost thirty years of alternative local history. We situate the concept of alternative local histories relative to literature examining heritage as the select framing of the past and discuss how heritage promotion has increasingly come together with displacement in recent decades. How Tskaltubo's history is now being selectively reframed through tourism promotion—and how IDPs are being faced with evictions in order to realize this vision—is then covered. Looking at community reactions to this process, we draw specific attention to how members of both Tskaltubo's local and IDP community hope to find prosperity through a return of the town's resort identity. Overall, we argue that the government's use of heritage tourism as a catalyst for urban recovery is erasing the town's unique past as a place of humanitarian shelter for Georgian IDPs. With the commodification of local heritage, Tskaltubo's complex layers of history are at risk of erasure.

The history of Georgia's largest balneological resort town

As the etymology of Tskaltubo suggests (*cqali*=water, *tubo*=warm),[3] the town has deep ties to its geographic location above natural radon-carbonated springs. While records dating back to the twelfth century describe the town's unique mineral water legacy, it was only after comprehensive scientific analyses by Russian chemist Genrykh Vasilyevich Struve in 1889 that the water's medical qualities were verified (Shavianidze, Shavianidze, & Shavianidze, 1990). As a result of Struve's findings, bathing in Tskltubo's spring water grew into an established form of medical self-treatment. Reports in the late-nineteenth century Georgian newspapers describe local travelers making their way to the area to relieve their arthritic pains, bathing in the natural hot springs (as mentioned in: Citlanadze, 1950).

In the first decade of the twentieth century, these practices expanded and Tskaltubo came to enjoy mild popularity among local holidaymakers. Purpose-built bath facilities, including the area's first four historic bathhouses were then developed in the 1910s (Shavianidze et al., 1990). In 1920, the government of the Georgian Democratic Republic (GDR)[4] declared Tskaltubo an officially designated resort location for hosting and treating wounded military servicemen (Lort'kip'aniże, 1920). Development of the resort sector was then accelerated by a 1931 decree of the Supreme Council of the Georgian Soviet Socialist Republic (GSSR) on establishing balneological resorts across Georgia (Babunashvili, 2018). Tskaltubo's 1936 development plan envisioned a centrally located, lush park space housing ten mineral water baths. Other medical facilities, including 22 sanatoria, were planned for the nearby hills (Figure 12.5).[5] In 1953, the status of Tskaltubo was elevated from "agricultural village" to "urban settlement," and it was awarded the title of an all-Union resort town[6] (National Agency for Cultural Heritage Preservation of Georgia, 2017).

The legacy of Tskaltubo as a spa resort town continued to expand throughout the second half of the twentieth century. Between 1950 and 1977, eleven new sanatoria were constructed, completing the town's master plan. In the 1980s, the town reached peak operations, welcoming some 125,000 guests annually (Shavianidze et al., 1990). There were 5,800 available guest beds and it was common to reach more than 80% occupancy (ibid.) The sanatoria's natural settings provided respite from work in factories, mines, or service in the military (Figure 12.6). Some of Tskaltubo's primary sanatoria were designed exclusively to serve Soviet ministries, trade unions, and members of the KGB (*Komitet Gosudarstvennoy Bezopasnosti*, or Committee for State Security). These sanatoria were administered through the all-Union institutions that coordinated employment in the USSR, with employees being assigned subsidized vacation passes (*putevki*) for access.

The local population also tremendously benefited from Tskaltubo's era of spa resort prosperity. Beyond employment opportunities and access to the town's lush parks, locals enjoyed the honor and prestige that came with their town being recognized across the Soviet Union.[7] According to the last Soviet census of 1989, Tskaltubo's population had reached 21,000—a fivefold increase in half a century—overwhelmingly linked to the booming spa resort industry. For residents, this period in the town's history is

Figure 12.5 Plan of Tskaltubo, showing the central park space with bathhouses, surrounded by peripheral sanatoria

recalled with great fondness as one of economic prosperity and international notoriety.[8]

The collapse of tourism and arrival of internally displaced persons

The collapse of the Soviet Union brought tremendous shocks to the political and economic functioning of Georgia, in turn greatly transforming Tskaltubo's resort industry. Seemingly overnight, guests stopped arriving and those in-residence went home. State employees of the resorts were no longer paid and ceased to come to work; the many bathhouses and sanatoria were closed and locked-up (Local focus group, Women, 45+; Babunashvili, 2018). The political, economic, and social instability of Georgia in the 1990s

Figure 12.6 Military servicemen outside of the Sanatoria of the USSR Ministry of Defense, (now Legends Tskaltubo Spa Resort), circa 1960s

ensured that tourism did not return in the same magnitude the town had seen previously. With all nine of the official bathhouses closed and all 22 of the sanatoria out of operation in the 1990s, it was no longer possible to receive professional balneological treatment. The minimal remaining tourist activity shifted toward small-scale guest houses, which were managed by residents who rented out spare rooms, generally to the few guests continuing to visit from within Georgia and neighboring Azerbaijan. By 2002, the population of the town had shrunk by 20% relative to 1989 (Bespiatov, 2018). Twelve years later, in 2014, the town's population was even lower at roughly 11,000 people, or only 46% of what it had been in 1989 (ibid.). Once a booming epicenter of Soviet tourism, Tskaltubo fell into decline, mirroring economic hardships in other former Soviet mono-towns.

Exacerbating the politico-economic struggles of the collapse of the Soviet Union were Georgia's geopolitical issues, experienced most acutely in its two secessionist regions of Abkhazia and South Ossetia, which unilaterally declared independence in 1991 and 1992, respectively. Ethnic conflict in Abkhazia between 1992 and 1993 produced some 251,000 IDPs, mainly ethnic Georgians (Kurshitashvili, 2012), roughly 8,000 of which arrived in Tskaltubo in search of temporary refuge in the now-vacant sanatoria. In the Fall of 1993, all but the highly securitized Georgian Ministry of Defense sanatorium had been allocated by the government to house IDPs: A total of

21 resort facilities.[9] The arrival of IDPs altered Tskaltubo's socio-economic composition and introduced new social hardships in the faltering post-independence economy. Although IDPs are culturally, ethnically, and linguistically identical to other Georgians, their loss of personal assets and resultant poverty, as well as their origins from a different geographic region, meant that they often faced discrimination (Brun, 2015, 2017; Kabachnik, Grabowska, Regulska, Mitchneck, & Mayorova, 2013; Mitchneck, Mayorova, & Regulska, 2009). IDPs were also physically isolated in the hillside sanatoria, originally designed to offer remoteness away from the town's central spaces. The IDPs' physical isolation ensured lower levels of daily contact with local residents and turned them into a distinct, socially-isolated community (Gogishvili & Harris-Brandts, 2019).[10]

Because the sanatoria housed IDPs, they were not privatized by the state during the widespread neoliberal market reforms that took place in the 1990s and 2000s.[11] Instead, they were converted into government owned and administered IDP Collective Centers.[12] In efforts to establish some semblance of a local life, IDPs began augmenting the sanatoria to support their needs—sometimes to dramatic degrees— including the addition or removal of walls, the enclosure of balconies, and the installation of utilities. Tremendous effort was required on behalf of the IDPs to render the sanatoria suitable for long-term residential inhabitation. Absent state resources in the first decade of Georgia's independence meant that IDPs were overwhelmingly left to rely on intermittent support from aid agencies, local residences, and their own community ingenuity. Many IDPs survived by taking up sustenance farming, a practice that transformed the once-formal sanatoria gardens into domestic agricultural spaces (Figure 12.7). Chickens and small livestock began roaming the parks formerly frequented by vacationing KGB servicemen.

Over the course of three decades (1992–present), Tskaltubo's IDPs constructed a new identity for the sanatoria, shaping them to meet their daily needs. While some humanitarian aid was available, it rarely addressed deteriorating building conditions, placing this burden on IDPs. Although officially the state claimed to be doing everything within its capacity to address the IDP crisis (Saakashvili, 2009b), keeping IDPs in ill-suited shelters had deeper political utility in that it bolstered government arguments for their right-of-return—a process linked to re-establishing Georgia's territorial integrity (Kabachnik, Mitchneck, & Regulska, 2015). During this time, the state focused more on highlighting the humanitarian crisis of IDPs for geopolitical reasons than on effectively integrating them into local society. Until the late 2000s, the removal of IDPs from collective centers was, therefore, a slow and unsystematic process (Brun, 2015, 2017; Gogishvili and Harris-Brandts, 2019; Kabachnik et al., 2015; Koch, 2015). Resettlement or monetary compensation was limited to buildings holding strong real estate value through state privatization. These efforts focused on property investment potential, rather than improving IDP welfare.

Figure 12.7 A pig grazing outside the former Savane sanatorium

In 2007, following much local and international criticism (Amnesty International, 2011; Brooks, 2011; Civil Georgia, 2011; UNHCR, 2009), the Georgian government finally developed an official state strategy for housing IDPs ("The State Strategy for Internally Displaced Persons–IDPs") (Government of Georgia, 2007). After several amendments (and complications caused by the arrival of new IDPs[13]), it began implementation in May 2010. Among other provisions, the strategy intended to abolish collective centers through privatization to IDPs, or by moving IDPs out of the collective centers "having special significance to the state" into new purpose-built residential buildings (Government of Georgia, 2007). The resettlement process was not always voluntary and, in some instances, collective centers were vacated by riot police using excessive force (Human Rights Watch, 2011, 2012).[14]

In December 2010, these resettlement efforts arrived in Tskaltubo. A wave of new apartment blocks were constructed in the town itself, as well as in the larger Georgian cities of Batumi and Poti (Gogidze and Ryan, 2011). Upon the completion of these apartments in 2011, 732 IDPs living at the Meshakhte, Tsiskari and Iveria Sanatoria (Civil Registry Agency, 2010) were evicted and resettled (Heinriech Boell Foundation Bureau in the South Caucasus, 2011). The residents largely vacated peacefully, their complacency being driven partially by the state's promises of better housing and the town's future economic prosperity.

At the same time, their resettlement caused much social impact on the IDPs community since they were so widely distributed across Georgia during such efforts. This detrimentally dismantled community relations, deteriorated decades of IDP social fabric, and severed existing IDP economic networks. For the town of Tskaltubo, the result was equally unfavorable in that many sanatoria simply stood empty, despite their privatization. Public spaces and schools also sat under-capacity and local shops lost revenues.[15] One local resident said: "if these sanatoria were going to just sit empty, then the government should have simply allowed IDPs to stay."[16]

(Re)constructing a "correct" heritage narrative

Various scholars have described the commodification of architectural heritage for tourism purposes, outlining how this process can result in the select re-telling of history (Alsayyad, 2013; Ashworth, Graham, and Tunbridge, 2007; Ashworth and Tunbridge, 2000; Ashworth and Larkham, 2013; Page and Mason, 2004; Sack, 1992; Waitt, 2000). AlSayyad notes that "in today's world, where the global heritage industry reigns supreme, the notion of authenticity has sometimes been cut completely loose from its moorings: the image of the thing may now actually replace the thing itself" (2013, p. 25). This is especially the case when there has been a long break in the timeline between when the site established its prominence and when it was acknowledged as heritage. When heritage is identified for its potential as an economic resource and catalyst for urban regeneration, buildings are assigned value not in accordance with their contemporary significance to local society, but on state politico-economic terms. The overwhelming use of heritage to bolster the economy means that rather than a testament of the past, "heritage is a contemporary commodity purposefully created to satisfy contemporary consumption" (G. Ashworth, 2013, p. 16).

Concerns over the impacts of heritage tourism on representations of local history stem not only from the risks of essentializing and fetishizing local cultures, but also from omitting minority narratives. The omission of such narratives becomes especially problematic when heritage is used to support a version of the past which is politicized (Timothy, 2007c). In reality, "[n]ot only does heritage have many uses but it also has multiple producers, both public–private, official–non-official and insider–outsider, each having varied and multiple objectives in the creation and management of heritage" (G. Ashworth et al., 2007, p. 2; see also, Graham, Ashworth, and Turnbridge, 2007). This dissonance in heritage should be understood as intrinsic to its production and market commodification, since "[e]xpressing one's heritage invariably means that another, different identification with the past is disinherited, excluded, or degraded" (Dwyer and Alderman, 2008, p. 172).

More than simply putting the past on display, the refurbishment of historic sites for tourism is therefore a practice, which is *place altering*, disrupting daily routines and reconstructing spaces in order to fabricate an historic image of an environment that is constantly changing. While heritage can provide familiarity with the past and function as a form of validation and belonging for communities feeling overlooked by society, it can also involve a problematic rejection of both the present and more immediate histories (Lowenthal, 1998, 2015). In this process, certain contemporary events are perceived as "flaws" or "anomalies" and their protagonists (such as the displaced) cast out from memory. Attempting to define what constitutes a "correct" heritage is typically the purview of society's powerful and wealthy, risking skewed accounts of history (Atkinson, 2005; Hobsbawm and Ranger, 1983; Timothy, 2007c). As place meanings are manipulated to cater to heritage tourism demands, economic profit dominates over historical accuracy, erasing important layers of collective memory.

The intrinsic subjectivity and normative nature of heritage designations has been well argued and demonstrated in a variety of global contexts (Alsayyad, 2013; Ashworth et al., 2007; Ashworth and Larkham, 2013; Graham, Ashworth, and Tunbridge, 2000; Page and Mason, 2004). That heritage should be considered only that which has historic provenience has further been contested, with authors demonstrating the value of other forms of cultural practice and commemoration (Eriksen, 2001; Smith, 2006; Smith and Waterton, 2012; Wiley, 2014). In the face of calls to see the definition of heritage expanded to recognize more intangible, lived, and hybrid cultural expressions, a growing body of critical scholarship has emerged since the 1990s (e.g. Eriksen, 2001; Graham et al., 2000; Keough, 2011; Lumley, 2004; Smith and Akagawa, 2008). In line with this work, urban scholars Alessandro Petti and Sandi Hilal have coined the term "Refugee Heritage," arguing for the acknowledgement of the unique forms of culture practiced by displaced communities in their places of humanitarian shelter (Petti and Hilal, 2017). Looking specifically at stateless Palestinian refugees, they raise important questions about whether a life lived in exile deserves commemoration as its own distinct form of cultural heritage—a question equally relevant for Tskaltubo's IDPs in the face of urban recovery.

In conditions of protracted displacement, communities may spend years or decades formulating histories in exile—histories that are deeply intertwined with their places of humanitarian shelter. Since much scholarly work has shown how place meanings are linked to the ongoing formation of collective identity (Dreier, Mollenkopf, and Swanstrom, 2001; Dwyer and Alderman, 2008; Gieseking, Mangold, Katz, Low, and Saegert, 2014; Low and Altman, 1992; Manzo and Devine-Wright, 2013; Rodman, 1992), it follows that displaced communities also have profound impacts on the environments of their host communities. In such instances, a recognition of "heritage from below"—a heritage that "can function as cultural resources for counter hegemonic expressions" (Robertson, 2016, p. 1)—can work to counter the social inequality of state-led heritage campaigns, and can draw attention to the alternative histories of displaced communities. In thinking about how to acknowledge and preserve cultural value through heritage tourism, such scholarship raises important questions about whose culture should be preserved and which periods of history carry priority.

Heritage promotion meets displacement

Heritage promotion has also been shown to have an uneasy relationship with displacement (Bloch, 2016; Chirikure, Manyanga, Ndoro, and Pwiti, 2010; Ghaidan, 2008; Herscher and Monk, 2015; Herzfeld, 2006; Shepherd, 2013). Bloch (2016) describes the community hardship that resulted from heritage promotion in Hampi, India, where the government evicted local residents in order to restore sites linked to the Vijayanagara Kingdom (1336–1646) in pursuit of United Nations Educational, Scientific and Cultural Organization (UNESCO) World Heritage status and greater tourism. In 2011, mass evictions began in Hampi, with properties being demolished to stage the historic glory of the fortifications, palaces, and places of worship that were once found at this medieval Hindu capital. Bloch argues that removing residents from their dwellings "was preceded by stigmatizing them in a heritage discourse shared by archaeologists, political authorities, religious leaders, and journalists" (Bloch, 2016, p. 574). The inhabitation of the site by villagers was thus described as "unnatural" and a deviation from India's "authentic" ancient history, standing in the way of the state's contemporary legitimacy and economy prosperity.

Along similar lines, Shepherd (2013) tells of how tourism to the sacred Buddhist site of Wutai Shan in China led to community displacement and the select re-telling of history. In efforts to resist eviction, the Taihuai residents of Wutai Shan put forward a petition, arguing for "their own place in this community as both the builders and defenders of religious sites" (Shepherd, 2013, p. 103). The petition asserts that "world heritage actually hurts the interests of common people and thus social development" (ibid.) Likewise, Caftanzoglou (2000, 2001) recounts evictions in Anafiotika below the Greek Acropolis, Arthurs (2013) describes Mussolini's destruction of the Marforio quarter in Rome, and Herzfeld (2006) shows the forced removal of residents from sites across contemporary Bangkok. Herzfeld argues that such evictions "are representative of an overall pattern whereby theme parks, partially made up of ancient materials but heavily restored and refurbished to suit modern ideas about the past, come to replace densely populated areas and in turn create growing zones of disaffected and displaced people" (Herzfeld, 2006, p. 132).

In a particularly stark example of heritage tourism driving displacement, Herscher and Monk (2015) describe the transformation of the walled citadel of Erbil, Iraq, which began conversion in 2006. This led to the removal of 840 Iraqi refugee and IDP families. Identified by UNESCO as "the longest continuously inhabited site in the world," the Kurdistan regional government hoped to extract tourism value from the regeneration of the citadel. The authors draw attention to the overwhelming irony of the initiative, in which "the citadel, which historically functioned as a place of refuge, could be institutionalized as historic architecture only by cleansing it of contemporary refugees" (Herscher and Monk, 2015, p. 71). That these individuals were twice-displaced underscores the vulnerability of IDP and refugee communities, which frequently lack (or are unaware of) their legal rights, including tenure security, and who are often forced to find temporary shelter in abandoned property. The global rise of conflict-induced displacement has meant that displacement is also becoming more urban and protracted (United Nations, 2018). Because of these trends, deteriorated properties (including vacant heritage sites) are more often coming to provide humanitarian shelter for vulnerable communities for periods of years and even decades. These groups then risk secondary displacement when local economies improve and governments identify untapped heritage value.

The reframing of Tskaltubo's history through urban recovery

A crucial component to facilitating urban renewal in Tskaltubo was the state's establishment of a new dominant heritage narrative. The town was presented not as a place undergoing multiple phases of change—each with equal import—but as long awaiting its return to a former spa resort glory. This nostalgic narrative had precedents elsewhere in Georgia. Prior to the government urban renewal of Tskaltubo, in the mid-2000s redevelopment projects had begun in Georgia's two largest cities of Tbilisi and Batumi. In these locations, IDPs had also been evicted as a part of urban renewal, under the banner of fostering Georgia's burgeoning hospitality sector, improving the country's international image, and supporting nation building. In Tbilisi, the centrally located Hotel Iveria and Hotel Adjara exemplified this process, where in the mid-2000s IDPs were forcibly removed. Following the renovation of the Hotel Iveria in 2009 (becoming the Radisson Blu), then-President Mikheil Saakashvili spoke at the inauguration, highlighting that the transformed property was "a symbol of new Georgia…[and of] fulfilled promises" (Saakashvili, 2009a). He sidestepped the contentiousness that had surrounded the IDP eviction process (Media Diversity Institute, 2004) while implicitly framing the presence of IDPs as unnatural and a hindrance to the country's prosperity (Manning, 2009).

Such rhetoric then became particularly effective in Tskaltubo, where the government promoted an urban renewal narrative of heritage tourism based on the glory of the town's past, while also downplaying the impact of IDPs on its more recent history. When mentioning Tskaltubo in speeches and news statements, Saakashvili framed the town as "the most famous tourist destination in the former Soviet Union" (Saakashvili, 2010, 2012a) and an untapped tourist hub for Europe. In a 2012 news interview outside the Meshakhte sanatorium, Saakashvili highlighted how the town's rehabilitation would bring new prosperity to the local community (Saakashvili, 2012b). Optimistically, he claimed that all Tskaltubo's sanatoria would soon have new owners and that the town would boast 100% employment (ibid.). The site of Meshakhte would be home to Eastern Europe's best sanatorium (ibid.). Television crews focused on unique architectural details and the president's promises that investors would "build the best hotels in the world," including an InterContinental (Saakashvili, 2012b). In other speeches, when IDPs were briefly mentioned by the president they were framed as aberrations in Georgia's "correct" history (Saakashvili, 2011). Through such rhetoric and selective heritage, both locals and IDPs were persuaded that the town's history could only be celebrated following the removal of IDPs, and that such a removal would bring back Tskaltubo's former prosperity and notoriety. IDPs were additionally promised improved social standing, better living conditions, and/or monetary compensation in their new purpose-built residential buildings.

In government efforts toward Tskaltubo's urban renewal, in 2014 a feasibility study was commissioned which resulted in the "Technical Proposal for the Tskaltubo Resort Development" (JSC Partnership Fund and Kohl & Partner Ltd., 2014), including market and location analyses, competition analyses, and a tourism strategy. In total, the Saakashvili government planned to allocate $90 million USD (200 million GEL) from the state budget for Tskaltubo's urban renewal (Sukhiashvili, 2015).[17] The decision to frame Tskaltubo as "Eastern Europe's premium spa destination" came out of this marketing research, which demonstrated that the town would have a hard time competing directly with Western Europe's top balneological locations (Danelia, 2018). Instead, it would more successfully garner nostalgia from former Soviet countries already familiar with such facilities (ibid.). The target tourist audience was thus those with fond memories of Tskaltubo during its zenith in the 1960–1980s. For urban renewal to be effective, government officials felt it was important that the town's branding play-up these positive memories while moving away from affiliations with more recent conflict and displacement, topics likely to deter tourists. Particularly for local Georgian tourists, reminders of the country's geopolitical instability and protracted internal displacement stood in contrast to notions of relaxation, restorative health, and wellness.

In fabricating an image of Tskaltubo's heritage tourism offerings, the state, tour operators, and property developers therefore carefully branded the town only in relation to its historic provenance as a Soviet-era resort destination, leaving out its IDP history. While it is rare for Georgia to celebrate aspects of its Soviet past, in this instance, the incredible former prominence of the town and the dramatic effect of the sanatoria's neo-classical buildings have led to an exception. Tourism advertising highlighted Tskaltubo's pedigree as a place once designed to house the prominent bathhouses of Joseph Stalin (Figure 12.8), as well as where one can access the best in the twenty-first-century balneological treatment (ArtMedia, 2018).

The government's two-decade, multi-phase initiative planned to transform Tskaltubo likewise focused on amplifying its resort identity (JSC Partnership Fund and Kohl & Partner Ltd., 2014). As with the president's

Figure 12.8 Interior of the Bathhouse #6 visited by Stalin

speeches, these proposed initiatives left out the history of IDPs and a sense of their formulation of a local community. The technical proposal recognizes that "[t]he problem with the IDPs has to be resolved" and that "[t]here are different possibilities in town or more outside of the resort" (JSC Partnership Fund and Kohl & Partner Ltd., 2014, p. 198). Yet, how specifically this will be accomplished is not divulged. IDPs were also not consulted during the technical proposal process. Framing IDPs as a "problem" further speaks to this selective framing of history.

Hope drives complacency: Garnering community support for IDP eviction

The effectiveness of government efforts toward establishing a new dominant heritage narrative for Tskaltubo were captured in the comments of our focus group respondents. Overwhelmingly, respondents from both IDP and resident groups viewed the promotion of heritage tourism as positive and believed that it would bring about significant prosperity. When asked if there was a risk of negative effects in the town's urban renewal, many respondents dismissed such concerns and called for tourism to begin immediately. In the words of one resident: "I support tourism with pleasure! If [the government] improves the town, it will create jobs, and future generations will live in a beautiful place. Of course, this is great for locals and IDPs. Both will have many job opportunities" (IDP focus group, Men, 18–45). Those in favor of heritage tourism felt that beyond personal gain and improvements to their own livelihood, urban renewal would bring back the town's prestige; "Tskaltubo will be restored" (IDP focus group, Women, age 45+).

For IDP respondents who had yet to be resettled, many felt that a new apartment would further afford them greater social respect and autonomy (IDP focus group, Women, age 45+). Moving into a new apartment would help them to shed the social stigma of displacement—a stigma to some extent developed by the government through its selective reframing of Tskaltubo's history.

While in general, focus group participants were optimistic toward the resettlement process, there were a number of informants who had already been resettled, and as a result, felt disillusioned. Although newly constructed apartments had provided them with security of tenure, these units still lacked basic utilities. Often, IDPs had to do maintenance work in their new apartments before moving in: "They [the government] let us in unfinished buildings" (IDP focus group, Men, age 18–45). "[When we moved] in the newly constructed collective center, we had to take our bedding from the sanatoria (IDP focus group, Men, 18–45)."

The state's claims of new employment opportunities also appeared to be overstated: "[prior to resettlement] they promised us jobs, although these were empty words. They moved us inside empty walls and we were given nothing in exchange... We [agreed on relocation to the seaside city of Poti] because of the new state project for a seaport. But there is ten times more poverty there. We needed to move to Poti for jobs, not for watching the sea on sunny days" (IDP focus group, Men, 18–45). In Tskaltubo, too, employment prospects have been overstated. Despite purchase clauses requiring buyers of the sanatoria to invest a certain amount of capital (~$1–6 million) and to employ a certain number of people (~60–100) (Agenda.ge, 2016a, 2016b), as of 2019, all three sold properties remain vacant and have yet to offer employment opportunities. The four other partially vacated sanatoria of Savane, Piliali, Megobroba, and Imereti have also not been redeveloped (Channel One News, 2019).

Another issue was the deterioration of community networks. The government's resettlement plans did not take into consideration existing community ties, and IDPs that formerly lived together were often assigned to entirely different localities (IDP focus group, Men, 18–45). Those IDPs evicted from Meshakhte sanatoria were allocated housing in the newly built Tamari Settlement near Batumi, 150 kilometers away from Tskaltubo. As a result of their resettlement, they have largely lost ties to the town. Focus group respondents highlighted how preserving their already formed communities would have made their lives easier and could have helped them to better adjust to their new realities. Several IDP households even left their new collective centers, either returning to the sanatoria of Tskaltubo or moving elsewhere in Georgia (IDP focus group, Men, 18–45). As a result of such experiences, some IDPs considered the resettlement process questionable and conducted solely because authorities were in a rush to lure new investors.

> Five to six sanatoria were resettled immediately... they (the government) were thinking that IDPs would be an obstacle for tourists...there was no systematic plan [of resettlement and sanatoria renovation], but some investor was interested in quickly taking over these properties. It is not because of IDPs occupying sanatoria that there is no tourism in Tskaltubo. When we arrived here [as the result of the war in Abkhazia],

> Tskaltubo was already malfunctioning... It is an illusion that the resettlement of IDPs will help to restore tourism in Tskaltubo.
>
> (IDP focus group, Women, 45+)

Feeling completely disillusioned with the government's resettlement process, one IDP stated: "How many times can we become displaced?! Resettlement is always the same [negative] situation for us (IDP focus group, Women, 18–45)."

Some locals, too, acknowledged that the removal of IDPs from the sanatoria had created a void in the town's collective identity: "Before [when IDPs were living in the sanatoria], you could hear them talking, there were lights on in the buildings. Now [the sanatoria] look like dungeons... Prior to resettlement, Tskaltubo's central marketplace was full of people, schools were full of children. Now the town is empty, all empty" (Local focus group, Women, 18–45). The stagnation of the town's promised urban renewal projects is therefore increasingly being met with suspicion, resentment, and even nostalgia for the former life lived alongside the previous sanatoria IDPs.

Conclusion

This chapter charted the complexities of heritage tourism development in Tskaltubo where Stalinist-era sanatoria are being vacated of IDPs in efforts toward urban recovery. The Georgian government's interest in Tskaltubo's formerly neglected historic buildings comes at a time when heritage preservation in the country is being undertaken in support of post-independence economic goals and nation building. The government's use of heritage for both economic and political purposes therefore has far-reaching consequences for Georgia's future development trajectory. As of July 2019, the resettlement of IDPs from Tskaltubo is still underway. Completed efforts point toward a number of shortcomings. Foremost has been the lack of community consultation and the high degree of uncertainty regarding how exactly IDPs will benefit from the town's urban renewal efforts. The selection process for resettlement has been driven by market forces, rather than by a comprehensive needs-based assessment of IDPs. This is disconcerting given the Georgian government's now three decades of dealing with IDPs.

Despite these ambitious plans to redevelop Tskaltubo, the government faces challenges in becoming Eastern Europe's premier wellness and spa destination. Aside from investment concerns, the redevelopment may eventually be faced with opposition from residents and IDPs that have yet to see the government's promises materialize into reality. With little concrete evidence of what the full return of tourists to Tskaltubo might look like or mean both culturally and economically, the problem is not tourism re-activation per se, but the way preparatory stages are being managed by the government—particularly with regards to IDP commemoration and resettlement.

As existing research has shown, the use of heritage tourism for urban renewal in areas inhabited by vulnerable populations can have detrimental impacts, from selectively re-narrating history to forced eviction and displacement. With the surge in both critical heritage and critical tourism scholarship since the 1990s, much has been done to advance our understandings of the role of heritage as a market commodity, and to investigate how heritage production may be detrimental to local communities. However, this literature has yet to fully intersect with scholarship showing the increased prevalence and protracted nature of conflict-induced displacement in heritage sites. In areas formerly experiencing (or in close proximity to) conflict, using heritage tourism as a means of urban recovery is showing itself to be uneasily associated with secondary displacement. More than just the exploitation of these communities, urban renewal builds up hopes of economic prosperity and social integration for isolated IDP communities.

In Tskaltubo, displaced individuals inhabiting the abandoned sanatoria were framed as anomalies in the town's formerly prosperous tourist history—a framing which greatly facilitated their eviction. Despite having lived in exile in Tskaltubo for over three decades, IDPs felt that they were perceived as outsiders by fellow Georgians and that the government framed their presence as "unnatural," even though they are Georgian citizens. They hoped that such perceptions would be ameliorated through Tskaltubo's urban renewal, which promised them greater employment and purpose-built apartments. Resettlement would then effectively erase their toll on society, returning Tskaltubo to its narrative of resort-era prosperity. The state's propaganda for the re-activation of tourism in Tskaltubo thus not only worked to attract new *outside* tourist audiences, but additionally built up optimism within local IDP and resident communities.

Notes

1. A *Sanatorium* (*sanatoria*, pl.) is a rest house for patients receiving treatment in bathhouses.
2. *Balneology* is bathing in natural spring water to ameliorate or treat physical ailments. The degree of its efficacy remains scientifically contested.
3. The meaning of *Tskaltubo* (*Tskaltubo* as the place) as the place for warm water has its origins in Mingrelian, a Kartvelian language, which has been spoken in the area for centuries.
4. The Georgian Democratic Republic (GDR) was a short-lived state entity, which emerged as the result of Russian Empire's retreat from the Caucasus region after World War I. In 1921, the GDR was then annexed by Soviet Russia and, in 1922, it was incorporated into the Soviet Union as the Georgian Soviet Socialist Republic (GSSR).
5. Tskaltubo's radial urban development was the brainchild of Tbilisi-based architect, Nikolay Pavlovich Severov. Severov's ideas were then influential to Tskaltubo's first comprehensive masterplan by Ioseb Zaalishvili and Vakhtang Kedia in 1950–1951. This later plan established new land use rules and divided the territory into balneological, sanitoria, and residential zones (Shavianidze,

Shavianidze, & Shavianidze, 1990). The impact of these divisions can still be felt today, with the sanatoria being largely isolated from the other spaces of the community.

6. The term all-Union resort (*Vsesoyuznaya Zdravnica*) denoted a resort complex, which hosted guests from all over the Soviet Union, and had sanatoriums allocated to the Union's central government entities, for instance, ministries or trade unions in Moscow.
7. As mentioned in two focus groups with local residents: Men 45+ and Women 45+.
8. As mentioned in two focus groups with local residents: Men 45+ and Women 45+
9. The remaining Georgian Ministry of Defense sanatorium was prevented from IDP use due to its state security affiliations.
10. Over the decades, this changed gradually and IDPs became more integrated with the local community. As of 1994, IDP children were transferred into the same schools and kindergartens as local children. IDPs also came to participate more in local employment and commercial activities, which ensured their further societal mixing. According to a survey conducted by Tbilisi State University in 2013, IDPs housed in sanatoria considered themselves as "integrated" and "adapted" with the locals, underscoring the change in their local place attachments over the last quarter century (see also: Salukvadze, Sichinava, & Gogishvili, 2013).
11. Unlike the sanatoria, the bathhouses in the center of Tskaltubo were privatized by the state following independence. From 1996, the majority of these facilities have been owned by the Georgian company, JSC Balneoservisi (National Public Registry of Georgia, 2016), although only bathhouses #1, 2, 3, and 6 are in operation. Bathhouse 6 was rehabilitated and re-opened in 1999, whereas Bathhouse 3 was re-opened in 2001 (Babunashvili, 2018).
12. A collective Centre in Georgia can be defined as a formerly residential or non-residential building where IDPs were settled temporarily (Salukvadze et al., 2013). There were over one thousand Collective Centres (CCs) for IDPs established across Georgia in the 1990s and 2000s (ibid.).
13. During the five-day Russo-Georgian War of August 2008, there was a renewal of hostilities in and around the border of South Ossetia and in the Kodori Gorge of Abkhazia, resulting in an increase in the number of IDPs by 26,000 (Amnesty International, 2010).
14. As of 2019, IDPs continue to be resettled across the country.
15. As mentioned in three focus groups with local residents: Men 18–45 and 45+, and Women 18–45.
16. As mentioned in the focus group with local residents: Women 18–45.
17. The process of Tskaltubo's rehabilitation was further supplemented by a new World Bank project, which was done through its cooperation with Georgia's Municipal Development Fund (MDF). In November 2012, the Municipal Development Fund of Georgia, with funding from the World Bank, started to implement its Second Regional Development Project, which intended to "improve infrastructural services…to support increased contribution of tourism in the local community" (World Bank, 2012).

References

Agenda.ge. (2016a, January 26). *Sanatorium in Georgia's Resort Tskaltubo Transforms Into Hotel.* http://agenda.ge/en/news/2016/207, accessed 30.6.2019.

Agenda.ge. (2016b, March 14). *New $5m Hotel for Georgia's Tskaltubo Resort.* https://www.agenda.ge/en/ news/2016/589, accessed 30.6. 2019.

Alsayyad, N. (2013). *Consuming Tradition, Manufacturing Heritage: Global Norms and Urban Forms in the Age of Tourism*. London and New York: Routledge.

Amnesty International. (2011). *Uprooted Again: Forced Evictions of the Internally Displaced Persons in Georgia* (No. EUR 56/005/2011). https://www.amnesty.org/en/documents/document/?indexNumber=EUR56% 2f005%2f2011&language=en, accessed 26.6.2020.

Arthurs, J. (2013). *Excavating Modernity: The Roman Past in Fascist Italy*. New York: Cornell University Press.

ArtMedia. (2018). *JSCo Balneoservisi*. http://tskaltuboresort.ge, accessed 2.12.2018.

Ashworth, G. (2013). From History to Heritage-From Heritage to Identity: In Search of Concepts and Models, in *Building a New Heritage: Tourism, Culture and Identity in the New Europe*, edited by G. Ashworth and P. Larkham. London and New York: Routledge.

Ashworth, G., Graham, B., and Tunbridge, J. (2007). *Pluralizing Pasts: Heritages, Societies and Place Identities*. London: Pluto.

Ashworth, G.J. and Tunbridge, J.E. (2000). *The Tourist-Historic City*. London and New York: Routledge.

Ashworth, G. and Larkham, P. (2013). *Building a New Heritage: Tourism, Culture and Identity in the New Europe*. London and New York: Routledge.

Atkinson, D. (2005). Heritage, in *Cultural Geography: A Critical Dictionary of Key Concepts*, edited by D. Atkinson, D. Jackson, N. Sibley, and N. Washbourne. London: I.B. Tauris, 141–147.

Babunashvili, A. (2018). *Personal interview*.

Bespiatov, T. (2018). *Cities and Towns of Georgia*. http://pop-stat.mashke.org/georgia-cities.htm, accessed 2.12.2018.

Bloch, N. (2016). Evicting Heritage: Spatial Cleansing and Cultural Legacy at the Hampi UNESCO Site in India. *Critical Asian Studies*, *48*(4), 556–578.

Brooks, C. (2011, February 13). Displaced Again: Georgian IDPs Face Eviction From Temporary Shelters. *RadioFreeEurope/RadioLiberty*. https://www.rferl.org/a/georgia_displaced_face_eviction/2307825.html, accessed 26.6.2020.

Brun, C. (2015). Home as a Critical Value: From Shelter to Home in Georgia. *Refuge: Canada's Journal on Refugees*, *31*(1), 43–54.

Brun, C. (2017). Living with Shifting Borders: Peripheralisation and the Production of Invisibility. *Geopolitics*, *0*(0), 1–18.

Caftanzoglou, R. (2000). The Sacred Rock and the Profane Settlement: Place, Memory and Identity under the Acropolis. *Journal of Oral History*, *28*, 43–51.

Caftanzoglou, R. (2001). Shadow of the Sacred Rock: Contrasting Discourses of Place under the Acropolis, in *Contested Landscapes: Movement, Exile and Place*, edited by B. Bender and M. Winer. Oxford, UK: Berg Publishers, 21–35.

Channel One News. (2019, May 24). [In Georgian] ts'q'alt'uboshi shesadzloa, khuti q'opili sanat'oriumis p'rivat'izeba mokhdes [In Tskaltubo, the Privatization of Five More Sanatoria is Expected]. *1TV*. s https://1tv.ge/news/wyaltuboshi-shesadzloa-khuti-yofili-sanatoriumis-privatizeba-mokhdes, accessed 26.6.2020.

Chirikure, S., Manyanga, M., Ndoro, W., and Pwiti, G. (2010). Unfulfilled Promises? Heritage Management and Community Participation at Some of Africa's Cultural Heritage Sites. *International Journal of Heritage Studies*, *16*(1–2), 30–44.

Citlanadze, G. (1950). [In Russian] *Tskhaltubo – Nauchno-populjarnyi ocherk* [Tskaltubo—Scientific-popular notes]. Tbilisi: Gruzmedgiz.

Civil Georgia. (2011, January 20). IDP Eviction Resumes. *Civil Georgia.* https://old.civil.ge/eng/article.php?id=23063, accessed 26.6.2020.

Civil Registry Agency. (2010). *Civil Registry Agency Data on Collective Centers in Georgia.* http://www.humanrights.ge/index.php?a=main&pid= 3009&lang=eng, accessed 26.6.2020.

Danelia, G. (2018). *Personal interview.*

Department of Cultural Heritage. (2018, February 7). [In Georgian] ts'q'altubos k'ult'uruli memk'vidreobis dzeglebi invent'arizacia [Inventory of Cultural Heritage Sites in Tskaltubo]. https://www.heritagesites.ge/ka/news_item/82, accessed 2.12.2018.

Dreier, P., Mollenkopf, J. H., and Swanstrom, T. (eds.). (2001). *Place Matters: Metropolitics for the Twenty-first Century.* Lawrence: University Press of Kansas.

Dwyer, O.J. and Alderman, D.H. (2008). Memorial Landscapes: Analytic Questions and Metaphors. *GeoJournal, 73*(3), 165.

Eriksen, T.H. (2001). Between Universalism and Relativism: A Critique of the UNESCO Concept of Culture, in *Culture and Rights: Anthropological Perspectives*, edited by J.K. Cowan, M.B. Dembour, and R.A. Wilson. Cambridge, UK: Cambridge University Press, 127–148.

Ghaidan, U. (2008). Damage to Iraq's Wider Heritage, in *The Destruction of Cultural Heritage in Iraq*, edited by P.G. Stone and J.F. Bajjaly. Suffolk: Boydell & Brewer Ltd, 85–96.

Gieseking, J.J., Mangold, W., Katz, C., Low, S., and Saegert, S. (eds.). (2014). *The People, Place, and Space Reader.* London and New York: Routledge.

Gogidze, L., and Ryan, C. (2011, February 3). Fact-checking the State of the Nation address: IDPs. http://www.humanrights.ge/index.php?a=main&pid=13009&lang, accessed 16.6.2019.

Gogishvili, D. and Harris-Brandts, S. (2019). The Social and Spatial Insularity of Internally Displaced Persons: "Neighbourhood Effects" in Georgia's Collective Centres. *Caucasus Survey, 7*(2), 134–156.

Government of Georgia. (2007). *Resolution #47: The State Strategy for Internally Displaced Persons – IDPs.*

Government of Georgia. (2007). [In Georgian] *gank'argykeba №47: idzulebit gadaadgilebul p'irta – devnilta mimart sakhelmts'ipo st'rat'egia* [Resolution #47: The State Strategy for Internally Displaced Persons – IDPs]. https://matsne.gov.ge/ka/document/view/2101072?publication=0, accessed 26.6.2020.

Graham, B., Ashworth, G.J., and Tunbridge, J.E. (2000). *A Geography of Heritage: Power, Culture and Economy.* London: Arnold.

Graham, B., Ashworth, G.J., and Turnbridge, J.E. (2007) The Uses and Abuses of Heritage, in *Heritage, Museums and Galleries: An Introductory Reader*, edited by G. Corsane. London and New York: Routledge, 26–37.

Heinriech Boell Foundation Bureau in the South Caucasus. (2011). [In Georgian] *sajaro disk'usia: devnilta satskhovrebeli p'irobebi da ts'q'alt'ubos samomavlo p'ersp'ekt'ivebi* [Public discussion: Living conditions of IDPs and future perspectives of Tskaltubo]. https://ge.boell.org/ka/2014/01/16/devnilta-sacxovrebeli-pirobebi-da-cqaltubos-samomavlo-perspektivebi, accessed 30.6.2020.

Herscher, A. and Monk, D.B. (2015). The New Universalism: Refuges and Refugees between Global History and Voucher Humanitarianism. *Grey Room, 61*, 70–80.

Herzfeld, M. (2006). Spatial Cleansing: Monumental Vacuity and the Idea of the West. *Journal of Material Culture, 11*(1–2), 127–149.

Hobsbawm, E. and Ranger, T. (1983). *The Invention of Tradition*. Cambridge, UK: Cambridge University Press.

Human Rights Watch. (2011, January 24). World Report 2011: Georgia. https://www.hrw.org/world-report/2011/country-chapters/georgia, accessed 16.6.2019.

Human Rights Watch. (2012, January 22). World Report 2012: Georgia. https://www.hrw.org/world-report/2012/country-chapters/georgia, accessed 16.6.2019.

JSC Partnership Fund and Kohl & Partner Ltd. (2014). *Technical Proposal for the Tskaltubo Spa Resort Development: Market & Technical Feasibility with Tskaltubo Vision Proposal*. www.enterprisegeorgia.gov.ge/ uploads/files/publications/5b4d-da3a83731-1.pdf, accessed 26.6.2020.

Kabachnik, P., Grabowska, M., Regulska, J., Mitchneck, B., and Mayorova, O.V. (2013). Traumatic Masculinities: The Gendered Geographies of Georgian IDPs from Abkhazia. *Gender, Place & Culture*, *20*(6), 773–793.

Kabachnik, P., Mitchneck, B.A., and Regulska, J. (2015). Return or Integration? Politicizing Displacement in Georgia. *Security, Democracy and Development in the Southern Caucasus and the Black Sea Region*, Bern: Peter Lang, 183–204.

Keough, E.B. (2011). Heritage in Peril: A Critique of UNESCO's World Heritage Program. *Washington University Global Studies Law Review*, *10*, 593.

Koch, E. (2015). Protracted Displacement in Georgia: Structural Vulnerability and "Existing not Living". *Human Organization*, *74*(2), 135–143.

Kurshitashvili, N. (2012). The Impact of Socially Irresponsible Resettlement on the Livelihoods of Internally Displaced Persons in Georgia. *Refugee Survey Quarterly*, *31*(2), 98–118.

Lort'kip'aniże, G. (1920, July 16). [In Georgian] *Brzaneba respublikis mt'avrobisa samxedro ucqebis mimart'* [Ruling of the Government of the Republic of Georgia to the Ministry of War]. Ert'oba.

Low, S.M., and Altman, I. (1992). Place Attachment: A Conceptual Inquiry, in *Place Attachment*, edited by I. Altman and S.M. Low. New York: Plenum Press, 1–12.

Lowenthal, D. (1998). *The Heritage Crusade and the Spoils of History*. Cambridge, UK: Cambridge University Press.

Lowenthal, D. (2015). *The Past is a Foreign Country—Revisited*. Cambridge, UK: Cambridge University Press.

Lumley, R. (2004). The Debate on Heritage Reviewed, in *Heritage, Museums and Galleries: An Introductory Reader*, edited G. Corsane. London and New York: Routledge, 15–25.

Manning, P. (2009). The Hotel/Refugee Camp Iveria: Symptom, Monster, Fetish, Home, in *City Culture and City Planning in Tbilisi: Where Europe and Asia Meet*, edited by K. Van Assche, J. Salukvadze, and N. Shavishvili. Lewiston: The Edwin Mellen Press, 319–349.

Manzo, L. and Devine-Wright, P. (2013). Introduction, in *Place Attachment: Advances in Theory, Methods and Applications*, edited by L. Manzo and P. Devine-Wrights. New York: Routledge, 1–9.

Media Diversity Institute. (2004, August). Residents Ordered to Check Out of the Iveria. *Media Diversity Institute (MDI)*, 1–2, http://www.media-diversity.org/additional-files/documents/Z%20Current%20MDI%20Resources/MDI%20special%20report%20-%20Refugees%20in%20Georgia.pdf, accessed 6.26.2020.

Mitchneck, B., Mayorova, O.V., and Regulska, J. (2009). "Post"-Conflict Displacement: Isolation and Integration in Georgia. *Annals of the Association of American Geographers*, *99*(5), 1022–1032.

National Agency for Cultural Heritage Preservation of Georgia. (2017). *Tsq'altubo* [Tskaltubo]. http://www.histowns.ge, accessed 6.30.2019.

National Public Registry of Georgia. (2016). *Excerpt from the National Registry of Georgia for the cadastrial parcel N 29.08.34.044.*

Page, M. and Mason, R. (2004). *Giving Preservation a History: Histories of Historic Preservation in the United States*. London and New York: Routledge.

Petti, A. and Hilal, S. (2017, February 22). Refugee Heritage. *E-Flux Architecture*, https://www.e-flux.com/architecture/refugee-heritage.

Robertson, I.J.M. (2016). Introduction: Heritage from Below, in *Heritage from Below*, edited by I. J. M. Robertson. London and New York: Routledge, 1–28.

Rodman, M.C. (1992). Empowering Place: Multilocality and Multivocality. *American Anthropologist*, *94*(3), 640–656.

Saakashvili, M. (2009a, February 9). [In Georgian] sakartvelos p'rezident'i mikheil saak'ashvili sast'umro "'redisonis'" gakhsnas daests'ro [President Mikheil Saakashvili attended the opening of the Radisson Blu hotel]. http://saakashviliarchive.info/ge/PressOffice/News/?p=2246, accessed 26.6.2020.

Saakashvili, M. (2009b, February). *Annual Report to Parliament*. Presented at the Annual Report to Parliament, Georgian Supreme Legislative Body. http://www.parliament.ge/en/media/axali-ambebi/mikheil-saakashvili-georgian-president-presented-annual-report-to-parliament-19590, accessed 26.6.2020.

Saakashvili, M. (2010, December 29). [In Georgian] sakartvelos p'rezident'i kutaisshi ganakhlebuli op'erisa da "miuzikholis" shenobis gakhsnis sazeimo ghonisdziebas daests'ro [President of Georgia attended the opening of renovated opera house and music hall in Kutaisi]. http://saakashviliarchive.info/ge/ PressOffice/News/?p=6001, accessed 26.6.2020.

Saakashvili, M. (2011, October 17). [In Georgian] mikheil saak'ashvili: "chven q'vela davbrundebit chven sakhlebshi, es aris chemi tskhovrebis mtavari amotsana" [Mikheil Saakashvili: We are all going back to our homes, this is my lifetime goal]. http://saakashviliarchive.info/ge/PressOffice/News?6941, accessed 26.6.2020.

Saakashvili, M. (2012a). [In Georgian] mikheil saak'ashvili ts'q'alt'uboshi [Mikheil Saakashvili in Tskaltubo]. https://www.youtube.com/watch?v=9K7byYsi_mM, accessed 26.6.2020.

Saakashvili, M. (2012b, December 1). [In Georgian] sakartvelos p'rezident'ma devnilebs batumis akhladashenebul binebshi akhalmosakhleoba miulotsa [President of Georgia congratulated IDPs with new apartments in Batumi]. http://www.saakashviliarchive.info/ge/PressOffice/News/Releases?p=7274&i=2, accessed 26.6.2020.

Sack, R.D. (1992). *Place, Modernity, and the Consumer's World: A Relational Framework for Geographical Analysis*. Baltimore: Johns Hopkins University Press.

Salukvadze, J., Sichinava, D., and Gogishvili, D. (2013). Socio-economic and Spatial Factors of Alienation and Segregation of Internally Displaced Persons in the Cities of Georgia. *Spatial Inequality and Cohesion*, *38*, 45–60.

Shavianidze, O., Shavianidze, M., and Shavianidze, G. (1990). [In Russian] *Kurort Cxaltubo i ego lechebny'e istochniki* [The Resort of Tskaltubo and its Medical Springs]. Tbilisi: Nauchno-medicinskoe obshchestvo kurortologov i fizioterapevtov Gruzinskoj SSR.

Shepherd, R.J. (2013). *Faith in Heritage: Displacement, Development, and Religious Tourism in Contemporary China* (1st edition). Walnut Creek, California: Routledge.

Smith, L. (2006). *Uses of Heritage*. London and New York: Routledge.

Smith, L. and Akagawa, N. (2008). *Intangible Heritage*. London and New York: Routledge.

Smith, L. and Waterton, E. (2012). Constrained by Commonsense: The Authorized Heritage Discourse in Contemporary Debates, in *The Oxford Handbook of Public Archaeology*, edited by R. Skeates, C. McDavid, and J. Carman. Oxford: Oxford University Press, 153–171.

Sukhiashvili, V. (2015, August 24). *We Were Building a Resort in Tskaltubo Much Like Baden-Baden… the Georgian Dream Coalition Killed this Idea*. https://factcheck.ge/en/story/19175-we-were-building-a-resort-in-tskaltubo-much-like-baden-baden-the-georgian-dream-coalition-killed-this-idea, accessed 26.6.2020.

Timothy, D.J. (2007a). *Managing Heritage and Cultural Tourism Resources: Critical Essays, Volume One*. Farnham: Ashgate.

Timothy, D.J. (2007b). *The Heritage Tourist Experience: Critical Essays, Volume Two*. Farnham: Ashgate.

Timothy, D.J. (ed.). (2007c). *The Political Nature of Cultural Heritage and Tourism | Critical Essays, Volume Three*. London and New York: Routledge.

UNHCR. (2009). *Protection of Internally Displaced Persons in Georgia: A Gap Analysis*. United Nations website: https://www.unhcr.org/4ad827f59.pd, accessed 26.6.2020.

United Nations. (2018). *Global Trends: Forced Displacement in 2018*. Retrieved from United Nations website: https://www.unhcr.org/5d08d7ee7.pdf, accessed 26.6.2020.

Waitt, G. (2000). Consuming Heritage: Perceived Historical Authenticity. *Annals of Tourism Research*, *27*(4), 835–862.

Wiley, T.G. (2014). The Problem of Defining Heritage and Community Languages and Their Speakers: On the Utility and Limitations of Definitional Constructs, in *Handbook of Heritage, Community, and Native American Languages in the United States*, edited by T.G. Wiley, J.K. Peyton, D. Christian, S.C.K. Moore, and N. Liu. London and New York: Routledge, 19–26.

World Bank. (2012). *Second Regional Project*. Retrieved from The World Bank website: http://projects.worldbank.org/P130421/regional-development-project-ii?lang=en&tab=documents&subTab=projectDocuments, accessed 26.6.2020.

13 Creative institutionalism

Statecraft beyond the state in Palestine

Chiara De Cesari

What role does the reconstruction of heritage play in enduring conditions of conflict, which never seem to give way to peace? When a state has either collapsed or never existed, can local civil societies establish or reinvent the functions of a state, whether in relation to cultural heritage or beyond? If so, how might they go about it? With heritage, urban regeneration, and museum projects of widely different scales proliferating across the Middle East, how can societies harness these developments to rethink 'urban recovery'? Might civil societies mobilize such tools of urban recovery to engender a measure of positive social change, even before a given conflict is over? This chapter sets out to explore these questions by way of a set of examples relating to Palestine, above all the West Bank. For Palestine offers a paradigmatic case of enduring conflict and contested heritages.

The story of the Palestinian state is one of interruptions and deferrals. It is a paradoxical story of a state which is not one. Post-Oslo developments in the region have raised fundamental questions as to whether establishing such a state in the West Bank and Gaza is viable or even desirable. Indeed, there may be alternatives that are both more feasible and more democratic. The Oslo Accords, which the PLO (Palestine Liberation Organization) agreed with the state of Israel in the mid-1990s, gave Palestinians the prospect of a state of their own. The reality on the ground was very different, however, and remains so today. Palestinians have a measure of autonomy and self-government, but only within a patchwork of walled-off areas in the West Bank and Gaza from which Israel has withdrawn. Under shifting colonial conditions, which many see as a form of apartheid (Yiftachel, 2012), Israel has maintained control over all borders and movement in and out of Palestinian areas. In this, Israel resembles South Africa's old Bantustans. In much of the West Bank, Israeli colonization has continued unabated after Oslo. According to the latest statistics, the West Bank now hosts over 600,000 settlers, a full 10% of Israel's Jewish population.[1]

The years following the Oslo Accords saw the failure of not only the so-called peace process, but the project of the Palestinian Authority (PA) too. Having been conceived initially as the vessel of a full-fledged Palestinian state, the PA was beset by territorial fragmentation and compromised sovereignty, not to mention corruption. In its inability to achieve full statehood, the PA has secured neither peace nor development for Palestinians; in fact, some critics have seen it as Israel's watchdog, administering the population and containing Palestinians (Stamatopoulou-Robbins, 2019). Despite the PA being unable to fulfil its promises, international donors—especially the European Union (EU) and European and Arab countries—have continued to provide it with essential support over the years. Although this international funding has indirectly subsidized the Israeli occupation and relentless dispossession of Palestinians (e.g., Le More, 2008), it has also kept Palestinian nongovernmental organizations (NGO) going. In the wake of the PA's many failures, these NGOs have stepped in to take up several governmental functions (Hammami, 2006; De Cesari, 2020). This process is evident in the field of heritage, which has historically played an important political role in the region.

Marred by their splintered geography, the Palestinian-controlled areas have experienced massive urbanization after Oslo. In an economy fueled by land speculation and real estate, people's relationship to the land has been thoroughly commodified (e.g., Bshara, 2015) while sprawling development has blurred the boundaries between the urban and the rural in these constrained territories. Barely endowed with local specificity, these new Palestinian urban sprawls exhibit a distinctly globalized aesthetics, in that they resemble Amman or Beirut. They contrast starkly, however, with other key urban forms: Refugee camps, but also newly restored and heritagized historic quarters. Derelict and largely abandoned in the mid-1990s, historic centers in the West Bank have been the object of restoration and rehabilitation campaigns ever since (Figures 13.1 and 13.3). Motivated by a urge to protect Palestinian vernacular heritage and architecture under threat from

Figure 13.1 Historic village of Beit Iksa; an Israeli settlement is visible in the background on the right. Photo by Michel Salameh

colonization and urbanization, these campaigns were driven by non- and semi-governmental Palestinian organizations, working with only a minimum of informal coordination with one another. In a context in which establishing the rule of law is difficult if not structurally impossible, these organizations have operated outside of existing legal frameworks and PA oversight. In fact, they have attempted to write the rules of the heritage game themselves.

Over the past two decades, urban regeneration has proliferated in the broader Middle East region and beyond (e.g., Daher and Maffi, 2014). Palestinian projects differ from equivalents in the wider region in that they are less dominated by a commodified and commodifying logic, which foregrounds the promotion of tourism and economic gains for the few. In Palestine, the main objective of urban regeneration is to stop the ongoing, unremitting erasure of the Palestinian historic landscape—the very Palestinianness of the land. Under such circumstances, regeneration means saving a distinctively Palestinian built environment from being drowned by the avalanche of concrete poured over it by Israeli settlements (and new Palestinian high-rise developments). For Palestinian NGOs in particular, it means regenerating both physical and social structures: (Re)building resilient communities from below.

The distinctive assemblage of actors and forces that are able to perform state-like functions in the West Bank—including the PA, Palestinian NGOs, international donors, the Israeli army, but also borders and walls—are less exceptional than it might seem at first glance. Much recent writing in political anthropology and critical state theory builds on the critique of the traditional Weberian model of the state as a monolithic entity separated from society. On this basis, this body of work has dramatically pluralized and expanded notions of what it means to be a 'state actor' (Cooper, 2019). In the strategic-relational approach, for example, the state is made up of social relations; as such, the state is conceived of as 'a contingent product of a changing balance of political forces located within and beyond the state ... as embedded in the wider political system and environing societal relations' (Jessop, 2008, p. 124). Reworking Foucault, anthropologists concerned to think beyond the Weberian state foreground the multiplicity of actors, institutions, practices, and technologies involved in the exercise of state power (Gupta and Sharma, 2006). They have also radically questioned the distinction between state and civil society, emphasizing how NGOs carry out the so-called 'state' functions across the world, especially under conditions of late liberal/neoliberal governmentality. International actors (that is, intergovernmental bodies such as the UN) and transnational actors (nongovernmental agencies such as Doctors Without Borders) have played a key role in this process (Mitchell, 1991; Ferguson and Gupta, 2002; Ferguson, 2004).

Alongside state and civil society, another questionable dichotomy is that between formality and informality. Scholars are increasingly recognizing how informal practices on the part of bureaucrats and others are key to the functioning of states. This holds true in both the Global South and North, and especially in conflict situations where state structures have collapsed (Gupta, 2013, 2015). Accordingly, urban studies scholars have insisted that informality is not positioned outside or against the state (that is, limited to the realm of civil society). Rather, the state is itself produced through a groundswell of informal practices and processes, in combination of course with formal ones (Davis, 2017; Roy, 2012). Indeed, the state is riven and constituted by trajectories of informalization and formalization. For Diane Davis, informality not only 'provides a largely unrecognized way of enabling governance, citizenship, and social order, no matter the developmental context' (2017, p. 316). What is more, it grounds an alternative way of conceiving how governance is enacted. This approach 'starts from a grounded understanding of how social order is produced, city services are distributed, spaces are utilized, citizens are mobilized, and local authorities are operating, instead of starting from modern state ideals' (2017, p. 320).

Mainstream political science has long seen informality as a problem characteristic of 'backward' postcolonial states that are 'permeated by society' (Gupta, 2015, p. 587; see also Migdal, 2001). Recent critical scholarship (and even policy debates), in contrast, has turned these presumptions about informality on their head, broadening the concept's analytic purview and showing its generative potential. Although Janus-faced, informality can facilitate service provision and even hold out 'avenues of participation' for those excluded by formal processes (Singerman, 1995). From the onset of neoliberalism onwards, policy debates have tended to present state reform as entailing a broader and deeper embrace of society. This is clear, for example, in the British Labour Party's recent calls for a 'relational state,' which would recognize the importance of 'human relationships' (Cooke and Muir, 2012).

What does this have to do with urban recovery? In this volume, Howayda al-Harithy conceives of it as an 'ongoing act of resilience that starts before a conflict or crisis ends.' Taking up this definition for my purposes of this essay, I approach urban recovery as an emergent, uneven process of socio-material reconstruction. Generated by civil society, this process involves mobilizing local social energies and material resources in unanticipated ways. It entails translocal and transnational alliances in what are imaginative, cunning responses to displacement and shattered lives, environments, and social worlds. As such, urban recovery represents an engine of resilience and even resistance in the face of often forbidding circumstances and configurations of power.

The Palestinian organizations that I discuss in this chapter do precisely this work of harnessing dormant socio-material resources. Arguably, they accomplish still more than that. Responding to the lack of basic state infrastructures in Palestine, the result of the failure of the PA's statehood and enduring conditions of conflict and colonization, these organizations have set up segments of an alternative infrastructure. As experimental as it is resourceful and astute, this work of 'statecraft' and institution building makes do with difficult circumstances. Although improvised and fragmented, this form of creative institutionalism is also deeply participatory. At times, it aligns closely with people's desires.

Experimenting with museums and biennials

The first Palestinian national museum after the Oslo Accords was arguably an artwork, Khalil Rabah's Palestinian Museum of Natural History and Humankind. The installation developed through both several international exhibitions, and local shows and initiatives undertaken by the museum itself. At the time of writing, Rabah has been working on this museum for at least twenty years (Stoughton, 2018). First exhibited at the 2005 Istanbul Biennial, it caused many visitors to wonder: 'But is it real?' The work was met with much the same reaction the following year when the museum was installed in the shadow of the Parthenon and the New Acropolis Museum in Athens. This location was significant indeed: It was as if Rabah's museum meant to 'absorb' institutional legitimacy through proximity with one of the quintessential European heritage sites. There were also clear political resonances between Rabah's project and the New Acropolis Museum, which is itself a monument to ongoing claims for the repatriation of cultural property (namely the Elgin Marbles) to its state of origin.

Although each iteration changes in response to its specific context, Rabah's installation usually consists of a museum-like display of artifacts, which are organized by theme in an array of cabinets. In addition, it puts on museumlike activities. Rabah's museum has organized lectures and auctions, runs a regular newsletter, and is publishing a catalogue as I write (Figure 13.2). Patently, this museum mocks the received model of the so-called universal museum, whose expansive collections span natural history, archaeology, and art. The first national museums in metropolitan centers, often born of colonial plunder, adhered to this model of museum. Like a natural history museum, Rabah's museum displays fossils, bones, and meteorites. Upon closer inspection, though, viewers discover that these are not natural specimen at all but rather human artifacts, which Rabah has crafted out of wood from the olive tree, a key symbol of Palestine and Palestinian nationalism. In some of its iterations, the museum's 'permanent collection' has the title 'Palestine before Palestine.'

Yet in Palestine, institutions such as national and universal museums are not there, physically. In such circumstances, the critique of established institutions takes the form of what I call creative or experimental institutionalism—the formation of counterinstitutions that do the state's work. Rabah's museum powerfully encapsulated the problem that Palestinians, as colonized nation, lack a space in which to represent themselves, narrative self-determination. But, the museum did more: It both called for a Palestinian national museum and laid some of the groundwork for a heritage-preservation infrastructure to be established in the future. Rabah's installation instanced an anticipatory mode of representation; in

THE PALESTINIAN MUSEUM OF NATURAL HISTORY AND HUMANKIND

NEWSLETTER

THE MUSEUM

EDITORIAL

Since its last newsletter issue The Palestinian Museum of Natural History and Humankind the Museum embarked on new projects and has proceeded with activities on already existing works. As the specialized teams of the Museum work on enlarging its Permanent Collection, expanding researches in its departments of the Museum, a new large and exciting project has been under construction literally. The Museum is building the New Palestinian Museum of Natural History and Humankind in Acropolis, Athens.

The staff of the Museum is to celebrate the opening of this event on July 17, 2006 and cordially invites you to this unique event.

ABOUT THE MUSEUM

THE PALESTINIAN MUSEUM OF NATURAL HISTORY AND HUMANKIND WAS ESTABLISHED IN 1936 TO INSPIRE WONDER, DISCOVERY AND KNOWLEDGE OF AMAZING PHENOMENA THAT PROVOKE CURIOSITY AND DEEPEN UNDERSTANDING OF OUR NATURAL AND CULTURAL WORLDS.

The growth and care of the collection, research and dissemination of knowledge through publications, exhibitions and educational projects remain the cornerstones of the Museum's mission today.

Our goal at The Palestinian Museum of Natural History and Humankind is to inform, connect, fascinate, amaze, delight, surprise and excite visitors' interest in the way the world works in the natural and cultural earth of Palestine.

The Palestinian Museum of Natural History and Humankind's activities are organized in four main departments: the Anthropological Department, the Botanical Department, Geology and Paleontology Department, Earth and Solar System Department.

Each section promotes its specific area of research through public events, expeditions, and co-operation with international organizations.

THE PALESTINIAN MUSEUM OF NATURAL HISTORY AND HUMANKIND

NEWSLETTER

THE MUSEUM

ABOUT THE MUSEUM

The Building of the New Palestinian Museum of Natural History and Humankind

Set in the shadow of the New Acropolis Museum in Athens, now the ***New Palestinian Museum of Natural History and Humankind*** *is being unveiled. Designed by PDF,* ***Palestinian Design Force*** *to house the Museum's Permanent Collection and special exhibitions, the architecture is deliberately designed to thwart expectation and blur boundaries between place non-place, container contained, displacement replacement and the right of return.*

18-20 Dionysiou Areopagitou Pedestrian Walkway. Acropolis, Athens Greece.

Figure 13.2 (a–c) The 2006 newsletter of the Palestinian Museum of Natural History and Humankind

that, it represented a Palestinian national museum before it was actually realized (De Cesari, 2012). Through its initiatives—both practical and symbolic—this art project contributed to a Palestinian museum and heritage infrastructure to come, as I explain below.

While exhibiting his museum worldwide, Khalil Rabah was simultaneously Creative Director of the main Palestinian heritage NGO, Riwaq. In this position, he punctuated Riwaq's campaign for the preservation of Palestinian historic vernacular with art installations and projects designed to magnify the organization's mission. In particular, he devoted a lot of effort to organizing the first (real) Palestinian biennials for art and architecture, which took place in 2005, 2007, and 2009 in various locations across

Figure 13.2 (Continued)

the Mediterranean. Whereas global events usually emphasizes a central art exhibition, Rabah's Riwaq biennial foregrounded performances, discussions, gatherings, assemblies, and walks across the Palestinian landscape. All of this was designed to reinvigorate and inject new energy into a fragile Palestinian commons and public sphere under severe threat. Crucially, the biennials were intended to bolster and expand Riwaq's work in the preservation and adaptive reuse of Palestinian historic heritage.

This objective came to the fore in the 2009 biennial, 'A Geography: 50 Villages.' It was named after Riwaq's most important national scheme, a long-term plan to restore the fifty most significant historic villages and towns of the West Bank that are being encroached upon by Israeli settlements. The idea for the biennial began life at the fifty-third Venice biennial, as part of the first ever Palestinian national pavilion in Venice, resonantly titled 'Palestine c/o Venice.' The pavilion included a one-day symposium and an exhibition of Palestinian artworks. These included Rabah's performance installation '50 Local Pavilions,' in which Riwaq's members and partners undertook their usual organizational activities (having all come to Venice

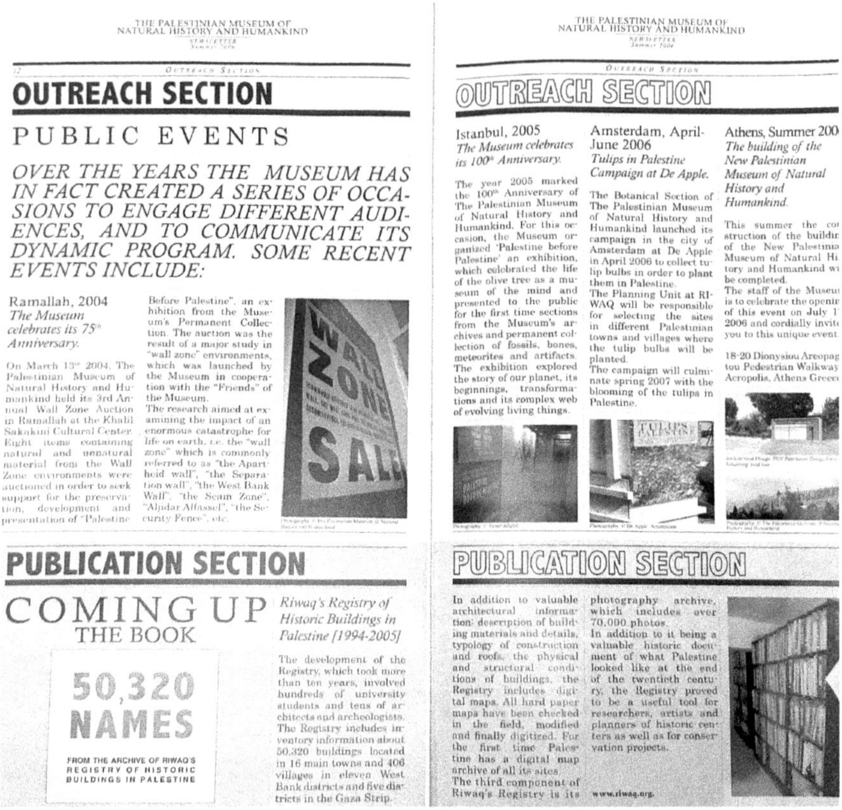

THE PALESTINIAN MUSEUM OF NATURAL HISTORY AND HUMANKIND

OUTREACH SECTION

PUBLIC EVENTS

OVER THE YEARS THE MUSEUM HAS IN FACT CREATED A SERIES OF OCCASIONS TO ENGAGE DIFFERENT AUDIENCES, AND TO COMMUNICATE ITS DYNAMIC PROGRAM. SOME RECENT EVENTS INCLUDE:

Ramallah, 2004
The Museum celebrates its 75th Anniversary.

On March 13th 2004, The Palestinian Museum of Natural History and Humankind held its 3rd Annual Wall Zone Auction in Ramallah at the Khalil Sakakini Cultural Center. Eight items containing natural and unnatural material from the Wall Zone environments were auctioned in order to seek support for the preservation, development and presentation of "Palestine Before Palestine", an exhibition from the Museum's Permanent Collection. The auction was the result of a major study in "wall zone" environments, which was launched by the Museum in cooperation with the "Friends" of the Museum.
The research aimed at examining the impact of an enormous catastrophe for life on earth, i.e. the "wall zone" which is commonly referred to as "the Apartheid wall", "the Separation wall", "the West Bank Wall", "the Seam Zone", "Aljidar Alfassel", "the Security Fence", etc.

PUBLICATION SECTION

COMING UP
THE BOOK

50,320 NAMES

FROM THE ARCHIVE OF RIWAQ'S REGISTRY OF HISTORIC BUILDINGS IN PALESTINE

Riwaq's Registry of Historic Buildings in Palestine [1994-2005]

The development of the Registry, which took more than ten years, involved hundreds of university students and tens of architects and archeologists. The Registry includes inventory information about 50,320 buildings located in 16 main towns and 406 villages in eleven West Bank districts and five districts in the Gaza Strip.

THE PALESTINIAN MUSEUM OF NATURAL HISTORY AND HUMANKIND

OUTREACH SECTION

Istanbul, 2005
The Museum celebrates its 100th Anniversary.

The year 2005 marked the 100th Anniversary of The Palestinian Museum of Natural History and Humankind. For this occasion, the Museum organized 'Palestine before Palestine' an exhibition, which celebrated the life of the olive tree as a museum of the mind and presented to the public for the first time sections from the Museum's archives and permanent collection of fossils, bones, meteorites and artifacts. The exhibition explored the story of our planet, its beginnings, transformations and its complex web of evolving living things.

Amsterdam, April-June 2006
Tulips in Palestine Campaign at De Apple.

The Botanical Section of The Palestinian Museum of Natural History and Humankind launched its campaign in the city of Amsterdam at De Apple in April 2006 to collect tulip bulbs in order to plant them in Palestine.
The Planning Unit at RIWAQ will be responsible for selecting the sites in different Palestinian towns and villages where the tulip bulbs will be planted.
The campaign will culminate spring 2007 with the blooming of the tulips in Palestine.

Athens, Summer 200
The building of the New Palestinian Museum of Natural History and Humankind.

This summer the co struction of the buildir of the New Palestinia Museum of Natural Hi tory and Humankind wi be completed.
The staff of the Museu is to celebrate the openir of this event on July 1 2006 and cordially invit you to this unique event.

18-20 Dionysiou Areopag tou Pedestrian Walkway Acropolis, Athens Greec

PUBLICATION SECTION

In addition to valuable architectural information: description of building materials and details, typology of construction and roofs, the physical and structural conditions of buildings, the Registry includes digital maps. All hard paper maps have been checked in the field, modified and finally digitized. For the first time Palestine has a digital map archive of all its sites.
The third component of Riwaq's Registry is its photography archive, which includes over 70,000 photos.
In addition to it being a valuable historic document of what Palestine looked like at the end of the twentieth century, the Registry proved to be a useful tool for researchers, artists and planners of historic centers as well as for conservation projects.

www.riwaq.org.

Figure 13.2 (Continued)

for the occasion). As Rabah put it, this was 'a biennale within a biennale, *an art work representing an institution*.'[2] In performing its institutional routine—the routine of heritage—Riwaq paralleled and mocked the biennial form, parodying, twisting, accentuating, and displacing global articulations of both heritage and contemporary art. And yet, Riwaq's people took all of this very seriously, for it represented a way of realizing projects. It is thanks to events such as the Palestinian biennials that the national scheme of village conservation got off the ground. Indeed, much of Riwaq's 2009 biennial after Venice took place in these villages in the hope that it would help stimulate the regeneration process.

At high-profile art biennials such as Venice's, national pavilions reflect state strength and national grandeur, put on by the countries' cultural institutions. The Palestinian intervention turned this convention on its head. Assembled by artists, civil society, and intellectuals, the Palestinian pavilion staged a parodic performance that subverted projects of national self-image. 'Palestine c/o Venice' was officially a collateral event, as opposed to a national pavilion proper (according to Venice biennale's policy, only states recognized by Rome can mount pavilions). Riwaq took the position that although the event was 'not supposed to be called the Palestinian pavilion … this is how we refer to it and it is important that we do so.' This stresses how the pavilion functioned performatively: Proceeding *as if* the exhibition were a full-fledged national pavilion represents a means of calling a future Palestinian state into existence. This is a practice of *anticipatory representation*, the performative instantiation of an institution that is yet to exist. Parody and parodic repetition are especially important in Rabah's and Riwaq's performative repertoire. Consider, to give just one example, how the mock 'Riwaq Biennale' passes imitated those distributed by the official biennial. Some visitors even managed to use them to gain access to the main Venice show.

Despite being a non-pavilion, organized by the civil society of a non-state, Riwaq's mock biennale fostered the development of state-like institutions. The performance functioned as a 'technology of enchantment' (Gell, 1992, title) in that it set out to realize Riwaq's '50 Villages' project. A heritage map of Palestine depicting the fifty villages that Riwaq was seeking to restore was distributed as part of the performance. It was produced in such a way as to look deceivingly like the so-called 'Oslo map' of the West Bank, which is marked by the mosaic of non-sovereign territories that emerged from the so-called peace process. Beneath its prosaic facade, Riwaq's heritage map is conceived of as 'a dream' in that it anticipates 'how we want to see the future of Palestine [by] creating the infrastructure for a national cultural project.'[3]

An alternative infrastructure of heritage preservation

A Palestinian Department of Antiquities was established in 1994 as part of the broader formation of the PA, which was expected to develop into an extensive state infrastructure within five years. The department's purpose was to take care of what are some of the most notable archaeological sites of the Middle East—such as the tumbling walls of Jericho. Although the failure of the peace process meant that the PA was never fully recognized and full statehood never arrived, the department continued working toward the preservation of Palestinian heritage. Its mandate and powers remained extremely limited, however, on account of the fragmentation of Palestinian sovereignty and inadequacy of the PA's legislative framework, which originated in the British colonial period (Kersel, 2015). Long staffed exclusively by archaeologists, the department was not legally mandated to protect either the historic built environment or more intangible forms of heritage. Indeed, the law protected only significant material remains dating back to before 1700 AD, leaving out all that came after (al-Jubeh, 2006; Maniscalco, 2005). The limited remit and outdated conception of heritage implicit in this received body of law reflected the traditional focus of archaeology in the region, which privileged the pre-Islamic, biblical periods, viewed as the cradle of (Western) civilization (Abu El-Haj, 2001; see also, e.g., Whitelam,

1996; Masalha, 2007). As such, archeology constituted an essentially orientalist field of study at the service of colonial agendas.

Owing to various micro- and macro-political factors, it took until 2018 to develop and ratify new legislation to protect the historic heritage. In the meantime, Palestinian civil society organizations began carrying out those tasks that fell outside of the PA Department of Antiquities' remit. Heritage organizations proliferated, for many Palestinians perceive the preservation of people's lived environment and traditional landscapes, whether urban or rural, as a matter of the utmost urgency and concern. It is also a national symbol of resistance. Encroached upon by colonization, urbanization, and land speculation, the traditional Palestinian landscape of domed white stone houses surrounded by terraces with olive trees is rapidly disappearing. Livelihoods, heritage, and a specific, distinctively Palestinian quality of place are vanishing with it. Traditional Palestinian architecture symbolizes historic Palestine, which was lost in the 1948 Nakba, or Palestinian catastrophe, in which 750,000 Palestinians were forced to leave what became the state of Israel and over 400 Palestinian villages and towns were destroyed and/or depopulated. Preserving this heritage means staying put, a chief Palestinian nationalist value. In staying close to the land, Palestinians are ultimately able to resist dispossession and cultural annihilation at the hands of settler colonialism (Figure 13.3).

Riwaq, the Centre for Architectural Preservation, was the first major Palestinian heritage organization. Initiated by militant intellectuals who, having participated in early Israeli–Palestinian peace negotiations, had a

Figure 13.3 Beit Iksa after restoration by Riwaq

foot in both culture and politics, Riwaq adopted a creative approach to heritage preservation from the outset. At the same time, it also assumed a clear sector- and institution-building role. Instructive in this regard is the fact that Riwaq's two founders, like many of these involved in Palestinian NGOs, did stints working for the PA in the 1990s only to then leave disenchanted. They would pursue their mission of creating democratic, viable, and vibrant Palestinian institutions outside the PA. In its nearly thirty years of activity since 1991, Riwaq has acted as the Palestinian shadow ministry of culture and cultural heritage. It was Riwaq, not the PA, that established the first national Palestinian biennial of art and architecture and used it to support heritage preservation in the West Bank. This points to the ingenious ways in which the organization has established quasi-national cultural institutions in a context where these either do not exist or do not function.

The early biennials are vivid instances of creative, experimental institutionalism. A few further examples help illuminate how Riwaq acts 'like a state' (Scott, 1998), though in distinctly inventive and collaborative ways. Using Swedish development funding and the help of hundreds of architecture students and professionals, it compiled the *Registry of Historic Buildings in Palestine* between 1994 and 2006 (Riwaq, 2006). (This is the only survey of its kind; the much less detailed PA inventory focused on archaeology.) Referred to as the 'National Registry' among Riwaq members, this comprehensive three-volume survey includes maps, photos, architectural, and historical information relating to approximately 50,000 Palestinian buildings and sites in the West Bank and Gaza that can be dated to before 1945. Rather unusually for a bureaucratic instrument, it is introduced by a poem.

Of the properties listed in this national registry, Riwaq prioritized what they saw as the fifty most significant and representative historic centers for rehabilitation in the years to follow. In this way, they effectively drew up a national plan for heritage conservation (though admittedly it was limited to Gaza and the West Bank for obvious reasons of accessibility). This is the '50 Villages' project, which Riwaq pushed forward through a variety of initiatives, including the biennials. Furthermore, in addition to compiling the national registry, Riwaq has played a major role in elaborating the new draft heritage legislation, which the PA ministry never approved because it deemed the proposed law too participatory and decentralizing (the ministry had its own law ratified but only in 2018). In the longstanding absence of a formal legal framework for the historic built environment and recent vernacular architecture, Riwaq managed to have some protections introduced through the planning system. What is more, it prepared heritage-led master plans for several West Bank towns and villages.[4] The national registry, '50 Villages' project, and these master plans show how in Palestine, supposedly 'nongovernmental' organizations, funded by international agencies, are conducting 'projects of legibility and simplification,' traditionally a key function of the state (Scott, 1998, pp. 9–83).

Riwaq collaborates with multiple actors across scales: From municipalities to local community associations, other Palestinian NGOs to international donors. Acting as a 'catalyst for ideas,' this NGO aims to 'spread the message' that Palestine's commons and historic environment must be cared for.[5] In this, they set an inspiring example for a kind of participatory urban regeneration that precipitates social change in Palestine and beyond. With community engagement at the core of its mission, Riwaq creates common spaces of self-support and possibility for marginalized communities. Therefore, the NGO targets local associations and disadvantaged groups within villages and small towns. Over the years, Riwaq has rehabilitated hundreds of historic buildings, largely houses and palaces from the Ottoman and Mamluk periods. Now marginalized groups can use them as social and cultural centers. But Riwaq also works closely with municipalities, within which it has trained personnel and established heritage protection units. Still, the organization's network reaches well beyond the local. It is by forging lasting partnerships with a diverse range of translocal and transnational actors, including other Palestinian organizations and international donors, which Riwaq has been able to significantly expand and magnify the scope of its activities.

What do Palestinians do in the absence of the state and a state infrastructure of the so-called cultural management? In a context in which state oversight of heritage preservation has long failed to protect or even record the recent vernacular past, NGOs and semi-governmental bodies have taken over the task, working largely independently of the PA. Indeed, Riwaq is far from being the only organization doing historic conservation and urban rehabilitation in Palestine. At the village level, it has collaborated with innumerable smaller organizations and community associations, which, if rather different from Riwaq in terms of the scale and scope of their activities, share its objectives of preserving Palestinian collective memory and identity as it is embodied in the built environment.

At the West Bank level, several large organizations focus on single cities, carrying out major schemes of urban regeneration in East Jerusalem, Bethlehem, and Hebron. Riwaq has often acted as consultant and coordinator for these organizations. All founded over the course of the 1990s, these NGOs share a recognizable common identity, discourse, practices, and formulas for development. They have restored large tracts of decaying, abandoned urban fabric, brought Palestinians back to live there under subsidized conditions (which include cheap rents, tax cuts, and free services), and implemented many social development activities. Restoring and repopulating Palestinian properties, these organizations have helped halt the expansion of Israeli settlements in old centers of cities such as Hebron and Jerusalem, which settlers have claimed as their own and partly expropriated. Given that the PA has no jurisdiction in these areas, organizations, such as the Hebron Rehabilitation Committee (HRC), have taken up some local municipal functions, providing key services to Palestinian residents (De Cesari, 2010, 2019a).

Riwaq is peculiar in how it both mobilizes art in the service of its agenda and acts as a mediator and translator linking multiple agencies. However, all of these organizations have far-reaching translocal and transnational connections and deep local roots. Their agency, their capacity to act effectively and introduce changes to people's lives, depends not only on their relations with donors, but on their profound knowledge of, and embeddedness in, local realities too. They often revitalize traditional customs and commoning practices, such as ancient forms of exchanging help and services. Most importantly, employees of the HRC or Riwaq often know residents and contractors by name and entertain personal relationships with them. In sharp contrast with the estranged indifference that characterizes Weberian state bureaucracies, these actors engage with citizens as humans and recognize their pain. This is true even of the organizations' directors. These NGOs have lived, intimate, and above all 'practical knowledge' of old cities. I mean this in James Scott's sense of an acquired intelligence that comes from the capacity to understand and respond promptly to a changing environment in a human, affective way, often from a position of closeness, familiarity, and confidence (1998, p. esp. 313ff). In the practice of the NGOs in question, this intimate form of knowledge is combined with the classic instruments of state bureaucracies (surveys, inventories, registers, and maps). As such, they can both engage personally and affectively with communities and draw on the kind of impersonal, standardized, and schematic knowledge through which states have governed territories and populations in modern times.

Sociologists and political scientists have criticized this combination of different techniques of governing the social as 'neopatrimonialism.' In blending bureaucratic rule with informal patron-client relationships, this argument runs, this mode of governance often reinforces structural inequalities (Banks and Thomas, 2010; Hinnebusch, 2015). In a less normative register, others scholars (anthropologists foremost among them) have pointed out that in adopting a 'family ethos,' social systems might well open up avenues for the poor to participate in civic projects and negotiate with authorities (Singerman, 1995; see also, e.g., Bayat, 2010). Certainly, inequalities and inequitable power relations endure in Palestinian villages and towns; the practices of Palestinian heritage organizations do sometimes reproduce these structures, if indirectly and unintentionally. Nonetheless, in a context in which infrastructure is lacking or rudimentary, this mixed mode of governing people, places, and things allows NGO to perform some form of 'state work.' Ultimately, this makes it possible to establish better living conditions for residents, the survival of the Palestinian socio-spatial fabric, and the reproduction of Palestinian life and livelihoods.

Networked associations: 'Cultural management' in Palestine

Over the years, a set of interlocking and continually shifting associations and partnerships both among Palestinian organizations, and between the NGOs and their donors, has produced an informal infrastructure akin to what elsewhere would be called cultural resource management. This infrastructure consists not of a fixed bureaucracy, but a loose lattice of provisional, contingent, and *ad hoc* achievements and alliances. This mode of cultural management is highly precarious; much more than stable institutions such as ministries and state heritage services, it is at risk of collapse and failure. Yet often, if unevenly, it allows for forms of experimentation. Gathering around matters of acute concern—Palestine's endangered landscape, culture, and livelihoods—NGOs and various cultural organizations join forces to realize national-scale events and institutions, even only for fleeting periods.

For example, many of these organizations have allied with each other to continue a Palestinian national biennial. These efforts build on varied early experiments, including the Riwaq biennial but also the Jerusalem Show, which the Al Ma'mal gallery and foundation for the arts and culture has run since 2007. Expanding on those precedents, a group of NGOs established Qalandiya International (QI), a larger-scale biennale that has taken place every two years since 2012. This is a diverse, transnational collection of organizations. It encompasses smaller art galleries and cultural community centers, such as Jerusalem's Al Hoash and Ramallah's Khalil Sakakini Center; much bigger NGOs and independent institutions, such as the Qattan Foundation and the Palestinian Museum; local municipalities; Palestinian cultural organizations in Gaza, Israel/1948, such as Haifa's Arab Cultural Association, and the diaspora, such as Beirut's Dar El-Nimer and Amman's Darat Al Funun, as well as agencies in Kuwait and London. In the biennial's mission statement, those who set up QI present it as 'an attempt to join forces and resources and form links across a fragmented geography—an innovative response to the need to find solutions that work for the collective rather than the individual institutions.'[6]

The choice of the name Qalandiya for the event is striking, for Qalandiya is probably the most infamous checkpoint separating East Jerusalem from Ramallah and the West Bank. Even though the UN recognizes East Jerusalem as occupied Palestinian territory, checkpoints such as Qalandiya block West Bank Palestinians from accessing the holy city. To visit this site, which is of paramount importance to them, Palestinians need Israeli permits that are very hard to come by. Qalandiya is also the location of a well-known refugee camp (also named Qalandiya), a historic village (which Riwaq is currently restoring), and Jerusalem airport. Now closed, this latter site had been envisioned as the Palestinian national airport in the early Oslo negotiations. Qalandiya, then, is a site of extended temporality, symbolically pointing toward both Palestinians' deep, indeed ancestral roots in the landscape and a potential future of self-determination. That future is now residual, perhaps, concealed as it is behind a rusty airport gate overgrown with weeds that stand for the interruption of the state-building phase inaugurated by the Oslo Accords. The latent possibilities lying behind that gate, however, might still be released.

Transnational alliances and the circulation of arts practice activate the energies dormant in Qalandiya's ruins and suspended futures. Here, Palestinian organizations together enact a kind of prefigurative politics, which involves 'experimenting with viable alter-presents—the shape of which can only be provisionally figured out in practice' (Rodríguez-Giralt, Marrero-Guillamón, and Milstein, 2018, p. 263). They make up what scholars in social movement studies, building on Latour, would call an 'actor-network' or 'assemblage.' That is to say, they constitute a 'precarious, contingent association between different actants' that remain diverse and autonomous, yet work together on the basis of horizontal collaboration and participatory democracy (Rodríguez-Giralt et al., 2018, p. 261). Working together, they enact practical, 'everyday utopias' (Cooper, 2013): Arrangements that exist in the present and yet open themselves up to future possibilities and imaginaries. Writing elsewhere in relation to Palestinian art, I have called it 'anticipatory representation' (De Cesari, 2012). This form of prefiguration entails an ongoing practice or set of experiments that both take place in the here and now and gesture toward an alternative, more just future. In this way, they open a clearing for possibilities and thinkabilities, instantiating 'pre-figurative lived models' of what the future could look like (Mundy, 2007).

The Palestinian Museum, which was inaugurated in 2016, mirrors this aggregate structure, despite being much more centralized. At least originally, its founders meant it to adopt a logic and aesthetic form that resembled QI. The largest Palestinian museum project ever begun; the museum was originally meant to commemorate the Nakba. It was initiated in the 1990s by the Welfare Association (WA), a major Palestinian NGO financed with expatriate Palestinian capital. Some of the most prominent Palestinian intellectuals served on its board. Back then, the PA was pursuing a parallel project to establish a memory museum. This collapsed, though, due to the onset of the Second Intifada and monumental costs that the project was expected to incur. Still, the Welfare project survived this and a series of other impasses. The museum would take on a new lease of life in the late-2000s. Since then art and culture have blossomed in nearby Ramallah, where a thriving, grassroots cultural scene has served as a counter-public sphere and a bastion of cultural resistance against both the occupation and the involution of Palestinian politics (Figure 13.4).

Growing through a process of institutionalization, the Palestinian Museum became an important player in this scene. In the 2000s, historian Beshara Doumani worked with philanthropist and filmmaker Omar al-Qattan to give shape to the museum. They developed a concept of a national museum that differed from the traditional monumental and monolithic format, which is 'built around a single chronological narrative.' Indeed, they wanted a 'mobilizing and interactive cultural project that can stitch together the fragmented Palestinian body politic by presenting a wide variety of narratives about the relationships of Palestinians to the land, to each other and to the wider world' (Doumani, 2010). In part, this approach made a virtue of necessity: Establishing a monumental museum with a major historical collection was and remains impossible under occupation, outside of the infrastructure of a sovereign state. Doumani and Al-Qattan, therefore, had to

Figure 13.4 Sculptural installation by Majdal Nateel, *Dream Is Possible*, in a restored historic building in Ramallah during Qalandiya International 2016

devise a different museum format befitting the Palestinian condition. They came up with an agile, cross-border institution made up of a 'hub' and many 'satellites' across Palestine and the diaspora, including several of the institutions involved in QI. A far cry from the traditional monumental site of memory presenting a hegemonic narrative, the diffuse structure of this model of museum resonates with what theorists of memory have described as 'noeuds de mémoire' or knots of memory (Rothberg, 2010). Omar al-Qattan has spoken explicitly about how this approach set out to counter established imaginations of the museum, as well as the initial plan for the Palestinian Museum as developed in the 1990s, which he criticized as 'conservative … and also very nationalist and patrician' (Al-Qattan, 2015).

The museum continues to work according to the hub-and-satellites model (its first exhibition took place in Beirut, for example). Nonetheless, the central institution does resemble other monumental museums elsewhere (Figure 13.5). Located next to Birzeit University in the hills north of Ramallah, the Palestinian Museum's main complex was eventually inaugurated in 2016. Over the years, it has developed into a full-fledged institution with its own bureaucracy. This come at the cost, perhaps, of being less agile, flexible, and inventive than was originally intended. Indeed, although the museum may have become a major player in Palestinian culture and heritage, some on its board have voiced concerns over the risks of bureaucratization. Still, amid these arguments over its form and function, mission, and program, the Palestinian Museum offers a

Figure 13.5 The Palestinian Museum

connective platform that links other Palestinian organizations and realities. Whatever its drawbacks, it holds out a space for experimenting with different cultural and exhibition formats and audience building.

Promises and pitfalls of informal institution building

A tension between institutionality and counterinstitutionality runs through the practices of the institutions set up by Palestinian cultural organizations over the last twenty years. As I have begun to suggest elsewhere, this tension may be the source of their promise (De Cesari, 2012, 2019b). They act as institutions and counterinstitutions simultaneously. They experiment with alternative organizational arrangements and forms of action that counter dominant global models, for they are propelled by an emancipatory ethos of resistance. These NGOs share the suspicion that traditional, major state institutions such as museums, with their embedded power relations and colonial heritage (e.g., Bennett, 1995), cannot but fail Palestinians and the powerless more broadly, for they were established precisely to reproduce the power of the state. At the same time, in the context of Palestine's weak, often failing, state institutions, civil society initiatives are usually the only game in town—the only functioning agencies in their domains. In the absence of state infrastructure, counterinstitutions then become institutions. They have managed to acquire some institutional stability, matching Western development aid with local and other, more controllable sources of funding. Still, both their own ethos and structures and the circumstances in which they operate have served to prevent them, at least so far, from ossifying into a 'stultified bureaucracy' (Al-Qattan, 2015). This ongoing dialectic of institutionalization and counterinstitutionalisation—that is, the push and pull of stabilization versus experimentation or disruption—shape Palestinian civil society initiatives.

Overall, the movement toward institutionalization is a source of much soul-searching. Initiatives becoming established, larger, more stable, and structured, and integrated into greater coordinated wholes provokes conflict and internal debate. The International Academy of Art Palestine is a good example of these tensions. As in the case of other Palestinian cultural institutions, this began as mobile art project. Initiated by artists, the academy offers training in the contemporary visual arts. One of the very few schools of its kind in the entire Middle East, much of its teaching and work reflects on the Palestinian political condition. As of 2017, the academy is an official faculty of Birzeit University, Palestine's main higher education establishment.

This merger was long in the making. The academy's founder and longtime director Khaled Hourani described how, having begun as an art project, the academy established itself as an institution.[7] This distinctive curve of development is fairly common in the Palestinian institutional landscape. The academy came into being during the Israeli reoccupation of the West Bank in 2002–2003. This was a time of heightened Israeli hostility toward—if not outright war upon—Palestinians and especially nascent Palestinian institutions, which the Israeli army was openly targeting. Against this backdrop, a group of important Palestinian artists set up mobile, experimental art workshops in a variety of places, including Gaza. In this way, they began 'experimenting with how to set up an [art] academy'. Necessarily, it had to be mobile so as to avoid attack and destruction. A Norwegian artist, who had been part of these efforts, then helped the group obtain Norwegian funding to bring the project to full fruition. Thanks to this funding, the academy officially opened its doors as a NGO in 2006. Over the following year, it transformed from being a project pursued by a grassroots group of artists to an institution, acquiring a building and offices, establishing a fixed organization and division of labor, and garnering external recognition. At first, the academy granted BA degrees in the contemporary visual arts under the aegis of the Norwegian Oslo National Academy of Arts. In the meantime, it was developing a curriculum that was both informed by global best practices and 'richly anchored' in Palestine.[8]

Between the academy's official opening in 2006 and its merger with Birzeit University in 2017, its members were embroiled in an internal debate about whether or not to seek accreditation with the Palestinian Ministry of Higher Education. Such accreditation was important in that it allowed the academy to give students official Palestinian degrees and continue paying employee salaries. In other words, it would confer stability and continuity on the organization. In the process of accreditation, however, the academy did not want to lose its own distinctive 'philosophy of … an alternative art institution,' that is, its freedom and 'flexibility.' 'To get certified you need a lot of procedures,' Khaled Hourani put it to me.[9] You need 'like … thousands of books in the library.' In becoming certified, he continued, 'we do not want to do the same as usual, our art academy is a platform for art dialogues, and we are trying to use all knowledge, all environment around us, even using streets as libraries and inviting farmers or policemen to talk to us.' The academy did not want to subordinate itself to the logic of state institutionalization and bureaucracy. In the end, it reached a compromise in the hope that it could maintain its alternative approach and philosophy. Although institutionalization secures survival, it entails risks too. Indeed, it can even threaten to disfigure or ossify a project's transformative energies. The Palestinian art academy resonates with what Davina Cooper has called 'activist statehood' or 'counterstates,' by which she means 'democratic, politically transformative formations that straddle and engage, even as they also oppose, formal state apparatuses' (Cooper, 2016, p. 454; see also 2016, 2017, 2019). Counterstates have features of both the state and social movements (namely an alternative philosophy and proposals for how societies might be better). Acting like social movements, projects such as the art academy enact the possibility of an emancipatory state in Palestine. They propose flexible and ultimately more feasible institutional formats that befit Palestinian conditions. Trajectories of institutionalization and counterinstitutionalization run in a variety of directions. Consider the Khalil Sakakini center, an important meeting ground for Palestinian culture and the arts, which took the opposite path to the International Academy of Art Palestine. Despite having been founded by the PA in the 1990s, it was soon transformed into an NGO because of the latter's chronic budget problems (Laïdi-Hanieh, 2006). As an NGO, it could take advantage of wider fundraising options. Regardless of this, NGOization did not solve the Sakakini's financial problems, which only worsened, reaching a crisis point in the mid-2010s. With international funding for Palestinian culture being cut across the board, the center's administrative costs were too high to be sustainable. It was at this juncture that artists on the Sakakini board decided to take a different tack, which would make the organization less dependent on international aid and more relevant at a local level.

The new iteration of the Sakakini had a less professionalized and more open and fluid organization, and was based on different principles of volunteering and crowdfunding. Yazan Khalil, an artist who served as its voluntary director during its transitional years, explained the challenges faced by the center (and indeed other cultural initiatives) in the following way:

> There is no more funding coming down, or it's very rare. And the cultural map—the funding and the grants—has changed, and the need for culture has changed as well. New and bigger institutions are opening like the [Palestinian] museum, like the new Qattan [Foundation] building. Qattan are shifting from being cultural funders to becoming producers of a huge public program.
>
> (Khalili, 2018)

Threatened with closure, the Sakakini transformed its practice in accordance with a new conception of public space and public service. It implemented a non-hierarchical organization, which moved beyond professionalization to involve different communities investing in it. In 'transforming our relationship with our audiences from passive to active—to "producive" audiences,' the center turned into an activating platform, in and through which a range of other individuals, artists, poets, groups, collectives, associations, initiatives, and institutions could collaborate and get their projects implemented (Khalili, 2018).

At the time of writing, the Sakakini functions as an energizing hub in a growing network. This is all thanks to what Khalili calls the 'Inreach' program:

> The 'Inreach' is a play on the outreach program. The outreach has been what we call the 'bread and butter' of most of the cultural NGOs in Palestine. It's all costly, all the NGO-ization of the cultural scene … we don't want to do any more huge projects outside the center. The Inreach that we do is at the center. … We no longer have to come up with, and fundraise for a project and then try to find audiences or producers for it. No. Every project we do is either one where we think of something and then we begin working with an individual or a collective to make it happen, or an individual or collective or whoever comes to the center and says, 'I have this project or this idea I want to do' and they want to do it with you. So in a way now all of our projects are in collaboration with our audience.
>
> (Khalili, 2018)

Clarifying the informality of arts management in this context, Khalili's discussion warrants quotation at length:

> if you want to do projects in the center, there is no big bureaucracy to do it. You don't need to apply. There's no written document … Sometimes you do a couple of things before you discover 'oh this idea wasn't good.' But what happens is that you become open to a huge spectrum of individuals and collectives that are thinking and working and looking for a space and looking for an institution to host them or to have them without having to invest so much time into administration and into bureaucracy to apply.
>
> (Khalili, 2018)

In this way, the artists on the board have precipitated a process whereby the Sakakini is becoming a place of dissensus (Stoler, 2018), a place in which active publics and counterpublics are fostered: A place, in short, of institutional experimentation.

As in other cases of informal institution building in Palestine, the stories of the Sakakini and the art academy are marked by resourcefulness and the mobilization of multiple cross-scalar connections. They are stories of crises been made to work to one's advantage, which speak of positive transformation and deep social inclusivity. The absence of an overarching framework of functioning rules in Palestine, however, poses a number of risks. Informality means freedom, flexibility, and horizontal organizations, certainly, but also often opacity and arbitrary procedures, which might conceal and reproduce asymmetric power relations. Still, I mean this only as a qualification to my emphasis on the potential of creative institutionalism: Many Palestinian institutional experiments are not sustaining social inequalities. Performed by artists, cultural workers, and grassroots groups amid a crumbling institutional landscape, such creative, experimental institutionalism might have implications and applications that point beyond Palestine.

Conclusion: Resilience and hope by way of creative institutionalism

Could Palestine serve as a model or at least an inspiring example of how to build social resilience elsewhere? A generalizable template for doing things otherwise? The problem of state failure, whether partial or full, is by no means limited to Palestine or other countries under conflict conditions. Forms of state failure can be observed in a range of contexts in which states exist but only under heavy strain. There are different declinations and forms: Arguably, the problem of state failure also affects 'strong states,' whether in the Global South or Global North. Indeed, it besets welfare and developmental states alike, both of which have massively failed in their foundational biopolitical promise to care for citizens, especially after neoliberal restructuring (Gupta, 2012). In building social resilience through experimental, participatory heritage institutions that heal and emancipate, could Palestinians be showing us one way in which people might respond to such failure?

State failure is no merely theoretical concern; nor is the problem limited to the Global South. The question of how to rethink the state in the wake of both the collapse or near collapse of many postcolonial developmental states in the Global South (Gupta, 2012) and neoliberal hollowing out in the Global North is of paramount significance for policy makers, NGOs, and actors on the ground (Cook and Muir, 2012; Cooper, 2019; Cooper, Dhawan, and Newman, 2019; De Cesari, 2020). Interestingly, some recent proposals for future states coming out of both critical theory and the most visionary policy circles share features with the Palestinian experiments I have described. These include the centrality of civil society; self-governing institutions; flexible forms of cooperative provision and 'co-creation,' sealed off from bureaucratic or market interference; and a recognition of the importance of lived experience and human relationships in government (Cook and Muir, 2012). For some, this means jettisoning clear or fixed visions of how the state should be in favor of embracing the creative, productive, and unforeseen qualities of processes of socio-institutional 'co-creation,' whose outcomes cannot be determined in advance (Stears, 2012).

Forced to cope with the volatile, stopgap temporality of conflict (and humanitarian aid and development), Palestinian heritage organizations have been deprived of possibilities for long-term thinking. These constraints have compelled them to 'develop a form of heritage and culture-based resilience that is focused on reinvigorating communities' capacities to care for each other and their lived environment, and to establish some control over spaces and conditions of everyday life (see Davis, 2012, p. 36). These organizations have come up with several creative solutions and experimental practices that gesture toward pluralist ways of rethinking the meaning of the state and cultural heritages, both in Palestine and beyond.

Notes

1. See https://www.btselem.org/settlements/statistics (accessed 27 November 2019). Since the 2007 split, Gaza has been administered separately by Hamas while the West Bank PA is run by Western-backed Fatah.
2. Introduction to the inaugural symposium, Venice, June 5, 2009. All quotations in this section come from my audio-recorded notes of the symposium.
3. See note 2 above.
4. Working with the Ramallah Municipality and Hebron Rehabilitation Committee, Riwaq has developed conservation master plans for Ramallah and Hebron respectively. Sixteen protection plans have been prepared for smaller towns and villages in cooperation with the various municipalities.
5. Riwaq has received numerous prestigious international awards for its work, including the Aga Khan Award for Architecture, the Netherland's Prince Claus Award, and UN-Habitat for the Best Worldwide Architectural Practices.
6. https://www.qalandiyainternational.org/why-qalandiya (accessed 17 December 2019).
7. The narrative that unfolds in this paragraph is largely based on the interview that I conducted with Khaled Hourani, then director of the academy, in Ramallah on September 29, 2011. If not otherwise noted, all direct quotes of the paragraph come from this interview.
8. See https://electronicintifada.net/content/launch-international-academy-art-palestine/687, accessed 17 January 2017.
9. See note 7 above.

References

Abu el-Haj, N. (2001). *Facts on the Ground: Archaeological Practice and Territorial Self-Fashioning in Israeli Society*. Chicago: University of Chicago Press.

al-Jubeh, N. (2006). Ay Qanun li-l-Turath al-Thaqafi Nurid? [What Kind of Cultural Heritage Law Do We Want?] Paper presented at the Conservation of Cultural Heritage Conference in Palestine, Jericho, February 20–22.

Al-Qattan, O. (2015). Full Interview between Omar Al-Qattan and Yasmine Eid-Sabbagh. Qattanfoundation.org, 31 March, http://qattanfoundation.org/en/qattan/media/news/full-interview-between-omar-al-qattan-yasmine-eid-sabbagh (accessed 17 April 2020).

Banks, A. and Thomas R. (2010). Neopatrimonialism in the Middle East and North Africa: Overview, Critique and Alternative Conceptualization. Paper presented at the workshop "'Neopatrimonialism in Various World Regions," GIGA

German Institute of Global and Area Studies, Hamburg, August 23. Available at: https://www.researchgate.net/publication/258325694_Neopatrimonialism_in_the_Middle_East_and_North_Africa_Overview_Critique_and_Alternative_Conceptualization (accessed December 30, 2019).

Bayat, A. (2010). *Life as Politics: How Ordinary People Change the Middle East.* Amsterdam: Amsterdam University Press.

Bennett, T. (1995). *The Birth of the Museum: History, Theory, Politics.* London: Routledge.

Bshara, K. (2015). Rural Urbanization: The Commodification of Land in Post-Oslo Palestine. In K. Bshara and S. Amiry (eds.), *Reclaiming Space: The 50 Village Project in Rural Palestine.* Ramallah: Riwaq, 93–103.

Cooke, G. and Muir, R. (eds.). (2012). *The Relational State: How Recognising the Importance of Human Relationships could Revolutionise the Role of the State.* London: Institute for Public Policy Research (IPPR).

Cooper, D. (2013). *Everyday Utopias: The Conceptual Life of Promising Spaces.* Durham, NC: Duke University Press.

Cooper, D. (2016). Enacting Counter-States Through Play. *Contemporary Political Theory* 15(4): 453–461.

Cooper, D. (2017). Prefiguring the State. *Antipode* 49(2): 335–356.

Cooper, D. (2019). *Feeling like a State.* Durham: Duke University Press.

Cooper, D., Dhawan, N., and Newman, J., (eds.). (2019). *Reimagining the State: Theoretical Challenges and Transformative Possibilities.* London: Routledge.

Daher, R., and Maffi, I. (2014). Introduction. In R. Daher and I. Maffi (eds.), *The Politics and Practices of Cultural Heritage in the Middle East*, London: Tauris, 1–52.

Davis, D. E. (2012). Urban Resilience in Situations of Chronic Violence. Final Report, May 2012. MIT https://www.wilsoncenter.org/sites/default/files/URVC%20Report.pdf

Davis, D. E. (2017). Informality and State Theory: Some Concluding Remarks. *Current Sociology Monograph 2017* 65(2): 315–324.

De Cesari, C. (2010). Creative Heritage: Palestinian Heritage NGOs and Defiant Arts of Government. *American Anthropologist* 112(4): 625–637.

De Cesari, C. (2012). Anticipatory Representation: Building the Palestinian Nation(-State) through Artistic Performance. *Studies in Ethnicity and Nationalism* 12(1): 82–100.

De Cesari, C. (2019a). *Heritage and the Cultural Struggle for Palestine.* Stanford: Stanford University Press.

De Cesari, C. (2019b). Anticipatory Representation: Thinking Art and Museums as Platforms of Resourceful Statecraft. In D. Cooper, N. Dhawan, and J. Newman (eds.), *Reimagining the State: Theoretical Challenges and Transformative Possibilities*, London: Routledge, 153–170.

De Cesari, C. (2020). Heritage beyond the Nation-State? Nongovernmental Organizations, Changing Cultural Policies, and the Discourse of Heritage as Development. *Current Anthropology* 61/1, to appear in February 2020.

Doumani, B. (2010). "A Post-Territorial Museum: Interview with Beshara Doumani." Interview by Ursula Biemann. *Arte East Quarterly* (Spring). https://arteeast.org/quarterly/a-post-territorial-museum/?issues_season=spring&issues_year=2010 (accessed 17 April 2020).

Ferguson, J. (2004). Power Topographies. In D. Nugent and J. Vincent *(eds.), A Companion to the Anthropology of Politics.* Malden, MA: Blackwell, 383–399.

Ferguson, J., and Gupta, A. (2002). Spatializing States: Toward an Ethnography of Neoliberal Governmentality. *American Ethnologist* 29(4): 981–1002.

Gell, A. (1992). The Technology of Enchantment and the Enchantment of Technology. In J. Coote, and A. Shelton (eds.), *Anthropology, Art and Aesthetics*. Oxford: Clarendon, 40–63.

Gupta, A. (2012). *Red Tape: Bureaucracy, Structural Violence, and Poverty in India*. Durham, NC: Duke University Press.

Gupta, A. (2013). Messy Bureaucracies. *HAU: Journal of Ethnographic Theory* 3(3): 435–440.

Gupta, A. (2015). State, Corruption, Postcoloniality: A Conversation with Akhil Gupta on the 20th Anniversary of 'Blurred Boundaries.' Interview by Nugent, D., and Sreenath, S., *American Ethnologist* 42(4): 581–591.

Gupta, A., and Sharma, A. (2006). Globalization and Postcolonial States. *Current Anthropology* 47(2): 277–307.

Hammami, R. (2006). Palestinian NGOs Since Oslo: From Politics to Social Movements? In J. Beinin and R. Stein (eds.), *The Struggle for Sovereignty: Palestine and Israel, 1993–2005*. Stanford, CA: Stanford University Press, 84–93.

Hinnebusch, R. (2015). Introduction: Understanding the Consequences of the Arab Uprisings—Starting Points and Divergent Trajectories. *Democratization* 22(2): 205–217.

Jessop, B. (2008). The State and State-Building. In S. A. Binder, R. A. W. Rhodes, and B. A. Rockman (eds.), *The Oxford Handbook of Political Institutions* Oxford: Oxford University Press, 111–130, http://www.oxfordhandbooks.com/view/10.1093/oxfordhb/9780199548460.001.0001/oxfordhb-9780199548460-e-7.

Kersel, M. (2015). Fractured Oversight: The ABCs of Cultural Heritage in Palestine After the Oslo Accords. *Journal of Social Archaeology* 15(1): 24–44.

Khalili, Y. (2018). Cultivating Creative Spaces in Palestine. Interviewed by Crasnow, S. *FIELD* 9(winter 2018). http://field-journal.com/issue-9/an-interview-with-yazan-khalili

Laïdi-Hanieh, A. (2006). Arts, Identity, and Survival: Building Cultural Practices in Palestine. *Journal of Palestine Studies* 35: 28–43.

Le More, A. (2008). *International Assistance to the Palestinians after Oslo: Political Guilt, Wasted Money*. London: Routledge.

Maniscalco, F. (2005). La legge n. 51/1929 e la tutela del patrimonio archeologico palestinese. In F. Maniscalco (ed.), *Tutela, conservazione e valorizzazione del patrimonio culturale della Palestina*. Naples: Massa Editore, 43–45.

Masalha, N. (2007). *The Bible and Zionism: Invented Traditions, Archaeology and Post-Colonialism in Israel-Palestine*. London: Zed.

Migdal, J. (2001). *State in Society: Studying How States and Societies Transform and Constitute One Another*. Cambridge: Cambridge University Press.

Mitchell, T. (1991). The Limits of the State: Beyond Statist Approaches and Their Critics. *American Political Science Review* 85: 77–96.

Mundy, J. (2007). Performing the Nation, Pre-figuring the State: The Western Saharan Refugees, Thirty Years Later. *Journal of Modern African Studies* 45(2): 275–297.

Riwaq. (2006). *Riwaq's Registry of Historic Buildings in Palestine*. Ramallah: Riwaq.

Rodríguez-Giralt, I., Marrero-Guillamón, I., and Milstein, D. (2018). Reassembling Activism, Activating Assemblages: An Introduction. *Social Movement Studies* 17(3): 257–268.

Rothberg, M. (2010). Between Memory and Memory: From Lieux de mémoire to Noeuds de mémoire. *Yale French Studies*, No. 118/119, Noeuds de mémoire: Multidirectional Memory in Postwar French and Francophone Culture: 3–12.

Roy, A. (2012). Urban Informality: The Production of Space and Practice of Planning. In R. Crane and R. Weber (eds.), *The Oxford Handbook of Urban Planning*, Oxford University Press. https://www.oxfordhandbooks.com/view/10.1093/oxfordhb/9780195374995.001.0001/oxfordhb-9780195374995-e-33

Scott, J. C. (1998). *Seeing Like a State*. New Haven, CT: Yale University Press.

Singerman, D. (1995). *Avenues of Participation: Family, Politics, and Networks in Urban Quarters of Cairo*. Princeton, NJ: Princeton University Press.

Stamatopoulou-Robbins, S. (2019). *Waste Siege: The Life of Infrastructure in Palestine*. Stanford: Stanford University Press.

Stears, M. (2012). The Case for a State that Supports Relationships, not a Relational State. In G. Cooke and R. Muir (eds.), *The Relational State: How Recognising the Importance of Human Relationships could Revolutionise the Role of the State*. London: Institute for Public Policy Research (IPPR), 35–44.

Stoler, A. L. (2018). On Archiving as Dissensus. *Comparative Studies of South Asia, Africa and the Middle East* 38(1): 43–56.

Stoughton, I. (2018). Palestinian Artist Khalil Rabah Explores 'Defeated Geographies'. *The National*, 26 February. https://www.thenational.ae/arts-culture/art/palestinian-artist-khalil-rabah-explores-defeated-geographies-1.707936

Whitelam, K. W. (1996). *The Invention of Ancient Israel: The Silencing of Palestinian History*. London: Routledge.

Yiftachel, O. (2012). Between Colonialism and Ethnocracy: 'Creeping Apartheid' in Israel/Palestine. In N. Jeena (ed.), *Pretending Democracy: Israel, an Ethnocratic State*. Pretoria: Afro-Middle East Centre AMEC, 95–113.

14 Souls of homes

Heritage as a manifestation of community relationships through space and time

Eva Ziedan

Paradigm for community-based living heritage

Important developments and constructive debates have taken place regarding the World Heritage Convention in 1972, in particular the Operational Guidelines for the Implementation of the World Heritage Convention, the Global Strategy for a Balanced, Credible and Representative World Heritage List in 1994, the 1994 Nara Conference and Document on Authenticity, the Intangible Heritage Convention in 2003, and their following revisions and updates.

However, the dynamics that accompanied these developments can be considered similar—they were top-down texts with differences in interpretation by member states, development of guidelines for explanation, and periodical updates and adaption. Some of these adaptations caused radical changes in the original texts, such as the Nara Document (which had an important impact on the European definition and preservation of heritage), and sometimes by changing certain terms, such as the cultural heritage criteria of "a unique artistic achievement" that "represent[s] a masterpiece of the human creative genius" in the 1994 Global Strategy text (Labadi, 2013, p. 60).

Despite all these efforts, there is still a gap between what UNESCO officially aims to achieve and what is practised on the ground, whether in the definition of heritage, its conservation, its listing criteria, or its connection to peace and sustainable development. According to Irina Bokova (2012), UNESCO Director General: "World Heritage is a building block for peace and sustainable development. It is a source of identity and dignity for local communities, a wellspring of knowledge and strength to be shared" (Bokova, 2012). In fact, the use of bottom-up terms is indicative of a participatory approach, in contexts that *de facto* use top-down practices that create confusion and ambiguity in understanding these texts.

According to Labadi, the nomination dossiers "include strategies employed by non-Europeans for coping with European ways of representing and interpreting the framework of the Convention" (Labadi, 2013, p. 21), and "People undertake to describe themselves in ways that engage with representations others have made of them" (Pratt, 1991, p. 35). In other words, this has a direct impact on societies identifying their heritage and, therefore, their identities and their culture became shaped inside imported boxes.

In this section of this chapter, I present another perspective on the concept of heritage, one that is rooted in grassroots practices by local communities as a form of resilience during times of crises. Thus, I propose here to adopt the expression of "living heritage." This living heritage is manifested by values and practices that ultimately produce the lifestyle and behaviour of local communities (David, 2004; Al-Kazzaz, 2010). Considering this definition, I focus on the dynamism of living heritage and its ability to adapt and develop itself.

This research for this chapter is in fact from the ground up and is based on direct observation and interaction with those active in local communities in Syria before and during the conflict. Local communities—and we focus here on the concept of "local" because they are not restricted to the elite and intellectual class, and not necessarily aim to work towards national objectives—they attempt to fill the vacuum created by the absence of state services. Accordingly, even during all its catastrophic effects, the conflict gives us an important opportunity to observe community behaviour. When some areas have limited access to State services, especially in countries with high centralization, like Syria. Here, people find themselves encouraged to use all their power to activate community resources and the sedimented legacy-based knowledge. This to find solutions to problems that directly threaten their lives. The viable solutions will then be collectively owned and used under collective responsibility. In fact, it is about interests and the satisfaction of local needs, without national or global dimensions. This entire process needs to evolve and change within the community itself. It is a societal behaviour that this chapter calls "living heritage" (Figure 14.1).

Case studies

These cases are distributed across four different contexts (Figure 14.2): a rural area that is part of a territory that has always been dependent on agriculture (the Ghab Plain); a historical city which has remained inhabited (Aleppo(; a city that has been re-settled (Raqqa(; and a rural area that has been re-settled (Tell Hasan Basha). Each location has its conditions; however, there are vivid commonalities portrayed in the concept of living heritage presented here: human needs, common interests, inherited legacies, social energy, and solidarity. This research is largely based on fieldwork and engagement by the author. Data was collected over six years based on observations of the continuous and collective work by local communities across Syria, in-depth interviews conducted with key actors in these four areas such as peace negotiators, professionals, historians, and youth activists in the civic associations.

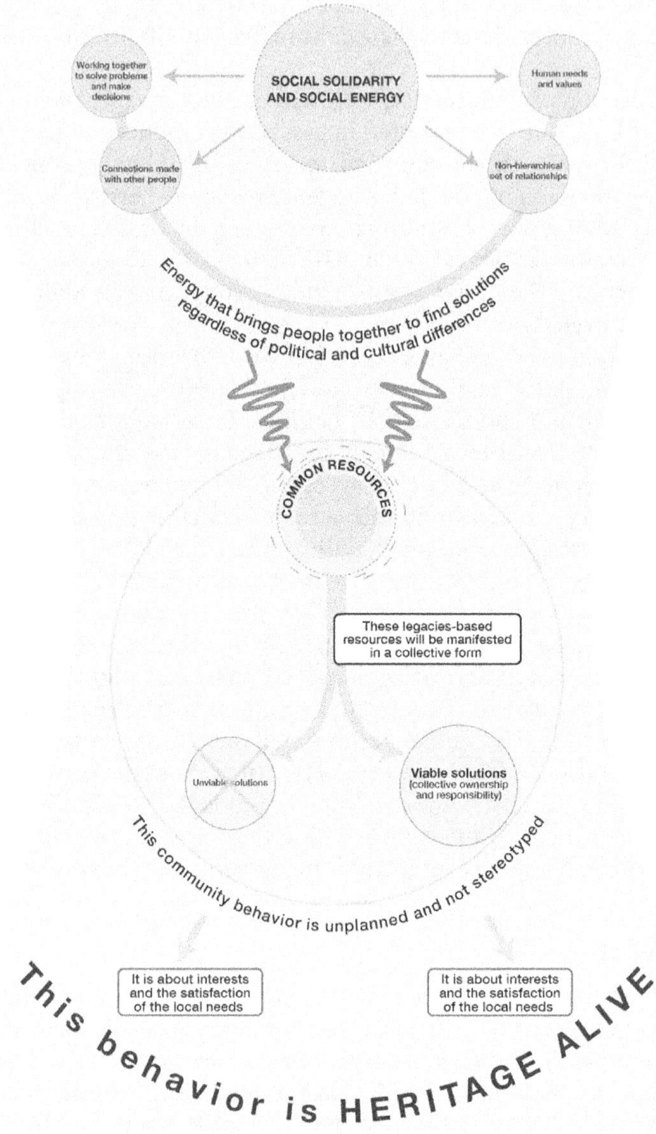

Figure 14.1 Community-based living heritage paradigm

Al Ghab plain

The Ghab plain is a fertile valley located in the middle Orontes basin, bordered by the Coastal Mountain Range to the west and Jabal al-Zawiya to the east and extends from Jisr al-Shughur in the north to the Masyaf to the

Figure 14.2 Location of the four case studies

south. It is 60–70 km long and 12–15 km wide (Al-Dbiyat, 2013, as mentioned in Dussaud, 2013). Archaeological studies have shown that for thousands of years, the area of the Orontes basin has been subject to intensive agricultural activity and, therefore, operations related to water management. Many sites flourished near the rivers, such as Tell An Nabi Mindu, Qatna, Homs, Hama, Tell Asharineh, Apamea, and Qalaat al–Madiq. As indicated in the map in Figure 14.3, during the Roman–Byzantine period, the area between Apamea, Rastan, and Salamiyeh had intensive and varied water supply systems such as irrigation channels,[1] qanats, and norias (waterwheels). Some of these ancient irrigation facilities have been rehabilitated and reused, while others have been discontinued (e.g. norias). Modern irrigation devices have also emerged (motor pumps) (Figure 14.4). Most of the lands in the case study area, al Ghab Plain, was covered by water until 1951 when the management of the al Ghab Project Institution started the draining, reclamation, and investment in the plain (Chehadat, 2013, as mentioned in Dussaud, 2013). Until now, water and its management have been

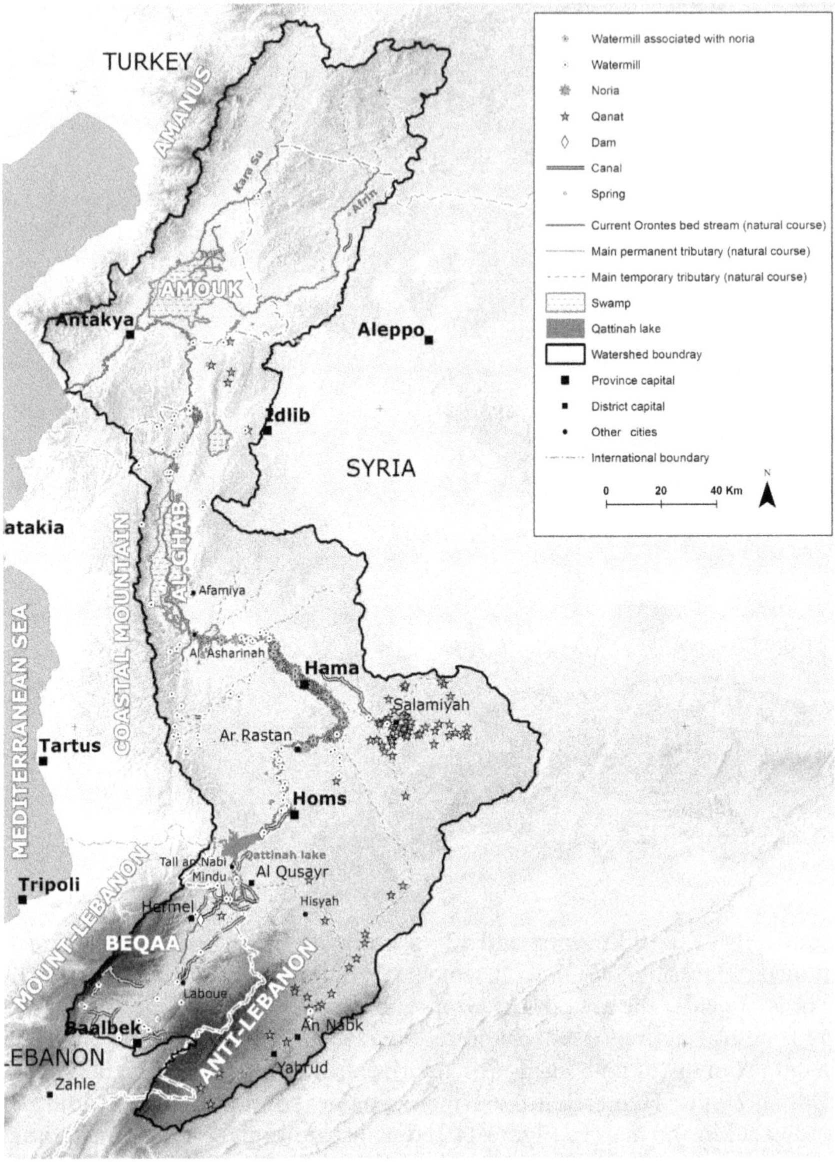

Figure 14.3 The ancient hydraulic structures in the Orontes River Basin

a central issue for the communities living in the area. They mainly relied on incomes generated by seasonal harvests; the Orontes basin, prior to the conflict, generated about 25 per cent of the total agricultural production in Syria (Haj Asaad and Shamaly, 2016). The towns and villages in the al Ghab area are historically characterized by a complex communal mixture

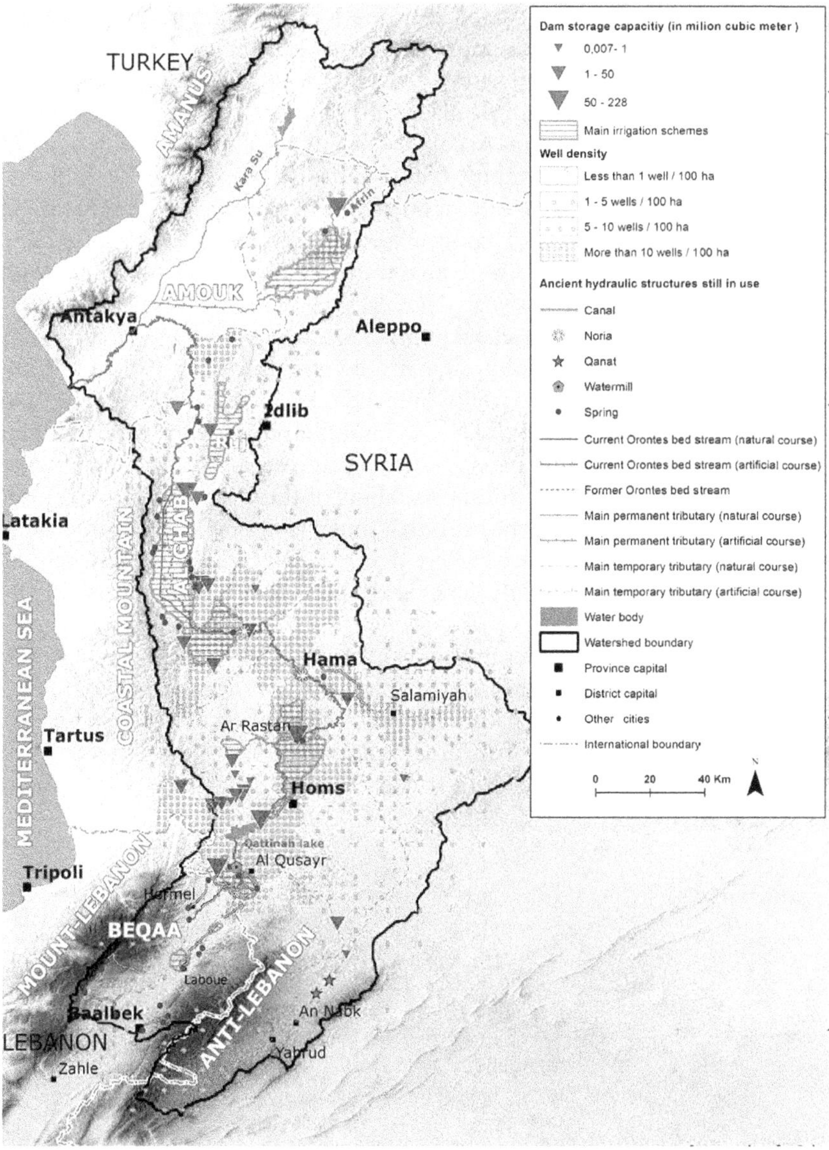

Figure 14.4 The contemporary hydraulic structures in the Orontes River Basin

of Alawites, Sunnis, Christians, and Murshidis. For centuries, the inhabitants have been sharing the same socio-economic environment.

However, as in many cases in other areas in Syria, deep roots have fuelled the on-going conflict, including economic factors: in 2010, most agricultural workers and their families suffered from poverty—the percentage of poverty in the agricultural sector is 72 per cent (Syrian Center for Policy Research, 2019).

The conflict in Syria impacted the area further. Starting from autumn 2012 the region was divided under the control of the Syrian central government and the armed opposition group; "About 1.6 million people from the Orontes basin have been displaced and have found refuge in Syria or neighbouring countries" (Haj Asaad and Shamaly, 2016, p. 12). The conflict border cut the irrigation channels causing drought due to the lack of fuel and electricity to power the water pumps, although it was known that these canals are the vital nerves that connect and sustain these areas (Figure 14.5).

In 2017, voluntary groups were able to negotiate and resolve issues with the parties in conflict for many civil issues, such as kidnapped persons, electricity disputes, robbery at checkpoints, and mutually important aspects of daily life. These informal groups are composed of key people in both opposition-held and government-controlled areas, some of whom could move between the two areas.[2] During my research, three main categories emerged in these voluntary groups: former or present teachers, drivers, and employees in government institutions. Many of them were able to carry out their work in the government-held area and then come back on the same day to the opposition-held area where they live. Officially, these are the only people who can pass through the checkpoints, with the exceptions of the various warlords.

They used to work together in public institutions such as the municipality. With the eruption of the conflict in 2011, members of the municipality were divided among local councils in opposition-held areas and others in government-held areas. Some of them stayed, however, in opposition to each other. Some of these people were trusted by their respective communities

Figure 14.5 An irrigation channel before maintenance works, Al Ghab Plain

and their colleagues at the local institutions. They were able to obtain permits from the local authorities (in both government- and opposition-held areas) to clean some of the irrigation channels in the contested areas, maintain the operational water pumps, and clear landfills across the conflict line (Figure 14.6). The common interest in the cleaning of these channels is the need to maintain and restore them so that irrigation water may be available to citizens in both areas that rely on agriculture. They tried to mobilize all the local heritage resources they share, including the archaeological site of Apamea (close to the conflict border) to attract attention and the interest of the Directorate General of Antiquities and Museums (DGAM) in facilitating the cleaning of the site. The positive results of the rehabilitation of one channel and the Apamea landfill led to other successful endeavours, an "upward spiral and success builds on success" (Flora et al., 2015, p. 21). People realized the importance of cultural and heritage resources, which are parts of their daily behaviours. The absence of state institutions made it possible for these communities to change the pattern of their relationship with local authorities. The relationship no longer involved central control; there was a need, and an opportunity, to make these institutions operate in other ways to allow communities to make decisions. However, it is crucial to emphasize that people did not intervene to protect the channels because of their heritage significance, but instead to protect their living practices. This presents an opportunity to rethink ways to protect heritage, which cannot be separated from development and socio-economic needs. The indirect achievement was facilitated by the temporary reduction of hostility between the parties in conflict during the execution of the work. This activity could

Figure 14.6 An irrigation channel after maintenance works, Al Ghab Plain

have significant repercussions on trade exchange and mitigate the effects of a war economy resulting from military barriers that, in case of a prolonged cease-fire, have no reason to exist. However, the continuation of the armed conflict, especially the escalation in 2018 which was directly linked to international and regional pressures, did not always allow for a follow-up of the results of this type of local negotiation based on common needs and socio-economic relations—in this case, living heritage (Figure 14.7).

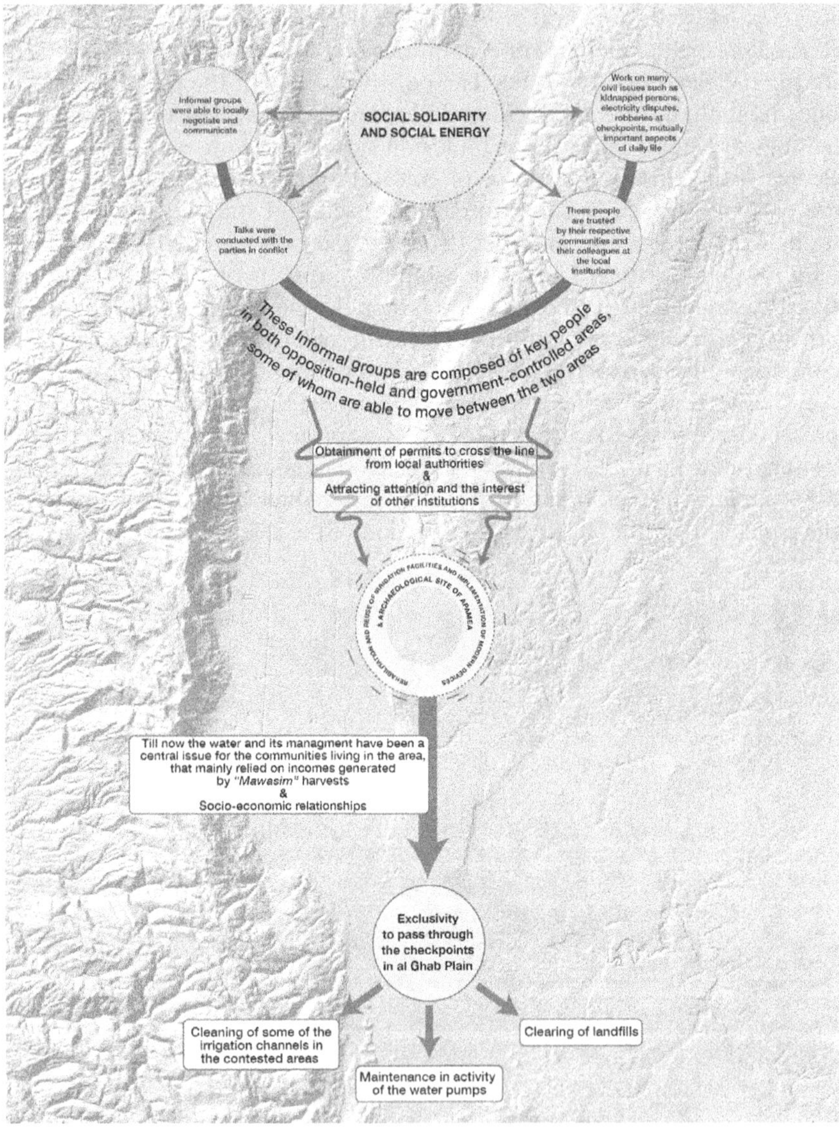

Figure 14.7 Al Ghab community-based living heritage

Aleppo

Aleppo is one of the oldest continuously inhabited cities in the world. For thousands of years at the crossroad of trade routes between Asia and the Mediterranean, Anatolia, and the Arabic Peninsula, Aleppo has been for decades one of the most significant industrial and commercial hubs of the entire Middle East. Nowadays, located in the northern region of Syria, the city is close to the Turkish frontier. Aleppo and its hinterland lie at the fringes of the Euphrates River to the east and the Orontes valley to the west, embracing the various urban-rural, mountainous, desert, and fluvial characteristics of central and northern Syria. It has been on the list of UNESCO heritage sites in Syria since 1986.

Many history books describe what wars and earthquakes have done to the city of Aleppo over hundreds of years, and how its people have been able to resist, restore, and renew their way of life over time (Al-Ghazzi, 1926; Sbeinati et al., 2005). The rules for the reconstruction and preservation were part of the Islamic traditions, customs (*'urf*), and values. However, the western concepts of heritage and authenticity have constrained the concept of living and its natural texture and restricted the freedom of people to deal with the surrounding space, that is, their everyday practices. Most importantly, these modern conservation practices came with top-down written values that have essentially different meanings than the same term in the Islamic context (Mahdy, 2019). For example, the *sabat* (one of the distinctive elements of Islamic architecture[3]) often expressed the integration of private rights and public rights, when there was a need to enlarge a house without harming the public right to use of a road. This traditional architecture reflected both needs and solidarity between people. Isolating one factor (architecture) from the other (needs and solidarity) will empty the traditional architecture of its dynamic and living criteria. Hence, the relations between these two factors, which identify the life pattern produced in this city, are not reflected in either state urban restrictions or in "outstanding values." Here, social energy and social solidarity between communities play an important role. These two terms were defined by Nembhard (2016):

> It's not just the money put in or the hours worked but, more importantly, it is about this energy—the connections made with other people, how we work together to solve problems, make decisions, and how and when we take leadership." "By solidarity system, I mean a non-hierarchical, non-exploitative, equitable set of economic relationships and activities geared toward the grassroots—that's of the people (people before profit), indigenous, participatory, based on human needs, humane values, and ecological sustainability.

In other words, it is the energy that brings people together in their shared space to find and create solutions in the face of political, social, and cultural differences.

The following example focuses on the importance of the solidarity system and social energy to living heritage. In 2012, the line of conflict passed among houses and markets, destroying two-thirds of the city of Aleppo and dividing it into two parts.[4] The governorate of Aleppo reported the highest rate of displacement within and outside Syria, with 1,632,838 fleeing outside Syria, 1,792,581 fleeing inside Syria (World Bank, 2017), and 34,487 persons killed (Figure 14.8) (Violation Documentation Center, 2019). The fighting has led to the destruction of infrastructure and all utilities and service stations; the most prominent states included the Al Khafsa station that provided water to Aleppo and was formerly under the control of the Islamic State. The entire eastern and western parts of the city were cut off, causing serious problems for residents who received only small quantities of water once every 10 days, and had to find alternative modes of supply.

Informal key actors, including the Syrian Arab Red Crescent (SARC) and The International Committee of the Red Cross (ICRC), and the initiative of the people of Aleppo *Mubadarat Ahali Halab*,[5] alongside other official and non-official associations, went into negotiations first with the opposing parties to get them to steer the battle away from areas with water pumps, working across city lines to repair water pipes and pumping stations, and then suggesting a group initiative to manage local resources in a sustainable manner.[6] The solution required digging wells across

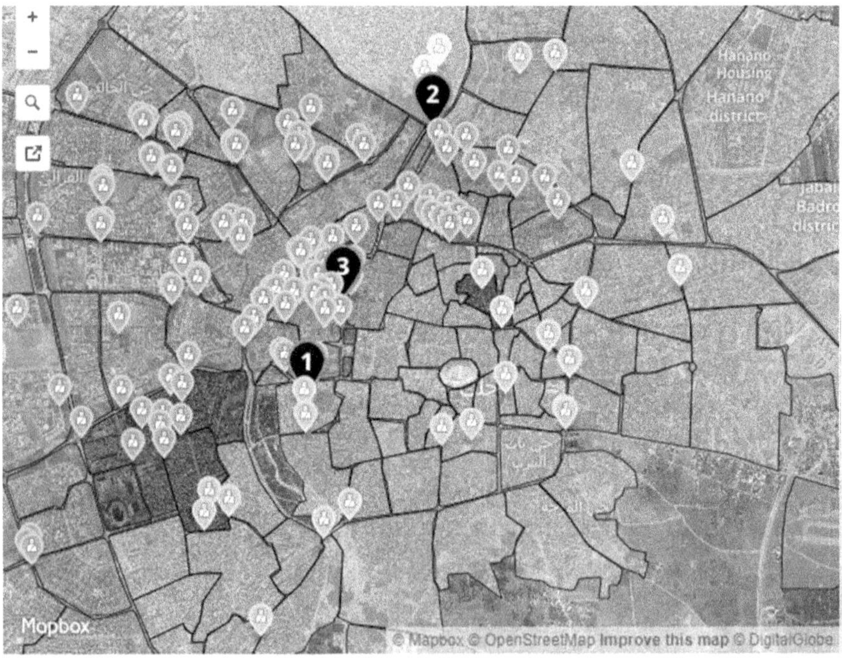

Figure 14.8 Military checkpoints, Aleppo 2013

the area of Aleppo to cover the needs of the residents and link the entire area to a water supply source. Restrictions and special conditions were set for digging the wells in public areas such as schools, public gardens, and government buildings to preserve and make sure they would not be monopolized by private groups. The funding came from some local CSOs, wealthy families, and Alepine diaspora, while the monitoring was done by the local community. Key actors were responsible for facilitating and negotiating the permits of government institutions (the Irrigation Institution because it is related to water provision, the Education Directorate because some wells were in schools, the General Organization for Housing and the governorate of Aleppo to supply excavators, etc.). More than 293 social media networks have been mapped in 2015 (Dayoub, 2015),[7] one of which is Dubara, which published the location of the wells and water-distribution points, asking its followers to take water to families that are not able to reach these locations. The social energy and the solidarity system are seen in their caring, in their enthusiasm and non-hierarchical spaces, and trust, allowing them to make decisions, take leadership, and innovate solutions to meet their needs (Figure 14.9). This posed two critical questions: how can conservation and reconstruction practices embrace and build on this energy and solidarity? How can we change the fact that many people in the half a century in Aleppo have *de facto* identified heritage as *athar* (antiquities) (Omar Abdul Aziz Hallaj quoted in Lamprakos, 2015, p. 236) and "moving away from the past and observing it" (David, 2004, p. 485).

Raqqa

> He died in the early 1940s after he created the foundations for the clan and the city together; for the clan through the construction of the Oda and the city through the construction of the main mosque, the main market and the municipality.
>
> (Al-Hassoun, 2017)

His name is Khalaf Al-Hassoun, one of the three brothers who came to Raqqa and in the 1860s re-established the city that we know today. The city, located around 200 kilometres east of Aleppo, in the northeast bank of the Euphrates, was re-established by the Ottomans in the second half of the eighteenth century as a garrison built in the south-east corner of the ancient Abbasid walls (Ababsa, 2009). The first thing Hassoun built was the *Oda* (Guests Room), also named *Madafa* (Ababsa, 2001), which became an essential part of every Arab house from the clans that inhabited the emerging city of Raqqa. In these structures, the inhabitants of Raqqa have been practising *Taalila*; *Taaliala* from the Arabic verb *'al-la-la* that means "imbibe"[8] during a gathering. Indeed, *Taalila* is a ritual that is followed at weddings and gatherings such as the wake after a death where

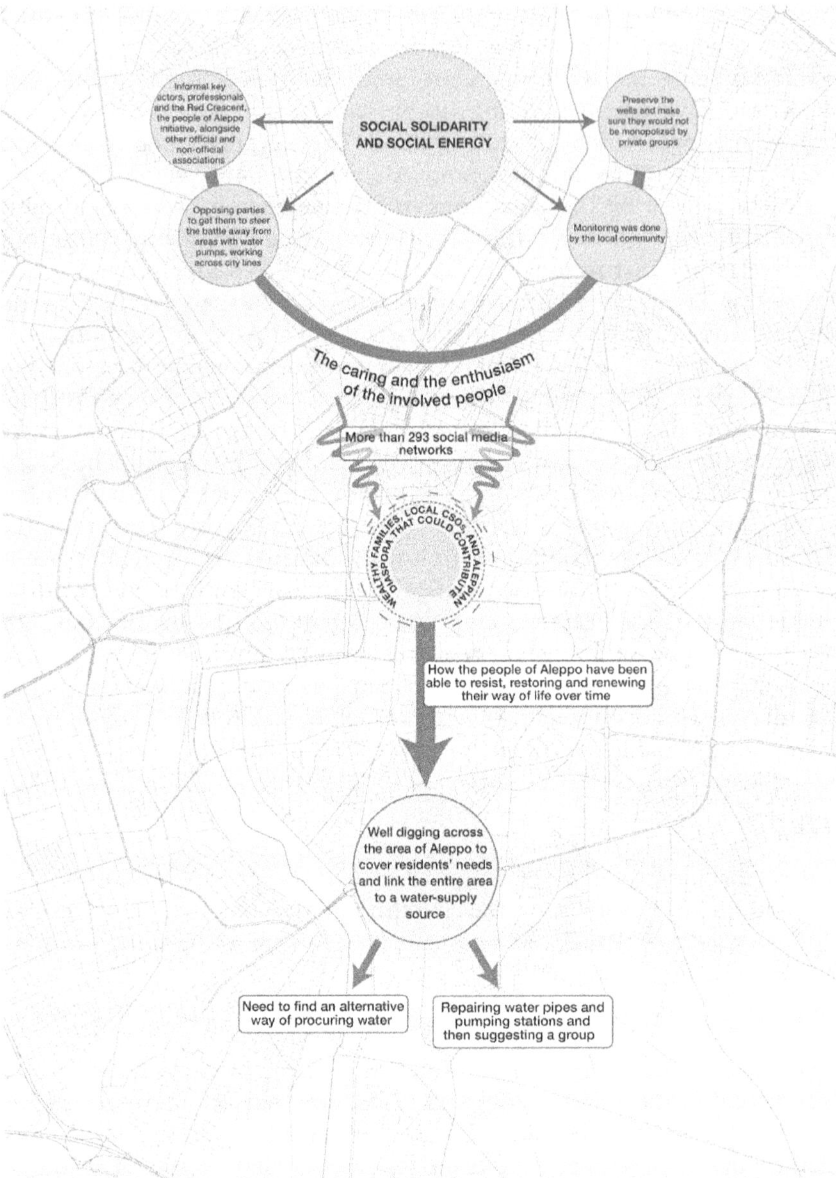

Figure 14.9 Aleppo community-based living heritage

the inhabitants and elders of an area come together. In Raqqa, *Taalila* is also a gathering to resolve issues faced by the community.[9] An example of that is when the intellectuals of Raqqa came together to resolve the problem of people from Deir az Zawr mistreating the students of Raqqa in 1953; there were no middle schools in Raqqa and students had to commute to

Deir az Zawr. During the gathering, the elites of the area decided to gather funds and open a school that they managed to get accredited.[10]

In Raqqa, the question "where is the *Taalila* tonight?" is a normal among people. Any well-off family can call people to their house where drinks, sweets, or sometimes dinner are served, especially on summer and long winter nights. "*Taalila* is a small parliament and an inherited tradition of pre-Islam," says Hassoun.[11] Before the conflict, *Taalila* was limited to representatives from the neighbourhoods[12] and those from neighbouring villages and passer-bys. It is about all kinds of public affairs—harvests, water from the Euphrates, marriage, social problems, poetry nights, and music (Hajj Saleh, 2017). As in the cases of the al-Ghab Plain and Aleppo, the conflict has shed light on the importance of this practice that has been used by the inhabitants of Raqqa to cope with the war. In the great battle of Raqqa in 2017, many inhabitants who had not previously left the city, even after it fell under the control of the Islamic State, had the chance to see their friends and relatives who had previously fled to the countryside.[13] After their reunion, they organized many meetings and *Taalila* that tackled the issue of the return to the city as soon as the shelling ceased. On 23 October, the shelling stopped, and large groups of people gathered and made their way back to the city of Raqqa. They were shocked by the state of the city, with the danger of unexploded devices and power and water cuts: it had turned into a ghost town, and the people made their way back to the countryside. During the *Taalila*, they decided to send groups to clean the area, men to repair houses, and people who had served in the army to remove the mines. They agreed that families would come back once the cleaning was done (Figure 14.10). There were several specialized international organizations assisting in these efforts; however, people had to wait for approvals for someone to come to their land remove the mines.

The role that women played in education was truly remarkable during the period of the crisis: many women taught children in secret during the time when the Islamic State had control. Some women even returned to the city to participate in local educational initiatives and asked retired teachers among the displaced people from Raqqa to help by providing online classes. Women thus became part of *Taaliala* that was alive and transformable. Diaspora originally from Raqqa started to send money to facilitate work, by either renovating houses or setting up schools. The importance of *Taalila* is not because it was threatened intangible heritage, but because it is essential due to the ability of its practitioners to achieve solutions based on their inherited knowledge (Figure 14.11).

Tall Hasan Basha

Tall Hasan Basha is a village in the Salamiya countryside. It was used as a settlement location by the Isma'ilis in the nineteenth century in the wider Ottoman policy to create a buffer zone to protect the fertile hinterland by

Figure 14.10 Locals from Raqqa; locals came back to clean their neighbourhood

the recurrent Bedouin attacks (Al-Dbiyat, 2004). In exchange for protection, the new settlers were granted land and were exempted from paying tributes. In this area, the Ismaili dignitaries led the resumption of the agricultural activities that had been conducted for over two thousand years in the area after being halted for many centuries. For the new settlers, Salamiye and its countryside were the motherland. They rehabilitated the Roman water channels for irrigation, which attracted inhabitants of the neighbouring villages, as well as other groups of various denominations and ethnicities who found employment opportunities in the Salamiya area with its Isma'ili majority. Neighbouring districts with farmland grew around the city (Al-Dbiyat, 2004).

Tall Hasan Basha is one of the villages that served as a home for Alawis, Turkmen, and Bedouins, all of whom worked in farming; the latter continued with this work after the agricultural reform. Even though they lived in the same village, the inhabitants were grouped into clearly discernible districts: the Bedouin neighbourhood, the Turkmen neighbourhood, and the Alawi neighbourhood. The inhabitants of Salamiya and the surrounding countryside used to sing rain chants during periods of drought, as was the custom in most agricultural areas in Syria (Abbas, 2018). They used to form a procession made entirely of children who held up a doll attached to a long stick and went from door to door collecting donations from the inhabitants.

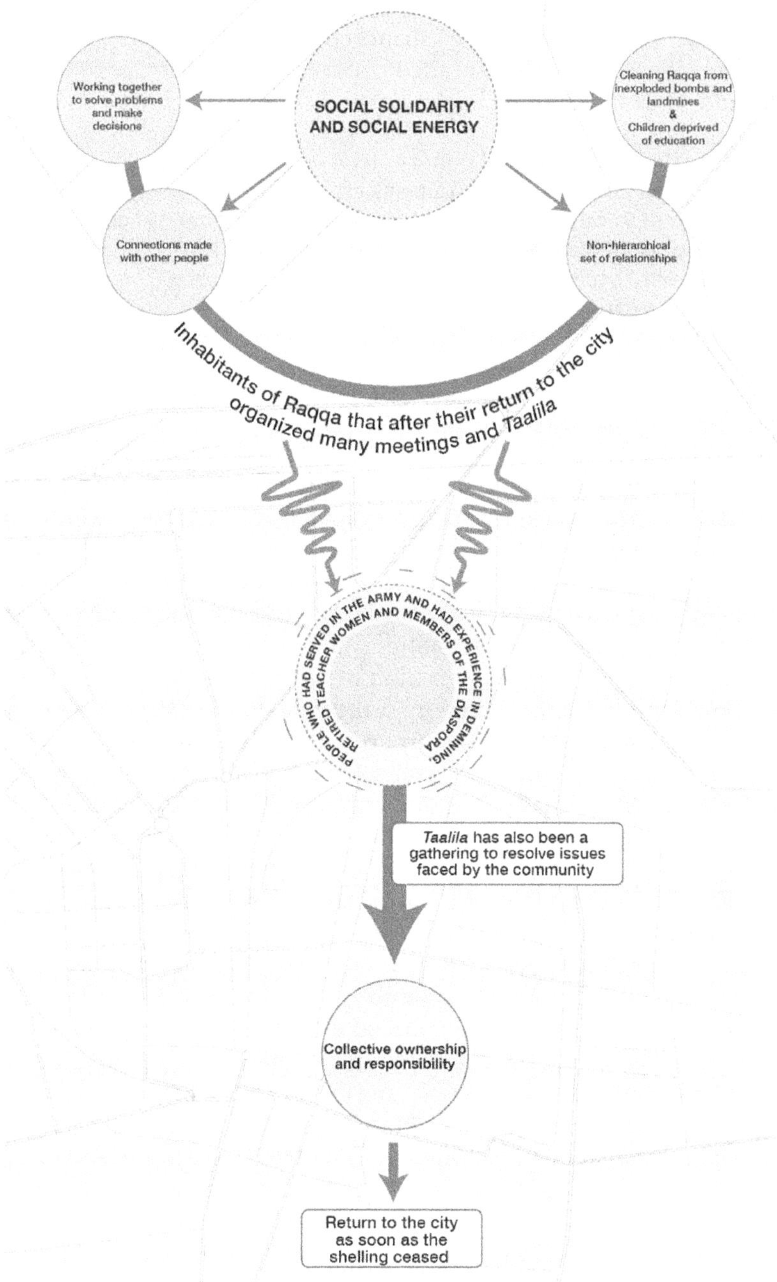

Figure 14.11 Raqqa community-based living heritage

Gifts included chickpeas, bulgur, and rice. The children's procession ended with women cooking all the ingredients collected and handing out food to the entire village. The doll was called Umm al Ghayth (Mother of torrential rain), and while the children held it up on their long sticks, they chanted:

> Oh, Mother of torrential rain, our river beds are empty,
> Come fill our pockets with rain aplenty,
> Allow us to wash ourselves and our garments, oh powerful deity,
> Allow us to clean the robes of our shepherd for eternity;
> Ever-our Shepherd, look upon our fields so scanty,
> Grow our crops, after years of drought so lengthy,
> Grant us your gracious rainfall, God Almighty,
> To water our crops, they're ever so thirsty,
> Grace our towns with rain and plenty,
> Grant us flowing torrents and rich bounty.

Since the eruption of violence in 2011, heightened religious and political differences deepened the gaps and further divided the diverse groups of people. This was the case in Tall Hasan Basha where people were either armed Bedouins or Alawis and, therefore, inherently pro-government. These stereotypes and differences became entrenched and, hence, the village was split.

After a long string of assassinations and accusations, each denomination in 2013 began to fear the reaction of other denominations or components, and they ended up fleeing from the village. The situation went back to its original state where every group was clustered in specific towns or villages and never mixed. After a state of relative calm in the area, people in 2014 began to return to their neighbourhoods but kept to themselves. Jawa Voluntary Team, a local voluntary group of activists, decided to include the countryside around Salamiya in its field visits; the visited Tall Basha and even met with teachers from the only school in the area, taking advantage of the meeting to speak about the team's peace-building endeavours.[14] In fact, the team includes people from the city of Salamiya, as well as internally displaced individuals from the cities of Homs and Al Rastan. One of the teachers took out a box with a doll and explained that all the inhabitants of the village came out and gathered during periods of drought, and the children marched in the streets and sang for rain while holding up the doll.[15] The team then asked the teacher whether it might a good idea to gather the children in the village and get them to chant their rain song, even though it was not a period of drought. As such, all the children gathered and walked the streets and neighbourhoods, growing in numbers with each house they passed, and successfully gathered chickpeas and bulgur (Figure 14.12).

They then headed to Al Khazzan square, the very square where they used to play together before the crisis and came across the members of their team, as well as women from the village who began to prepare the food they had brought. Everyone decided that the doll should be christened the Peace Doll. Its story meant that common traditions allow people to overcome

Figure 14.12 Families around Al Khazzan Square; Umm al Ghayth has returned. Tell Hasan Basha 2015

difference; they bring communities together to meet basic human needs. The same procedure was used to perform another mission, which successfully brought together the area's inhabitants. Although such practices are not considered intangible heritage, they are a reliable basis for new beginnings (Figure 14.13).

This community behaviour is unplanned and not stereotyped; it does not want to list or preserve heritage just for the sake of preserving it. It does not want to compete with other heritages. In this case, living heritage was evoked to initiate peace building during the time of conflict.

Perspectives

This research is an open-ended process that discusses the community-based living heritage paradigm in times of conflict. This does not mean that we do not tackle situations that do not involve wars. On the contrary, understanding the nature of societal behaviour in the context of armed conflict and the absence of government institutions is an important basis for recognizing the role of the government (before and after the armed conflict) in understanding, managing, politicizing, and reassessing heritage. This paradigm needs to be contextualized; indeed, understanding the specific conditions

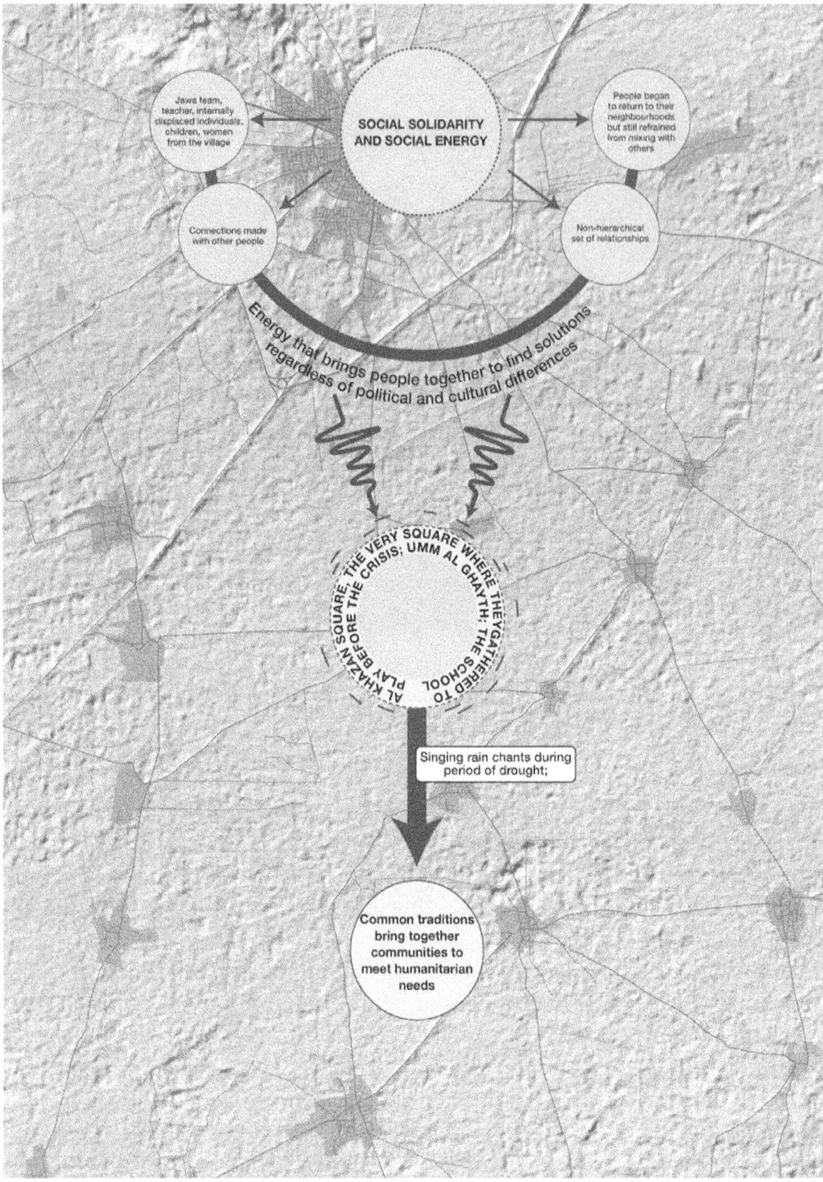

Figure 14.13 Tell Hasan Basha community-based living heritage

of each area is crucial. For instance, the cases described in this chapter are current examples affected by the war in Syria after 2011. Yet, to understand them well, we need a thorough and deep understanding of the roots of the conflict before 2011, which have gradually influenced the behaviours of the communities—or rather their living heritage.

For example, we will not be able to capture the social energy and social solidarity in Aleppo if we do not investigate the degree of socio-economic conflict between the countryside and the city. It will be difficult to contextualize *Taalila* as living heritage if we do not acknowledge that Raqqa has been neglected for decades. We will not be able to trace the extent of to which farmers in the Orontes and the villages around it suffer from poverty.

Accordingly, local communities are the vital nerve of living heritage. Criticism of top-down laws does not mean always obeying the traditions of the past; it is rather an opportunity to generate new knowledge and practices based on understanding the region. Here, in addition to the community itself, intellectuals play a vital role as risk evaluators, development experts, and peacemakers. These various multi-level solidarity connections mean that living heritage constitutes entry points for those who want to build peace and recovery in real, sustainable, and practical terms. Local communities will keep trying to protect their own achievements by using all their resources and relying on their legacies and knowledge—their living heritage. For instance, in Qalaat al Madiq people were able to stop the snipers from both sides of the parties in the conflict under the pretext of cleaning the water channel. Stopping the snipers facilitated the movement of goods without taxes and reduced the role of warlords in controlling the road that links the two areas. Consequently, there are great responsibilities for the organizations concerned with heritage, development, and reconstruction, including UNESCO. In this case, the intervention needs to be able to find harmonious solutions and practices for building collective memories, common cultural identities, and living heritage that is meaningful to everyone without bias and rejection of diversity.

With these kinds of interventions, heritage naturally maintains itself without the direct need to be fixed between tangible and intangible, between outstanding and non-outstanding values, or by the state. Hence, none of these communities will place a high priority on preserving a castle or preserving the craft of mosaic making, only to have it labelled as heritage. In fact, living heritage is about answering the needs of communities and protecting the interactions between people.

Acknowledgements

I express my profound gratitude to all my connections in Syria that allowed me to enrich my understanding of the research topic. I want also to thank COSV, an international non-governmental organization working to support the development of local communities in Syria and the Middle East. Through my work with COSV, I was able to widen my network in Syria. Special thanks also go to Omar Abdulaziz Hallaj and the enlightenment Howayda Al-Harithy for their support and encouragement.

Notes

1. "Dams and irrigation channels are difficult to date, although they have been documented in the Near East from the Bronze Age onwards (3600–1200 BC)" (Chambrade et al., 2014, p. 4).
2. Personal communication with Abdul Rahman Al Yahia, Director of Syrian Centre for Heritage Protection in Maarat al Numaan.
3. The configuration and possibility of bridging a public right-of-way emanates from the concept of the fina. It is a device that allows for the creation of additional space attached to a building. Codes written by Muslim jurists clearly stipulate the legal rights associated with constructing sabats (Hakim, 2008).
4. Using satellite photographs taken between 2013 and 2017, UNOSAT has assessed the damage in Aleppo as 32.7 per cent of the total count of damaged structure (https://towardsdatascience.com/damage-caused-by-the-syrian-civil-war-what-the-data-say-ebad5796fca8).
5. *Mubadarat Ahali Halab* (The people of Aleppo initiative) is a civil organization that worked across city lines during the military division in Aleppo to restore public services.
6. Personal communication with Mahmud Kusa, Muhammad Uthman and Mahmud Abdul Rahman, civil society activists in Aleppo, May-July 2019.
7. Mapped by the Syrian Initiative, part of the Common Space Initiative. The Common Space Initiative is a Beirut-based NGO working on supporting the creation and consolidation of peace assets, particularly in the civil society. It develops common grounds for dialogue among Syrian stakeholders at all levels and strengthens local and international advocacy for peace.
8. Usually wine (personal communication with Maabad Al Hassoun, June 2019), but it could also include tea (personal communication with Ibrahim Al Khalil, July 2019). Another interpretation includes the Arabic verb *'al-la-la* that means "to find a solution."
9. Personal communication with Maabad Al Hassoun, June 2019.
10. From the Facebook page of Humsi Farhan Al Hammada, an intellectual from Raqqa, www.facebook.com/www.Homsialhamada/photos/a.1542883059257617/1659997190879536/?type=3&theater, accessed July 19, 2019.
11. Personal communication with Maabad Al Hassoun, June 2019.
12. "For some reason, the people of Raqqa replaced the names of their clans and places of their residence with neighbourhoods. This may reflect the preparations and tendencies to replace the clan with the city, replacing the clan's connotations with the city's connotations" (Al Hassoun, 2017).
13. Personal communication with Hassan Shuayb, civil society activist from Raqqa, June 2019.
14. From the Facebook page of Jawa Voluntary Team. https://www.facebook.com/pg/jawa.voluntary.team/photos/?tab=album&album_id=859954504047649&ref=page_internal, July 2019.
15. Personal communication with Eman Nasser and Khaled Al Agha from Jawa Voluntary team, June 2019.

References

Ababsa, M. (2001). La mad^afa `a Raqqa: Mutation d'un lieu de sociabilité tribale en attribut de notabilité citadine. *Géographies et Cultures*, 2001, 17–36.

Ababsa, M. (2009) *Raqqa: Territoires et Pratiques Sociales d'une Ville Syrienne*. Beirut, IFPO.

Abbas, H. (2018). *Traditional Music on Syria* (in Arabic). Beirut, UNESCO.

Al-Dbiyat, M. (2004). Syrian Urban History and the Resurgence of Small Cities in Syria, in *La ville en Syrie et ses territoires, héritage et mutations* (in Arabic). Edited by David J. C., and Al-Dbiyat, M., Damascus, Dar Al Jundi, 421–433.

Al-Dbiyat, M. (2013). Footnote number 3, p. 233 in Dussaud, R., *Topographie historique de la Syrie antique et medieval* (in Arabic), Beirut, IFPO.

Al-Ghazzi, K. (1926). Nahr al-Dhahab fi Tarikh Halab (The River of Gold in the History of Aleppo) (in Arabic), Part 3, Aleppo, Dar Al Qalam, 329–334.

Al-Hassoun, M. (2017). Raqqa, from the Tribe to the City: The Ottoman Police Station, Transient States. Baath and Jihadists (in Arabic). http://alaalam.org/ar/society-and-culture-ar/item/538-655160617, accessed April 2020.

Al-Kazzaz, T. (2010). Urban Transformations: The City is a Tapestry of Changing Relationships in Time and Space, in *Beirut Heritage in conservation and preservation* (in Arabic). Edited by Sarraj, N., Beirut, Beirut Urban Observatory and Arab Scientific Publishers, Inc., 29–47.

Bokova, I. (2012). Celebrating 40 years of World Heritage, UNESCO Director-General. www.unesco.org/new/en/beirut/singleview/news/celebrating_40_years_of_world_heritage, accessed July 14, 2019.

Chambrade, M. L., Al-Dbiyat, M., and Geyer, D. (2014). Ancient Water Infrastructure. www.water-security.org/article/ancient-water-infrastructure, accessed July 2019.

Chehadat, I. (2013). Comments in *Topographie historique de la Syrie antique et medieval.* Edited by Dussaud, R. (1927), Beirut, IFPO,, 233.

David, J. C. (2004). Presence of the Past: The Formation of Urban Heritage, in *La ville en Syrie et ses territoires, héritage et mutations* (in Arabic). Edited by David J. C., and Al-Dbiyat, M.,, Damascus, Dar Al Jundi, 467–488.

Dayoub, D. (2015). Wartime Coping and Its Re- Configuration of the Urban Space. The Case of Aleppo City during the Protracted Syrian Conflict. A Thesis submitted in the Partial Fulfillment for the Requirement of the Degree of Master of Science in Integrated Urbanism and Sustainable Design, University of Stuttgart.

Dussaud, R. 2013). *Topographie historique de la Syrie antique et medieval (in Arabic).* Beirut, IFPO.

Flora, C., Flora, J., and Gasteyer, S. (2015). *Rural Communities: Legacy + Change.* Edition 5, Boulder, Colorado, Westview Press.

Haj Asaad, A., & Shamaly, O. (2016). The Effects of the Conflict in the Orontes River Basin in Syria, in *Water Resources Management in the Lower Asi-Orontes River Basin: Issues and Opportunities.* Geneva & MEF University, Istanbul, Graduate Institute of International and Development Studies.

Hajj Saleh, Y. (2017). Religion and Personal Experience: Testimony, in *al Jumhuriya*, September 22, 2017, Beirut (in Arabic). www.aljumhuriya.net/ar/content, accessed July 2019.

Hakim, B. 2008. Mediterranean Urban and Building Codes: Origins, Content, Impact, and Lessons. *Urban Design International*, 13, 21–40.

Labadi, S. (2013). *UNESCO, Cultural Heritage, and Outstanding Universal Value: Value-based Analyses of the World Heritage and Intangible Heritage Conventions*, London, AltaMira Press, 60.

Lamprakos, M. (2015). *Building a World Heritage City: Sanaa, Yaman.* London-New York, Routledge.

Mahdy, H. (2019). Is Conservation of Cultural Heritage Halal? Perspectives on Heritage Values Rooted in Arabic-Islamic Traditions, in *Values in Heritage Management*. Edited by Avrami, E., Macdonald, S., Mason, R., and Myers, D., Los Angeles, The Getty Conservation Institute, 127–140.

Nembhard, J. G. (2016). Building a Cooperative Solidarity Commonwealth, in *The Next System*. https://thenextsystem.org/building-cooperative-solidarity-commonwealth, accessed July 2019.

Pratt, M. L. (1991). Arts of the Contact Zone. *Modern Language Association*, 33–40.

Sbeinati, M., Darawcheh, R., and Mouty, M. (2005). The Historical Earthquakes of Syrian Analysis of Large and Moderate Earthquakes from 1365 B.C. to 1900 A.D. *Annals of Geophysics*, *48*(3).

Smith. L., and Campbell, G. (2017). The Tautology of "Intangible Values" and the Misrecognition of Intangible Cultural Heritage, *Heritage & Society*, 10:1, 26–44, DOI: 10.1080/2159032X.2017.1423225.

Syrian Center for Policy Research (2019). Food Security and Conflict in Syria, p. 29, https://www.scpr-syria.org/launch-of-food-security-conflict-in-syria-report/, accessed July 17, 2019.

World Bank (2017). The Tool of War: The Economic and Social Consequences of the Conflict in Syria, https://www.worldbank.org/en/country/syria/publication/the-toll-of-war-the-economic-and-social-consequences-of-the-conflict-in-syria, accessed July 16, 2019.

Violation Documentation Center (VDC) (2019). http://vdc-sy.net, accessed July 16, 2019.

Part IV

Space and imaginaries in framing post-crisis recovery(s)

15 Transient city – steadfast camp

Re-construction of ancient Rome and present Dheisheh

Elena Isayev

What does it mean to perceive of place and one's relationship to it – which some term belonging – from the perspective of an inhabitant, not of a city, but of a camp – not as a citizen, but as a refugee? Such a perspective does not come easily to those who approach the world as members of nation-states, the existence of which depends on that crucial, and seemingly self-evident relationship between belonging, territory and citizenship.[1] A way of thinking into that other space beyond the nation-state is by considering historical alternatives that pre-date it.[2] One such historical context is 2000 years ago, when the people of ancient Italy were formulating what it meant to be Roman in relation to the site of Rome itself.

Part of that discourse is captured by the ancient historian Livy, through the imagining of what would happen if Rome was moved or exiled instead of being rebuilt after it was sacked. Could it still be Rome in its new location? Would those inhabiting it be Romans? Does the land on which it sits matter to what it is? This projection allowed for a re-configuration of the subject – Rome – from the outside. The power of such an external perspective is equally revealed in the discourse between Dheisheh refugee camp, in the West Bank, and 'its' suburb, whose inhabitants lobbied to be part of the camp. Both the ancient and the contemporary scenarios seem fantastical, but perhaps no more so than the organisation of the world into nation-states, that invests the soil with magical powers. This quality of the soil was subverted by Cildo Mereiles in his quiet work – *Geographical Mutations: The Rio–São Paulo Border,* created during the military dictatorship in Brazil:[3]

> Rio De Janeiro State shares a border with the state of São Paolo. In 1969, Cildo Mereiles stood on the Rio side of that border. He was carrying a suitcase specially constructed for the occasion. A leather divider separated the inside of the case into two compartments. Mereiles also had a shovel. He proceeded to dig a hole, filling one of the case's compartments with the dirt of Rio De Janeiro State. Mereiles then closed the case and stepped across the border to São Paolo State. There he dug another hole, filling the second compartment with dirt from that state. So, the two were combined in one case.

The two cases, the one of Dheisheh camp and the other of the city Rome, are by no means comparative in terms of their circumstances, and that is not the point of examining them together. Rather, it is to consider alternative constructions of the imagined community and their ambivalent relationship to the land and fabric of any site. In each case, the formation and persistence of collective identity is articulated through a spatio-temporal juncture, but with important differences. The rootedness of the city to its site (*sede* in Latin, literally meaning seat) – a seemingly permanent feature of being Rome – is destabilised by Livy in his projection of its absence and removal to elsewhere (Livy *Ab Urbe Condita,* 5.51.2). What at first appears an aberration, or a mere flight of fancy, as we will see, exposes the paradoxes of territorial thinking in a world demarcated neither by nation-states nor spatially contained citizenship. Conversely, the case of Dheisheh and its suburb subverts the condition of temporariness that emerges out of exile and displacement, which defines the camp and refugeeness as a liminal existence – an absence. Instead, it exposes the possibilities for constructions of exceptional place, politics, and forms of collective identity, deemed impossible in a world demarcated by pockets of spatially contained citizenship. Setting the camp (Dheisheh) and the city (Rome) in contrast with the territorial nation-state is to destabilise the relationship between territoriality and political subjectivity. Both force a reconfiguration of the essence of place and its relation to identity. Both the ancient and future discourse expose the doublethink behind the reliance on territoriality in identity formation, as being simultaneously enduring and ephemeral.

The camp in Palestine shares some of the challenges that Livy presents in shifting between the ephemerality of the city, in defining what Rome is beyond the physical site, while trying to articulate why the land that hosts Rome too is meaningful. Livy draws, as we will see, on the rituals and practices through which Rome is created – personal family rites and public customs with jurisdictions that are delineated territorially – asking whether these would still have significance if de-coupled from their locations. There is no discussion about land ownership or territory in any abstract sense. Rather, he points to Rome's location as chosen by divine will and crafts it as a mosaic of sites of memory, which make up identity and belonging. Thus, he sketches imaginaries of Rome, within which its future presence is dependent on remaining in its place. How then to read the presence of a refugee camp, the existence of which is meant to be temporary, negating any ties with any land it is placed on, or that beyond it. Qasmiyeh from his deep experience of the camp articulates this in his poem: *There it is: the camp that is yet to be born*, from which come the following insights (Qasmiyeh, 2019):

> Find me a place whose meaning is that of its absence.
> Find me a place where nothing is not exactly nothing but its equivalent.

The camp in its existence signifies absence – absence of people from their place of origin, and from their land of belonging, as that embodied in Palestine. Furthermore, it challenges the absence of place itself, for it is precisely through its spatiality and the practices in, and of, the camp that there remains significance in the locations of the sites of origin and exile, which the camp may represent, but not replace. The mosaic of sites of memory are recreated through the camps whether in the West Bank, Jordan or Lebanon, and thus so is Palestine.[4] This allows for a particular way of belonging to the camp as it references the original place, with its entitlement to citizenship and to nation. It subverts the seemingly self-evident relationship between belonging, territory and citizenship by becoming a representation of the original place. 'The Camp is a translation of something beyond itself'.[5]

To exist in a refugee camp is to inhabit a paradox: from an outside perspective, ostensibly the true place of belonging is to be imagined elsewhere while the camp – where one lives and may have even been born – is not. The camp, demarcated by temporality, is seemingly denied the possibility of being a place and having the potential for future memory that is projected to another somewhere and somewhen. The reality, however, is not that of temporariness. The increasing permanence of the camp means that it is becoming a heritage site (Petti, 2017a,b; Hillal and Petti, 2018). The camp is no longer made up of tents, and although it continues to be a site of national exclusion, it is not a site of marginalisation, poverty and political subjugation. It has intricate structures that make up its fabric: the walls and rooves that form into buildings, houses, shops, which delimit passageways – some almost yards – an infrastructure of weaving streets, also conducting water, electricity and stories. All these, often in a perpetual phase of reconstruction give the appearance of place (Figure 15.1). Architect and educator Alessandro Petti observed that 'the prolonged exceptional temporality of this site has paradoxically created the condition for its transformation: from a pure humanitarian space to an active political space' (Abu Aker and Al Lahham, 2013, p. 9). This transformation comes to the fore in numerous acts, which are at times materialised in the reconfiguration of the camp, for example, when some of its inhabitants leave to construct a nearby suburb. Such a relocation disrupts the meaning of the camp and any collective political identity of refugees. It brings into question the status of the new site. In what capacity are its inhabitants still refugees?

These issues are explored first-hand with the creation of the suburb Dahiyyat Al-Shuhada adjoining the Dheisheh Refugee Camp in Palestine. Its residents have lobbied the camp to allow the suburb to be part of it.[6] This particular case, outlined below, exposes the complex nature of the discourse on membership and identity in relation to the refugee status and the position of the camp. Here, there is a seemingly inverse relationship between space and political status. In Dheisheh's case, the suburb's inclusion in the camp constitutes a bestowal, rather than a removal, of political existence. Correspondingly, inclusion in 'the norm' – the existing municipalities – in

Figure 15.1 Looking across Dheisheh Refugee Camp, West Bank, Palestine

this instance would be depoliticising. The broader narrative in the study of Dheisheh, which underpins the politics and the political imagination and thus shapes people's actions, is that of displacement, exile and the hope for return. At stake is not only the politics of the camp, but one that challenges and dialogues with the nation-state's sanctioning of territorial space as a condition of political existence. It also intersects with the politics of international and refugee organisations, and that of Palestine. Thus, encompassed in the case of Dheisheh, are several political fields that impinge on how identity is positioned; some are spatially delineated, but others not, even if at times they take on a territorial guise.

The diverse forms of being political has been intensely scrutinised in relation to the Camp context of today, with debates spurred on by the work of Schmitt and its development by Agamben (Schmitt, [1950] 2003; Agamben, 2003). The critical reactions to the exceptionalism that they ascribe to the camp are indicative of a more profound understanding of such contexts, but also of the transformation in the nature of camps over these last decades (see examples in Qasmiyeh, 2019, Hillal and Petti, 2018; Agier, 2014; Feldman, 2015; Sigona, 2015; Turner, 2016). The new perspectives bring into question the analytical capacity of Agamben's influential conceptualisation of the camp as a space of exception. At the core of these new understandings of the nature of camps and camp-like institutions is the way that, the camp is a space of human agency for those who inhabit them despite their exceptionalism.[7] These expose the variety of interactions and social relations, within the camp and between the camp and its surrounding settlements, the city and the state. They articulate the strategies that people who inhabit them develop in their everyday lives to claim rights and membership. Sigona draws on such findings and brings them together with Isin's work on the relationship between citizenship and being political, in proposing

> the concept of 'campzenship' to capture the specific and situated form of membership produced in and by the camp, the complex and ambivalent relationship of its inhabitants with the camp and the ways the camp shapes the relationship of its inhabitants with the state and their capacity and modes of being political.
>
> (Sigona, 2015; Isin, 2002)

Numerous studies have unsettled the categories of state and citizenship (Chaleshtoori, forthcoming; Isin, 2009; Harvey, 2009; Malkki, 1996). These new understandings are about recognising the camp for the life that has been lived in it over generations, while still challenging normalisation – *tawtin*. They are about seeing the camp as part of a continuum of forms of settlement that look to the future – a fragment of a city yet to come: 'Today, refugees are re-inventing social and political practices that improve their everyday lives without undermining the exceptionality of the camp. Camps have become semiautonomous zones where different social, political, and spatial structures have emerged; a fragment of a city yet to come' (Hillal and Petti, 2018, p. 33).

The camp and the suburb

The land that became Al-Shuhada suburb lies on the edge of Dheisheh (Figure 15.2). Its complex and forgotten ownership history allowed for its development into a new neighbourhood from 1995, primarily by refugees from Dheisheh who today make up almost 90% of its populace. The residents are not owners but hold a document that authorises the 'right to use' the land. The 'edge' of Dheisheh adjoining the suburb is the boundary of the original camp set up in 1949 by the United Nations (UN), on a half-a-kilometre square piece of land leased for 99 years. Within the edges of the camp the United Nations Relief and Works Agency for Palestine Refugees (UNRWA) has the responsibility to provide protection and services for the refugees living there, whose population has grown from 3,000 to 13,000. Across the street from the North edge of the camp is the municipality of Doha, the population of which is also largely made up of refugees. Beyond the South edge of the camp is the village of Irtas, with its own autonomous governing structures. Al-Shuhada too could have become a municipality or joined an already existing one outside the camp; in 2003, Irtas had hoped to annex it, but the suburb's residents refused. Instead, in 2012, the Local

Figure 15.2 Aerial view of Dheisheh Refugee Camp and adjacent Suburb

Committee of Al-Shuhada made a request to the Popular Committee of the Dheisheh refugee camp to include the suburb in its jurisdiction. By implication, it would mean expanding the initial camp boundaries. Such a territorial inclusion and acceptance of the suburb as part of the camp, would allow the camp's services, including water and electricity, to be extended to the suburb. But this is not just about services, as similar services would have been provided if they joined Irtas. A major motivation for joining Dheisheh is political. It would mean that the suburb's residents would be under the auspices of the Popular Committee – the camp's representative, although not technically governing, body run by volunteers. It would allow them to be represented along with Dheisheh's residents *as refugees* in discussions with external organisations, in particular, the UN and the Palestinian Authority (PA). Officially, the request cannot be granted because technically the Popular Committee in conjunction with UNRWA cannot operate outside the physical boundaries of the camp. Nevertheless, the Popular Committee is keen to support the suburb and include them within their remit.

The representation *as refugees*, and not simply as Palestinian residents of a municipality such as Doha, or Irtas, is crucial for the inhabitants of Al-Shuhada. In terms of identity and status, several interrelated aspects come into play, thus interleaving past, present and future: 1) the birthplace of many of the suburb's residents, which is Dheisheh; 2) the wider identity of being Palestinian; 3) the status of being a refugee in exile; 4) the belonging linked to the place of origin before exile and 5) the physical place of residence in a suburb. Citizenship does not seem a relevant category here, and we may ask how it would sit alongside these, whether by replacing or encompassing them. Technically, the status of refugee remains whether one lives in the camp or not, since it is dependent on the condition of being displaced, not residence. Hence, one can still live in Bethlehem, Ramallah or overseas, and remain a refugee. Crucial to the residents of Al-Shuhada is the ability to maintain that status not only individually, but collectively, and to be represented politically on the strongest possible platform – the camp. It would allow the Popular Committee of the camp to represent refugees of the suburb as if they were part of Dheisheh. Although these committees were meant to be only service providers, in practice they have taken on the responsibility to politically represent the camp at the local level (Abu Aker and Al Lahham, 2013). It is they who hold the main authority, not UNRWA, nor Fatah, which retains control of the PA in the West Bank through its president. The Committee's successful lobbying has transgressed camp boundaries, effectively acting to support surrounding municipalities, such as Doha. The issue is that, although there is the existence of a Department of Refugee Affairs (DORA) inside the Palestine Liberation Organizatio (PLO), there is no special body that represents the Palestinian refugees politically on a national level, as the PA is the representative of all Palestinians, without distinction of whether they are refugees or not.

Al-Shuhada's popular request to be part of the Dheisheh camp, despite the lack of any official political framework, challenges the relationship between community, land and representation. It brings to the fore the deception of temporariness and the question of what happens to the camp once the 99-year lease is up. It is an act against the static idea of normalisation of the exceptional condition of the camp, and also a claim of a place beyond, which the camp has come to represent, the place of origin and the wider Palestine. At the same time, it is a recognition of the need for an imaginary of the future that is not invested solely in a single-event return from exile or an arrival. Rather, the return is to be re-imagined through a multiplicity of smaller acts, no longer through a dependence on an outside force or resolution, but through internal action and through individual, splintered acts, which reach towards a future incrementally (Hillal and Petti, 2018). There is an ambivalence as to how the two-state solution may have resolved these contradictions. The suburb's request to be part of the camp pressures UNRWA to acknowledge its responsibility to the expanded population of refugees, and to recognise the suburb as a natural extension of the camp. The paradox is that to have representation as refugees, UNRWA needs to extend its support geographically to include the suburb, yet the argument is centred on the non-territorial identity and practices of being a refugee.

Through such a discourse, the camp is acknowledged not simply as a humanitarian space but as a dynamic site of socio-political interaction, a platform for negotiation, and the intersection of memories and practices. It is not just a camp – but a place, and a representation of place of origin beyond it. It unsettles the city as a model against which all other practices are measured. It also disturbs the city's perceived static nature, exposing the dynamic forces and flows on which it rests, and through which it is constantly remade – at times violently – and re-imagined. As articulated by geographer Doreen Massey, 'place… changes us, not through some visceral belonging (some barely changing rootedness, as many would have it) but through the practising of place, the negotiation of intersecting trajectories; place as an arena where negotiation is forced upon us'.[8] Through such a definition, can the two sites be brought into dialogue about the relationship of such seemingly ephemeral practices and the materiality and spatiality of sites? How does the delineation, construction, destruction, removal, extension, reconstruction or even refusal of a site's fabric and territory affect its placeness – the meaning for those who inhabit it in body or in the imagination? Both can thus be reconstructions of an 'other' place.

Mapping the mobile city

Is any city, even Rome, ever more than the mobile or metaphorical city of Kandinsky's dream – perpetually changed through experience?[9] According to a legend, the site of ancient Rome was first chosen as an asylum and a refuge for the de-placed, whether those brought together by Romulus or Aeneas. As the ancient city grew, it encompassed a dispersed and transient community made coherent through ties to a specific point in space and in the imagination. Its physical being and site were imbued with mythic significance. Yet ancient conceptions of space made it difficult to map and physically define Rome.[10] It was, therefore, a challenge for Livy to make its material fabric and its spatial location meaningful to those who called themselves Roman. Livy's history of Rome, written in the first-century BCE, tried to achieve this in the context of a highly mobile Mediterranean, a world full of cities but without nation-states. There were no regional boundaries, nor maps to scale on which to plot them. It is not that the capabilities for mapping were not there. Euclid's work already in the Hellenistic period, from the fourth-century BCE, recognised global space mathematically, but it did not translate to a global geography. In other words, the abstract was not deployed for cartographic representation despite the fact that the tools were available (Lefebvre, [1974]1991). The most accurate cartographic depictions were employed by the *agrimensores* (land surveyors) for demarcating private land allotments and laying out settlements, but this site-plotting did not translate to depictions of regional spaces. Nor were they employed in the demarcation of state-borders. The boundaries which separated outsiders from community members in the context of the ancient Mediterranean were those of status, and these could be much better policed by enforcing cultural norms than by scrutinising city-ports or gateways.

Ancient (and medieval) cartographic creations were not interested in the depiction of static space. Rather, such itineraries as the Tabula Peutingeriana, from the fourth-century CE, celebrated the intricate road network of paths and highways that gave an impression of a coherently bound together Late Roman world (which by then was rapidly unravelling).[11] With the Renaissance, a new spatiality was conceived: the medieval maps that marked out itineraries and pilgrimage routes, with distances calculated in hours or days, gave way to Early Modern maps that erased the itineraries and colonised space. Within this new mapping – the foundation of our mapping today – there is no spatiality that is not organised by the determination of frontiers (De Certeau, 1984). Many works expose the proliferation of maps as tools of conquest, as for example Padrón's, which charts the sixteenth-century collusion of cartography and empire in the early modern Hispanic world, and the way that America was created rather than discovered (Padrón, 2004). He shows how the authority of such maps was used to strip native forms of territorial imagination and contributed to de-territorialisation of the First Nations Peoples of America. There was growing interest in geometric and mathematical – Euclidian – representation of space, which, as Lefebvre has shown, privileges a particular kind of spatiality at the expense of others (Lefebvre, [1974] 1991). With its focus on abstraction, territorialisation becomes its primary concern; instating an authority over abstract space, while supplanting with representation that of the perceived and the lived. These cartographic practices were unlike the ancient depictions of the known world that reveal an interest, not in the landmass nor any abstract Euclidian space, but the dynamism of the journey, connectivities and place-moments. They do not provide a spatial framework – like the gridded maps of today – with empty spaces onto which new locations can be plotted. There was no way of mapping a community on a two-dimensional plane, as citizenship and territory did not overlap. This was especially true in the case of Rome, whose membership was given to communities and individuals beyond it.

Against this background, Livy's 142 volume work – *Ab Urbe Condita* - *From the Founding of the City* – and especially the speech of Camillus in Book 5, allows for an exploration of how place-specific belonging is generated in people's historic experience. It was written just as Rome established itself as the imperial centre of the Mediterranean in the reign of Augustus. The city was the perfect protagonist even for a historian who originated well outside it. Livy the city's greatest biographer was from Northern Italy, near Padua. His Cisalpine home only received full Roman citizenship when he was about ten years old, decades after the rest of Italy was enfranchised in 89 BC.[12] With the extension of citizenship across the peninsula came the reconfiguration of the meaning of Rome – a capital city, yet still a city-state – especially for those who lived outside it. Their allegiance and loyalty needed to be re-inscribed. For the first time, there were attempts to articulate why the physical site of Rome was meaningful. Livy achieves this through the lens of events that took place four centuries before his time, evoking the aftermath of the Sack of Rome by the Gauls in 387 BCE (Livy, 5.51–54. Full text and translation in Appendix). He describes how on their return from refuge the inhabitants find their city devastated, and gather to decide what to do next. A debate ensues of whether they should rebuild the ruins of Rome or move to the nearby city of Veii.

The argument against abandoning Rome, Livy puts into the speech delivered by the general Camillus. He appeals to the emotional and practical sensibilities of his audience, including ideas of homeland, exile, sacredness and political process, to construct the city as a site of communal memory and practice. The external structures may be temporary, but the household gods, sacred sites and practices have their appointed places. Every bit of soil has meaning, which he exemplifies through the episode of sacrifice by the young C. Fabius (5.52.4), part of the infamous Fabii clan.[13] Camillus recounts how this youth risked his life by leaving the protection of the citadel during the Gallic Sack. Making his way through enemy lines, he climbs the Quirinal hill to perform his family's annual sacred rites at the appointed place and time. Livy attempts to transpose such respect for private obligations to those of the citizens' duties to Rome. In light of these could Rome exist anywhere, or was its essence rooted to its site? And if its people moved, would they still be Romans? Did Camillus convince the assembled Romans not to move? Livy's comment, to our surprise, is not quite, although they stayed anyway. His final observations are that to him Rome looks more like a settlement we would associate with a camp rather than an organised city (Livy, 5.55.5). To understand the meaning of Rome at the height of empire, and seemingly unlimited power, Livy chooses a moment of emergence from a position of exile and constraint. To what extent is such a moment of emergence precisely the one that a camp like Dheisheh finds itself in now?

Livy's construction of Rome does not present it as a closed or bounded community. Neither do his contemporaries. Roman statesman and lawyer Cicero, when defining community and citizenship, privileges the laws as binding, not any notion of territory (Cicero *De Re Publica,* 1.39, 49; 2.61). Roman identity divorced citizenship from ethnicity and location, allowing for inclusivity which, some argue, was fundamental to its success.[14] Romans, at the beginning of empire, from the third-century BCE onward, in their origin stories and foundation myths imagined their community as the product of migrations. Rome's beginnings, according to one legendary strand, show how the city originated from the creation of an 'asylum' by Romulus who invited others to join him at the site, including vagrants and refugees (see Isayev, 2017a; Dench, 2005). Rome's other mythical strand, recounted in Vergil's *Aeneid*, also envisions refugees as ancestors of the original populace of the city, but here specifically from Troy, thus allowing them to be tied into the most prolific Mediterranean myths of the time, with the Trojan War at their centre.

Extensions of citizenship to individuals and communities beyond Rome, and the settlement of Romans outside of the city and its immediate territory, made Rome potentially limitless and perpetually expandable. This latter extension of Rome, well beyond any container, has echoes with what is witnessed in the camps, into which the meaning of Palestine and place of origin have stretched, although through a dispersal due to exile, rather than imperial expansion. Still, in the case of ancient Rome, boundary-making is not absent from the speech of Camillus. He refers to the *pomerium*, a boundary that did not encompass all that was Rome, nor did it coincide with the city walls.[15] Rather, it delimited the space for performing certain rites and rituals associated with political activity and delineated the jurisdiction of civic and military authority. For example, the dead were to be buried beyond the *pomerium*, while auspices needed to be taken within it.[16] The *pomerium* confined the jurisdiction of the consuls, which could only be exercised within it. Physically and conceptually, it was a line that acted as a sacred, political and social divider. In the context of Camillus's speech, the *pomerium* is presented relationally – as a set of customs and practices – not as a cordon that circumscribed Rome as a coherent space. Livy does not use the *pomerium* as an opportunity to argue for a bounded territory, and there is little reference to bordering practices overall, outside of those that concern the sacred precincts (Livy, 5.50.2). The evidence we have of Roman mapping practices shows that there was less concern in circumscribing a territory with a cordon, than an interest in the 'line', or a series of lines stretching from a given point or *terminus*. Rome's power and influence spread from just such a point.

Livy presents us with one imaginary of place in the last decades of the first-century BCE. It confronts the expectation of attachment to a particular site and the emotions that may be invested in it. By building a definition of Rome through a contemplation of its absence, Livy explores the multiple meanings of community and its relationship with the natural and human landscape. In so doing, he epitomises the process by which Rome is recognised as more than simply a 'site' – a physical location and its urban fabric – but as 'the place'. That Rome was an intersection of practices and memories is unquestionable; it was a meshwork of pathways and the central node from which Empire spread. But, was it immoveable in the same way as a *domus* (home), or a *templum* (sacred precinct), the two key models that Livy employs to articulate the importance of a physical site? It was a *place* not because of where it was located, but because it housed the stories and the *monimenta* (monuments or physical markers) associated with them,[17] which Livy's writings helped to create, entrench and maintain in living memory. The Rome of Livy's Camillus speech cannot be easily located on a conventional map. Even the *pomerium* is an elusive boundary when it comes to delineating what Rome is spatially. The speech of Camillus exposes the way that the place of Rome is made central to the imagined community that saw itself as Roman.

Reconstruction of memory in place

Through interleaving ancient and contemporary spheres, which exist outside of nation-state paradigms, there has been an attempt to understand the interdependence between practices, community, politics, spatialities and materialities, as they are exposed and subverted in contexts of exile, displacement, destruction and return. What convergences allow or deny the positioning of oneself in relation to community, place and time? Unlike the ancient exiles who returned to 'their' Rome, destroyed as it was, for the refugees in the twenty-first century who continue to be displaced, the positioning is with reference to the place of origin, to which return is pending. How then is that identity reconstructed in relation to sites that are not that place of reference. Dheisheh is a discourse on how new politics is made and meaning is given to such sites, a meaning which can encompass within it the original place of reference and the right of return. The paradox that this entails, which is both a recognition of, and resistance to, an externally fuelled regime of permanent temporariness, can be considered against an ancient, differently enigmatic, dialogue about the meaning of Rome as *urbs* – city and *patria* – fatherland, at the time that Livy was writing his histories.

In today's world, the term *patria* is mostly reserved for the nation-state and is unlikely to be used in reference to a city. But in ancient writings, its use was much less consistent even within individual texts.[18] In exploring the distinction in Latin literature between sentiment towards a local *patria* (for example a birthplace with 'natural' attachment), and the civic or political *patria*, Bonjour found that even expressions of attachment to birthplace are inconsistent (Bonjour, 1975). Livy and ancient politician Cicero, who lived slightly earlier in the same period, evoke different forms of the *patria* in reference to the city. For Cicero, the *res publica* can be presented as the *patria*, which is intangible and exists through practice.[19] The site of Rome is important to him for what it facilitates. But, unlike his birthplace in Arpinum, it is not imbued with emotional attachment. Cicero's conception of the city of Rome and *patria* displays a more relational approach than we find in Livy's speech of Camillus, which, in its argument for fixity, moves closer to a conception of absolute space. Cicero in his *De Legibus* – *On the Laws* (2.1–6) – written about a decade before Livy began his work – creates a fictional dialogue exploring the links between place, citizenship and belonging as part of a wider discussion of natural law.[20]

The conversation is set in Cicero's birthplace of Arpinum some 100 km South-West of Rome. It takes place between himself and his Roman-born friend Atticus, who asks him whether he has one *patria* or two, in Arpinum and Rome. The answer is not clear cut, as Cicero is loath to choose between places dear to him. He concludes by referring to Rome as his adoptive *patria* that holds Arpinum the other *patria* within it. He specifies that the former is *patria* by law (*iuris*) and the latter by location (*loci*), presumably of birth (Cicero *De Legibus* 2.5: *habuit alteram loci patriam, alteram iuris*). His dialogue identifies the versatile and multiple ways in which *patria* could be conceived, and one's relationship with it. *Patria* is not used to mean simply or exclusively the place where one is born; nor is there a direct relationship to the citizenship that one holds. It can simultaneously refer to a physical site such as Arpinum, or to a wider concept implied by Rome. It can also be used to encompass as broad an entity as Italy.

As Atticus and Cicero continue their walk, the conversation moves to Athens, Atticus's current abode. Atticus explicitly states that what attracts him to Athens is not its great buildings, nor its ancient art works, but the memories of the great men who passed through them, and the public memories that cluster there.[21] It is Cicero's revelation of such memories that now drive Atticus's attachment to Arpinum.[22] As he states, they are meaningful because 'we are affected in some mysterious way by places about which cluster memories of those whom we love and admire' (Cicero *De Legibus,* 2.4). Here, we recognise the profound impact when such sites of deep memory are destroyed or built over – even if they were never monumental. It is this emotion which Livy's speech of Camillus struggles to elicit for the whole of Rome, for it to be such a site of memory. The chronological depth in the layering of this memory, whether in Athens or Arpinum, also reveals the difficulty of reestablishing such connections to place elsewhere, especially when the new site of habitation is one created out of a moment of hostility and exile.

When Cicero's dialogue with Atticus turns to consider Rome as *patria*, the discourse is framed differently to that of Arpinum. There is no appeal made to comparable Roman sites of memory as those in Arpinum. Rome is primarily discussed in relation to the institution of citizenship; it is presented in the abstract and appears intangible. The meaning of Rome as the *patria communis,* for which he is willing to die, is not derived from its structures or location (Cicero *De Legibus,* 2.5). Rome's physicality serves as the meeting point for action and institutions to facilitate the existence of the *res publica.* But, as Cicero argues in his writings after his own return from exile (he was exiled from Rome in 58 BC by rival politicians), even that could be moved elsewhere if not in the right hands. The communal memories which Cicero presents tend to be centred on the institutions or interactions, the fabric of the city is rarely invoked.[23] For Cicero, Rome is a contemporary place, and it is 'practiced' through his own activities and life there. By his own acknowledgement, it takes the writer Varro's revelation of the stories about Rome for it to be made into a site of communal memory: '"You are right, Varro", I replied. "For in our own city we were like foreigners wandering and drifting but your books brought us home, so that we might recognise who and where we were..."' (Cicero *Academica Posteriora* 9 (1.3))

In appealing to the memory scape, ancient authors exemplify the same process that Tilley had already noticed in the pre-historic monumental landscape. He observed that

> If stories are linked with regularly repeated spatial practices, they become mutually supportive, and when a story becomes sedimented into the landscape, the story and the place dialectically help to construct and reproduce each other. Places help to recall stories that are associated with them, and places only exist (as named locales) by virtue of their employment in a narrative.
>
> (Tilley, 1994, p. 33)

There is no place without narrative and hence no place without memory.[24] We may add that there is also no place without the potential of future memory, and it is this that the work of architects Alessandro Petti and Sandi Hillal from within camps confronts directly. Petti addresses this in the following dialogue, responding to the observation by Eyal Weizman on their work.

> WEIZMAN: … But you battled against this situation of a receding future, and against the intellectual and architectural challenges of dealing with that. There is always an immediacy in your built architecture, in Shu'fat, Fawwar, or The Concrete Tent, but the future that you speculate on is a long-term one.
>
> …
>
> PETTI: I'm against this discourse about the future, because we've found a much more effective way to think political transformation than messianic Marxism. We understood decolonization as an endless struggle, one that is happening right now, right here. There are already fragments of futures in the present. You imagine something, and at the same time live it. It is liberating to understand political transformation without being trapped in the idea that one day everything will be solved, and we will all live happily. The work that we have been doing in refugee camps is already the future; it is already something that deals with people that live outside the nation state. Working within and against the condition of permanent temporariness means opposing two fronts at the same time: the perpetuation of the status quo, that imposes an unbearable condition of precarity on people, an normalization, trying to put all the broken pieces of the nation state back into its box.
>
> (Hillal and Petti, 2018, pp. 141–2)

Twinning the city and the camp

In the construction of the imagined community of the Dheisheh Camp and the City of Rome, there is an ambivalent relationship with territory as a way of articulating what a community is, whether that of Palestinian refugees or Romans. There is an acceptance that place-based rites and practices are the catalysts for memory and myth-making that inspire belonging and attachment. These make land more than just a site of refuge or asylum; there is the potential to convert it into a home, transforming inhabitants into a collective through shared practices. For Palestinian refugees, this brings with it concerns of normalising the exceptionality of the camp. Yet, there is also a potential for allowing it to be 'a translation of something beyond itself'. In both cases, despite the very different contexts, the condition of exile has a conspicuous place in giving coherence to the community, rather than for example citizenship, which seems extraneous. Interestingly, neither invokes images of the 'other' to achieve such coherence.

In both the city and the camp, there is an acknowledgement of the role of boundaries that delineate the official jurisdiction of politics and services, even if in practice, they stretch outside them. In their very different positions of strength, both Dheisheh, through the workings of its Popular Committee, and Rome, in its imperial capacity, are both representative of, and have influence on, communities beyond these boundaries (whether the 1949 UN cordon, or the *pomerium*). Their influence spreads not in a blanket way but in patches from a central point, a metaphoric *terminus*. In Rome, this was represented by the deity *Terminus* – the god of beginnings and endings – epitomised in a single boundary stone, which stood on the Capitol. In Dheisheh, the Al Feniq Cultural Centre, which has become the hub of socio-political activity, may act in the same capacity, especially through its name – meaning Phoenix – also a symbol of endings and beginnings (Figure 15.3).[25] They are also a link to that place of origin as not just a point of belonging in the past, but into the future, through the present of the camp. These symbols embody the ambivalence between steadfastness and transience – the immobility of physical boundaries and the temporality of

Figure 15.3 The courtyard of Al Feniq Cultural Centre, Dheisheh Refugee Camp, the West Bank, Palestine. A gathering for a workshop: Reading Fanon in Palestine Today

the camp or the city fabric. Such ambivalence is problematic for the nation-state, predicated on being a fixed political, civic, cultural and geographic unity. In the case of Dheisheh and Rome, despite the very different historical circumstances of their formation and their constituency – one a seemingly homogenous community of refugees and their descendants, the other a cosmopolitan capital – there is an acceptance of the co-existence between such fixity and mobility in time and space. The shared condition of the city and the camp may be symbolised in the grave marker of the French activist Fernand Tuil, in the gardens of Al Feniq, commemorating the twinning of his Montataire with Dheisheh.

Acknowledgements

This chapter is a result of discussions at distant intersections between city, campus and camp. I am grateful to those who generously shared ideas and imaginations, in 2015 in Dheisheh, Palestine, with *Campus in Camps,* in 2016 at the *Topographies of Citizenship Seminar* in Cambridge, and in 2019 at *City Debates,* AUB in Beirut. The ideas in this chapter developed from the possibility offered by Alessandro Petti and Sandi Hillal's hospitality at *DAAR* in Beit Sahour, Isshaq Al-Barbary and Diego Segatto's introduction to the *Suburb,* begun during our early explorations, and reading ancient insights together in Ramallah and Dheisheh: *Place, Heritage and Belonging: Livy and Cicero,* 17 November 2015 (http://www.campusincamps.ps/projects/place-heritage-and-belonging-livy-and-cicero/). I am indebted to Charles Tripp's astute response to an early version of this chapter, and Howayda Al-Harithy, Boano Camillo's, Elena Fiddian-Qasmiyeh's, Samar Maqusi and Yousif M. Qasmiyeh's expansive insights on rethinking the City and the Camp.

Notes

1. These ideas were first formulated in discussions during the workshop '*Al Masha'* or the Space of the Common', curated by Sandi Hilal and Alessandro Petti, as part of *A Journey of Ideas Across: In Dialogue with Edward Said.* Haus der Kulturen der Welt, Studio Armin Linke, ECLA of Bard, Liberal Arts University in Berlin, October 29–November 2, 2013.
2. See, for example, the discourse in the volume *Displacement and the Humanities,* especially between Gray (2018) and Boano (2019); Padilla Peralta (2019); Isayev (forthcoming).
3. The work as described in the writing between ancient and modern by Husseini, Isayev and Magee (2019, pp. 295–6). For details of the art by Cildo Mereiles work, see discussion in Brett et al. (1990, p. 40).
4. For a discourse on memories and meanings of camps, see Fiddian-Qasmiyeh (2020).
5. I am grateful to the patient discussions with Howayda Al-Harithy in developing these ideas and pushing my understanding of the relationship between city and camp.

6. For discussions on the suburb, much of the material has been drawn from conversations with members of Dheisheh refugee camp, participants of Campus in Camps, and the publication: Abu Aker and Al Lahham (2013). *The Suburb: Transgressing Boundaires*, Campus in Camps, Dheisheh Refugee Camp.
7. For a longue durée perspective on refugee agency, see Isayev (2017b).
8. Massey (2005, pp. 154, 158). Developing Lefebvre ([1974] 1991, esp. 10–12), she goes on to note that at any moment negotiations of place take place on the move, between identities which are on the move.
9. Kandinsky (1969, p. 57): 'A great city built according to all the rules of architecture and then suddenly shaken by an unpredictable and incalculable force'. The city is changed through movement and practices, creating under the surface a metaphorical or mobile city within the one that is planned. In this work, Kandinsky conceives of spatial experience mainly in psychoanalytic terms, and traces it to the experience of space in childhood. In relation to practised space, De Certeau (1984, p. 110).
10. These issues and Livy's speech of Camillus, especially in the context of ancient mobility, as well as ancient mapping practices (or rather countercartographies) are discussed in detail in: Isayev (2017a, Chapters 10 and 11).
11. Discussion of roads and the Roman mindset, suggesting that even the Tabula Peutingeriana was less of a practical itinerary than a showpiece: Talbert (2010); Talbert (2012, pp. 248–50); Salway (2005).
12. The Cisalpine region was probably enfranchised in 49 BC: Sherwin-White (1973, pp. 157–59); Williams (2001, pp. 15–16); Isayev (2017a, Chapter 9), for the enfranchisement of Italy in 89 BC.
13. The myths surrounding the near destruction of the Fabii clan (*gens*) and on the history of the clan within Roman society: Smith (2007); Richardson (2012).
14. Eckstein (2006, pp. 311–12), notes how the Roman system of legal status-groups replaced ethnicity and geographical location as the basis of membership in the polity.
15. Livy 5.52.15. Varro describes the etymology of the *pomerium*, and its practice: Varro *De Lingua Latina* 5.143; Varro, *De Re Rustica* 2.1.9–10, and Cato *Origines* F66 (= Chassignet 1.18, Peter F18; Servius *ad Verg. Aen.* 5.755), describes the related Etruscan rite of city foundation, with overview of city foundation rituals and their origin.
16. The Twelve Tables outlined the burial laws in *Tabula* X = *Roman Statutes* 40, (Vol. II, 704–11). A more detailed account of such laws survives as part of the *Lex Coloniae Genetivae*, Chapters 73–74 = *Roman Statutes* Law 25, (Vol. I, 393, esp. 403–4).
17. Varro *de Lingua Latina,* 6.49 for a definition of monimenta and its connection to memory.
18. The fundamental work on Roman *patria* is Bonjour (1975). For a summary of the scholarship on *patria*: Farrell (2001, p. 18). A more recent collection of studies on *patria* and political and ethnic unity: Urso (2008). For the development of the concept in later periods: Eichenberger (1991); Gasser (1999).
19. For definitions of the *res publica,* especially in reference to Cicero *De Re Publica* 1.39.1: Schofield (1995); Hillard (2005); Ando (1999, p. 14).
20. *De Legibus* is a philosophical treatise written alongside his other work *De Re Publica* and uses paradigms from Plato. The dialogue, which opens the second book of the *De Legibus* is part of a wider discussion on the constitution, religion and its connection to politics as well as the conception of law and nature. For Cicero's aims in the *De Legibus*: Dyck (2004, pp. 238–63). For the meaning of *patria* in this part of the dialogue, see Salmon (1972); Farrell (2001, pp. 18–27). For the rhetorical aspects of this work, see Steel (2001); Connolly (2007).

21. This theme is taken up again in Cicero's *De Finibus* 5.1–2, as Cicero, Piso and their friends make their way through Athens pointing out places that move them, evoking stories, both real and mythical.
22. In presenting this dichotomy between natural and cultural attachment, Farrell (2001, pp. 18–27), notes that it takes Cicero to point out to Atticus that, since he is Roman born, he has no sense of a natural *patria* as distinct from an adoptive one.
23. For the way that Cicero assigns meanings to the material world (for example, in the *Verrines*), and produces the city in different forms, whether as the centre of all vice or culture, depending on his need (as in the *pro Roscio* or the *pro Caelio*): Vasaly (1993, Chapters 3 and 5).
24. Rowlands (1993, p. 144) articulates the way material culture, through asserting its own memory, links past, present and future.
25. The Al Feniq Cultural Centre is a hub for diverse activities and gatherings, Campus in Camps classroom, Edward Said library, a wedding venue and a hostel for families whose relatives are in nearby hospitals. The image here in Figure 15.3 depicts the gathering for a workshop: *Reading Fanon in Palestine Today* 19 November 2015. (http://www.campusincamps.ps/projects/reading-fanon-in-palestine-today/).

References

Abu Aker, Q. and Al Lahham, A. (2013). *The Suburb: Transgressing Boundaires.* Campus in Camps, Dheisheh Refugee Camp, www.campusincamps.ps/projects/05-the-suburb/.

Agamben, G. (2003). *State of Exception.* Chicago, IL: University of Chicago Press.

Agier, M. (ed.) (2014). *Un Monde de Camps.* Paris: La De´couverte.

Ando, C. (1999). Was Rome a Polis? *Classical Antiquity* 18 (1), 5–34.

Boano, C. (2019). From Exclusion to Inhabitation: Response to Gray, Benjamin. Citizenship as Barrier and Opportunity for Ancient Greek and Modern Refugees, in *Displacement and the Humanities* (Special Issue of *Humanities* 6/7/8.3), edited by E. Isayev, and E. Jewell, https://doi.org/10.3390/h8030125 - 16 Jul 2019.

Bonjour, M. (1975). *Terre Natale: Études sur une Composante Affective du Patriotisme Romain.* Paris: Les Belles Lettres.

Brett, G., I. Gallery, and Gallery, C. (1990). *Transcontinental: An Investigation of Reality: Nine Latin American Artists.* London: Verso.

Chassignet, M. (2002). *Caton: Les Origines, Fragments.* Paris: Les Belles Lettres.

Chaleshtoori, S. (forthcoming). 'Deconstructing Citizenship: Subjects, Status and States'.

Connolly, J. (2007). *The State of Speech: Rhetoric and Political Thought in Ancient Rome.* Princeton – Oxford: Princeton University Press

Crawford, M.H. (1996). *Roman Statutes.* London: Institute of Classical Studies.

De Certeau, M. (1984). *The Practice of Everyday Life* (trans. by S. Rendall). Berkeley: University of California Press.

Dench, E. (2005). *Romulus' Asylum: Roman Identities from the Age of Alexander to the Age of Hadrian.* Oxford: Oxford University Press.

Dyck, A.R. (2004). *A Commentary on Cicero, De Legibus.* Ann Arbor: University of Michigan Press.

Eckstein, A.M. (2006). *Mediterranean Anarchy, Interstate War, and the Rise of Rome.* Berkeley: University of California Press.

Eichenberger, T. (1991). *Patria: Studien zur Bedeutung des Wortes im Mittelalter (6.-12. Jahrhundert).* Sigmaringen: J. Thorbecke.

Farrell, J.A. (2001). *Latin language and Latin Culture: From Ancient to Modern Times*. Cambridge – New York: Cambridge University Press.

Feldman, I. (2015). Looking for Humanitarian Purpose: Endurance and the Value of Lives in a Palestinian Refugee Camp, *Public Culture* 27 (3), 427–447.

Fiddian-Qasmiyeh, E. (2020). Memories and Meanings of Camps (and more-than-camps), in *Refugee Imaginaries: Contemporary Research Across the Humanities*, edited by D. Farrier, A. Woolley, L. Stonebridge, S. Durrant, and E. Cox. Edinburgh: Edinburgh University Press.

Gasser, F. (1999). *Germana Patria*. Leipzig: B. G. Teubner.

Gray, B. (2018). Citizenship as Barrier and Opportunity for Ancient Greek and Modern Refugees, in *Displacement and the Humanities* (Special Issue of *Humanities* 6/7/8.3), edited by E. Isayev, and E. Jewell, https://doi.org/10.3390/h7030072 – 19 Jul 2018.

Harvey, D. (2009). *Cosmopolitanism and the Geographies of Freedom*. New York: Columbia University Press.

Hillal, S. and Petti, A. (2018). *Permanent Temporariness*. Stockholm: Art and Theory Publishing.

Hillard, T.W. (2005). Res Publica in Theory and Practice, in *Roman Crossings: Theory and Practice in the Roman Republic*, edited by K. Welch and T.W. Hillard. Swansea: Classical Press of Wales, 1–48.

Hussaini, A., Isayev, E., and Magee, P. (2019). 'The Sky is Hidden': On the Opening up of Language and National Borders, *GeoHumanities* 5 (1), 295–311. DOI: 10.1080/2373566X.2019.1586560.

Isayev, E. (2017a). *Migration, Mobility and Place in Ancient Italy*. Cambridge: Cambridge University Press.

Isayev, E. (2017b). Between Hospitality and Asylum: A Historical Perspective on Agency, *International Review of the Red Cross, Migration and Displacement* 99 (1), 75–98.

Isayev, E. (forthcoming). Ancient Wandering and Permanent Temporariness – The Turning Point, *Displacement and the Humanities* (Special Issue of *Humanities* 6/7/8.3), edited by E. Isayev, and E. Jewell.

Isayev, E. and Jewell, E. (eds.) (2017–2020 forthcoming). *Displacement and the Humanities* (Special Issue of the *Humanities* 6/7/8.3). https://www.mdpi.com/journal/humanities/special_issues/Manifestos_Ancient_Present

Isin, E. (2002). *Being Political: Genealogies of Citizenship*. Minneapolis: University of Minnesota Press.

Isin, E.F. (2009). Citizenship in Flux: The Figure of the Activist Citizen, *Subjectivity* 29 (1), 367–388.

Kandinsky, V. (1969). *Du spirituel dans L'art*. Paris: Denoël.

Lefebvre, H. ([1974] 1991). *The Production of Space*. (*Production de l'espace*, Paris, trans. by D. Nicholson-Smith), Oxford – Cambridge Ma.: Blackwell.

Malkki, L. (1996). National Geographic: The Rooting of Peoples and the Territorialization of National Identity Among Scholars and Refugees, in *Becoming National*, edited by G. Eley, and R. G. Suny. Oxford: Oxford University Press, 434–453.

Massey, D. (2005). *For Space*. London: Sage.

Padilla Peralta, D. (2019). Citizenship's Insular Cases, from Ancient Greece and Rome to Puerto Rico, in *Displacement and the Humanities* (Special Issue of *Humanities* 6/7/8.3), edited by E. Isayev, and E. Jewell, https://doi.org/10.3390/h8030134 - 08 Aug 2019.

Padrón, R. (2004). *The Spacious Word: Cartography, Literature, and Empire in Early Modern Spain*. Chicago: University of Chicago Press.

Petti, A. (2017a). Refugee Heritage – Part 1 – Identification of the Property, *e-flux architecture*: https://www.e-flux.com/architecture/refugee-heritage/100502/identification-of-the-property/.

Petti, A. (2017b). Refugee Heritage. Part III Justification for Inscription, in *Displacement and the Humanities* (Special Issue of *Humanities* 6/7/8.3), edited by E. Isayev, and E. Jewell, https://doi.org/10.3390/h6030066

Qasmiyeh, Y.M. (2019). There It Is: The Camp that Is yet to Be Born, *Refugee Hosts*, 5 August 2019, https://refugeehosts.org/2019/08/05/there-it-is-the-camp-that-is-yet-to-be-born/.

Richardson, J.H. (2012). *The Fabii and the Gauls. Studies in Historical Thought and Historiography in Republican Rome*. (Historia Einzelschriften, 222). Stuttgart.

Rowlands, M. (1993). The Role of Memory in the Transmission of Culture, in *Conceptions of Time and Ancient Society (World Archaeology)* 25 (2), 141–51.

Salmon, E.T. (1972). Cicero, Romanus an Italicus Anceps, in *Cicero and Virgil, Studies in Honour or Harold Hunt*, edited by J.R.C. Martyn. Amsterdam: Hakkert, 75–86.

Salway, B. (2005). The Nature and Genesis of the Peutinger Map, *Imago Mundi* 57 (2), 119–35.

Schmitt, C. ([1950] 2003). *The Nomos of the Earth in the International Law of the Jus Publicum Europaeum* (trans. by G.L. Ulmen). Candor, NY: Telos Press.

Schofield, M. (1995). Cicero's Definition of Res Publica, in *Cicero the Philosopher: Twelve Papers*, edited by J.G.F. Powell. Oxford: Clarendon Press, 63–83.

Sherwin-White, A.N. (1973). *The Roman Citizenship*, 2nd edn. Oxford: Oxford University Press.

Sigona, N. (2015). Campzenship: Reimagining the Camp as a Social and Political Space, *Citizenship Studies* 19, 1–15.

Smith, C. (2007). *The Roman Clan: The Gens from Ancient Ideology to Modern Anthropology*. Cambridge: Cambridge University Press.

Steel, C.E.W. (2001). *Cicero, Rhetoric, and Empire*. Oxford: Oxford University Press.

Talbert, R.J.A. (2010). *Rome's World: The Peutinger Map Reconsidered*. Cambridge: Cambridge University Press.

Talbert, R.J.A. (2012). Roads not Featured: A Roman Failure to Communicate? in *Highways, Byways, and Road Systems in the Pre-modern World*, edited by S.E. Alcock, J. Bodel, and R.J. Talbert. Chichester: Wiley-Blackwell, 235–254.

Tilley, C. (1994). *A Phenomenology of Landscape, Places, Paths and Monuments*. Oxford – Providence: Berg.

Turner, S. (2016). What Is a Refugee Camp? Explorations of the Limits and Effects of the Camp, *Journal of Refugee Studies* 29 (2) (Special issue: *What is Camp?*), 139–148.

Urso, G. (ed.) (2008). *Patria Diversis Gentibus Una? Unità Politica e Identità Etniche nell'Italia Antica*. Pisa: ETS.

Vasaly, A. (1993). *Representations: Images of the World in Ciceronian Oratory*. Berkeley: University of California Press.

Williams, J.H.C. (2001). *Beyond the Rubicon: Romans and Gauls in Republican Italy*. Oxford: Oxford University Press.

Appendix

Livy *Ab Urbe Condita*
Book 5.50 - 55
Camillus' Speech on the move to Veii

From the Founding of the City
Translation and text based on various sources, including J. Bayet and G. Baillet 1954, *Belles Lettres*, and B.O. Foster 1924, *Loeb* editions.

[49]
Line 7
Dictator reciperata ex hostibus patria triumphans in urbem redit, interque iocos militares, quos inconditos iaciunt, Romulus ac parens patriae conditorque alter urbis haud vanis ladibus appellabatur.

IXL

The dictator, having recovered his country from her enemies, returned in triumph to the city; and between the rough jests uttered by the soldiers, was hailed in no unmeaning terms of praise as a Romulus and Father of his Country and a second Founder of the City.

8
Servatam deinde bello patriam iterum in pace haud dubie servavit cum prohibuit migrari Veios, et tribunis rem intentius agentibus post incensam urbem et per se inclinata magis plebe ad id consilium; eaque causa fuit non abdicandae post triumphum dictaturae senatu obsecrante ne rem publicam in incerto relinqueret statu.

His country, which he had saved in war, he then indubitably saved a second time, now that peace was won, by preventing the migration to Veii: though the tribunes were more zealous for the plan than ever, now that the City lay in ashes, and the plebs were of themselves more inclined to favour it. This was the reason of his not resigning the dictatorship after his triumph, for the senate besought him not to desert the state in its hour of uncertainty.

[50]
Lines 1–3
Omnium primum, ut erat diligentissimus religionum cultor, quae ad deos immortales pertinebant rettulit et senatus consultum facit: fana omnia, quoad ea hostis possedisset, restituerentur terminarentur expiarenturque, expiatioque eorum in libris per duumviros quaereretur; cum Caeretibus hospitium publice fieret quod sacra populi Romani ac sacerdotes recepissent beneficioque eius populi non intermissus honos deum immortalium esset;

L.

His first act in line with his scrupulous attention to religious obligations was to lay before the senate those matters relating to the immortal gods, and to obtain the passage of a decree that all shrines, in so far as they had been in possession of the enemy, were to be restored, their boundaries established and the rites of purification celebrated and the duumvirs should search the (Sybelline) Books for the proper rites; that a covenant of hospitality would be entered into by the state with the people of Caere, because they had received the sacred things of the Roman People and its priests and thanks to their good deeds, worship of the immortal gods had not been interrupted;

(continued)

4–5

Ludi Capitolini fierent quod Iuppiter optimus maximus suam sedem atque arcem populi Romani in re trepida tutatus esset; collegiumque ad eam rem M. Furius dictator constitueret ex iis qui in Capitolio atque arce habitarent. Expiandae etiam vocis nocturnae quae nuntia cladis ante bellum Gallicum audita neglectaque esset mentio inlata, iussumque templum in Nova via Aio Locutio fieri.

That Capitoline Games should be held, because Jupiter Optimus Maximus had protected his home/place/abode and the Citadel of the Roman People in its time of danger; and that Marcus Furius, the dictator, should to that end constitute a college of men from amongst those who were living on the Capitol and in the Citadel. A proposal was also made for propitiating the voice which was heard in the night to foretell disaster before the Gallic War, and was disregarded, and orders were given for a temple to be built in the Nova Via to Aius Locutius.

6

Aurum quod Gallis ereptum erat quodque ex aliis templis inter trepidationem in Iouis cellam conlatum cum in quae referri oporteret confusa memoria esset, sacrum omne iudicatum et sub Iouis sella poni iussum.

The gold which had been carried away from the Gauls and that which during the alarm had been collected from the other temples, and carried into the shrine of Jupiter, since there was no clear recollection where it ought to be returned, the whole was adjudged to be sacred, and ordered to be deposited under the throne of Jupiter.

7

Iam ante in eo religio ciuitatis apparuerat quod cum in publico deesset aurum ex quo summa pactae mercedis Gallis confieret, a matronis conlatum acceperant ut sacro auro abstineretur. matronis gratiae actae honosque additus ut earum sicut uirorum post mortem sollemnis laudatio esset.

Even before this the religious feeling of the citizens had already been apparent in the fact that when there was not sufficient gold in the public coffers to make up to the Gauls the stipulated sum, they had accepted what the matrons got together, to avoid touching the sacred gold. For this, a vote of thanks was given to the matrons, and they were granted the honour of having eulogies pronounced at their funerals, as in the case of the men.

8

His peractis quae ad deos pertinebant quaeque per senatum agi poterant, tum demum agitantibus tribunis plebem adsiduis contionibus ut relictis ruinis in urbem paratam Veios transmigrarent, in contionem uniuerso senatu prosequente escendit atque ita uerba fecit.

It was not until after those measures which concerned the gods, and lay within the competence of the senate, had been enacted that, heeding the importunity of the tribunes, who were urging the plebs unceasingly to quit the ruins and emigrate to a city ready to their hand at Veii, Camillus went up into the assembly, attended by the entire senate and discoursed as follows:

(continued)

[51] Lines 1–2	**LI.**
"Adeo mihi acerbae sunt, Quirites, contentiones cum tribunis plebis, ut nec tristissimi exsilii solacium aliud habuerim, quoad Ardeae uixi, quam quod procul ab his certaminibus eram, et ob eadem haec non si miliens senatus consulto populique iussu reuocaretis, redituras unquam fuerim. Nec nunc me ut redirem mea uoluntas mutata sed uestra fortuna perpulit; quippe ut in sua sede maneret patria, id agebatur, non ut ego utique in patria essem. Et nunc quiescerem ac tacerem libenter nisi haec quoque pro patria dimicatio esset; cui deesse, quoad uita suppetat, aliis turpe, Camillo etiam nefas est.	"So painful to me, Quirites, are controversies with the tribunes of the plebs, that my most bitter exile knew no other solace but this, all the time that I lived at Ardea, that I was far away from these contentions. And they are likewise the cause that though you had a thousand times recalled me by resolution of the senate and the people's vote, I intended never to return. Nor have I now been induced to do so by any change in my desires but by the alteration in your fortunes. For the issue was whether my country was to remain in her seat, not whether I was to be in my country at any cost. Even now I would gladly remain quiet and hold my peace, were not this too the battle for my country; whom to fail while life endures is in other men disgraceful, but in Camillus impious.
3–5	
Quid enim repetiimus, quid obsessam ex hostium manibus eripuimus, si reciperatam ipsi deserimus? Et cum uictoribus Gallis capta tota urbe Capitolium tamen atque arcem dique et homines Romani tenuerint, uictoribus Romanis reciperata urbe arx quoque et Capitolium deseretur et plus uastitatis huic urbi secunda nostra fortuna faciet quam aduersa fecit? Equidem si nobis cum urbe simul positae traditaeque per manus religiones nullae essent, tamen tam euidens numen hac tempestate rebus adfuit Romanis ut omnem neglegentiam diuini cultus exemptam hominibus putem. Intuemini enim horum deinceps annorum uel secundas res uel aduersas; inuenietis omnia prospera euenisse sequentibus deos, aduersa spernentibus.	For why did we seek to win her back, why rescue her, when besieged, from the hands of the enemy, if now that she is recovered, we desert her? And although, while the Gauls were victorious and in possession of the entire City, the Capitol nevertheless, and the Citadel were held by the gods and men of Rome, and now that the Romans are victorious and the City recovered, are the Citadel and Capitol to be deserted? Shall our prosperity make Rome more desolate than our adversity has done? Indeed, if we had no religious rites established with the founding of the City and by tradition handed down, yet so manifest has at this time the divine purpose been in the affairs of the Romans, that I for one should suppose it no longer possible for men to neglect the worship of the gods. For consider these past few years in order, with their successes and reverses; you will find that all things turned out well when we obeyed the gods, and ill when we spurned them.

(continued)

6–7	
Iam omnium primum, Veiens bellum—per quot annos, quanto labore gestum.—non ante cepit finem, quam monitu deorum aqua ex lacu Albano emissa est. Quid haec tandem urbis nostrae clades noua? Num ante exorta est quam spreta uox caelo emissa de aduentu Gallorum, quam gentium ius ab legatis nostris uiolatum, quam a nobis cum uindicari deberet eadem neglegentia deorum praetermissum?	First of all the war with Veii. How many years we fought, and with what painful exertion! And the end came not, until, admonished by the gods, we drew the water off from the Alban Lake. What I beseech you, of this unparalleled disaster that lately overwhelmed our City? Did it come before we disregarded the voice from heaven announcing the approach of the Gauls? before the law of nations was violated by our envoys? before we, that ought to have punished their fault, had passed it by, with the same indifference towards the gods?
8–9	
Igitur uicti captique ac redempti tantum poenarum dis hominibusque dedimus ut terrarum orbi documento essemus. Aduersae deinde res admonuerunt religionum. Confugimus in Capitolium ad deos, ad sedem Iouis optimi maximi; sacra in ruina rerum nostrarum alia terra celauimus, alia auecta in finitimas urbes amouimus ab hostium oculis; deorum cultum deserti ab dis hominibusque tamen non intermisimus.	And so it was that, defeated, captured, ransomed, we received such punishment at the hands of gods and men that we were a warning to the whole world. Adversity then turned our thoughts upon religion. We fled for refuge to the Capitol and its gods, to the seat of Jupiter Optimus Maximus; of our holy things, some, in the ruin of our fortunes, we concealed in the earth, others we removed to neighbouring cities out of sight of our enemies; in the worship of the gods, albeit forsaken of gods and men, yet were we unceasing.
10	
Reddidere igitur patriam et uictoriam et antiquum belli decus amissum, et in hostes qui caeci auaritia in pondere auri foedus ac fidem fefellerunt, uerterunt terrorem fugamque et caedem.	Therefore they have restored to us our country again, and victory, and our ancient renown in war that we had forfeited; and against our enemies, who, blinded with greed, broke treaty and troth in the weighing of the gold, have they turned dismay and rout and slaughter.

(continued)

[52] Lines 1–2 Haec culti neglectique numinis tanta monumenta in rebus humanis cernentes ecquid sentitis, Quirites, quantum uixdum e naufragiis prioris culpae cladisque emergentes paremus nefas? Urbem auspicato inauguratoque conditam habemus; nullus locus in ea non religionum deorumque est plenus; sacrificiis sollemnibus non dies magis stati quam loca sunt in quibus fiant.	**LII.** " As you consider these momentous effects upon the affairs of men, of serving or neglecting the gods, do you begin Quirites, to perceive how though yet scarce clear of the wreckage of our former guilt and calamity, we are headed towards a grievous sin? We have a City founded with due observance of auspice and augury; no site in it is not permeated by religion and the gods; for our annual sacrifices, their places are no less fixed than are their days.
3–4 Hos omnes deos publicos priuatosque, Quirites, deserturi estis? Quam par uestrum factum [ei] est quod in obsidione nuper in egregio adulescente, C. Fabio, non minore hostium admiratione quam uestra conspectum est, cum inter Gallica tela degressus ex arce sollemne Fabiae gentis in colle Quirinali obiit? An gentilicia sacra ne in bello quidem intermitti, publica sacra et Romanos deos etiam in pace deseri placet, et pontifices flaminesque neglegentiores publicarum religionum esse quam priuatus in sollemni gentis fuerit?	Are you, Quirites, going to desert all these gods both of state and family? How far does your action come up to that of the noble youth C. Fabius in the recent siege, which the enemy beheld with no less astonishment and admiration than yourselves, when he descended from the Citadel through the missiles of the Gauls and offered the annual sacrifice of the Fabian clan on the Quirinal Hill? What? Would you suffer no interruption, even in war, of family rites, but desert the public rites and the gods of Rome in time of peace? Would you have the pontiffs and the flamens less careful of the ceremonies of the state religion than a private citizen has been of the anniversary of his clan?
5–7 Forsitan aliquis dicat aut Veiis ea nos facturos aut huc inde missuros sacerdotes nostros qui faciant; quorum neutrum fieri saluis caerimoniis potest. Et ne omnia generatim sacra omnesque percenseam deos, in Iouis epulo num alibi quam in Capitolio puluinar suscipi potest? Quid de aeternis Vestae ignibus signoque quod imperii pignus custodia eius templi tenetur loquar? Quid de ancilibus uestris, Mars Gradiue tuque, Quirine pater? Haec omnia in profano deseri placet sacra, aequalia urbi, quaedam uetustiora origine urbis?	Perhaps someone may say that we shall either do these things at Veii, or thence dispatch our priests to Rome to do them; but of these courses neither can be followed without violation of the sacred rites. For not to enumerate all the kinds of rites and all the gods, is it possible at the feast of Jupiter that the couch should be spread elsewhere than in the Capitol? Why need I speak of Vesta's eternal fires, and the image which is preserved as a pledge of empire in her temple? or of your sacred shields, O Mars Gradivus and Quirinus our Father? All these would you leave behind on unconsecrated ground – things coeval with the City, and some more ancient than the City's origin?

(continued)

8–10

Et uidete quid inter nos ac maiores intersit. Illi sacra quaedam in monte Albano Lauiniique nobis facienda tradiderunt. An ex hostium urbibus Romam ad nos transferri sacra religiosum fuit, hinc sine piaculo in hostium urbem Veios transferemus? Recordamini, agite dum, quotiens sacra instaurentur, quia aliquid ex patrio ritu neglegentia casuue praetermissum est. Modo quae res post prodigium Albani lacus nisi instauratio sacrorum auspiciorumque renouatio adfectae Veienti bello rei publicae remedio fuit? At etiam, tamquam ueterum religionum memores, et peregrinos deos transtulimus Romam et instituimus nouos. Iuno regina transuecta a Veiis nuper in Auentino quam insigni ob excellens matronarum studium celebrique dedicata est die".	See, too, how great the difference between us and our ancestors. They handed to us certain rites to be solemnized on the Alban Mount and in Lavinium. But if we scrupled to transfer sacred rites from hostile cities to ourselves in Rome, can we shift them without sin from Rome to Veii, city of our enemies? Call to mind, I pray you, how often sacrifices are renewed because some point of antique ritual has been, through carelessness or accident, omitted. What was it, a while ago, after the portent of the Alban Lake, that brought relief to the commonwealth – then in the throes of war with Veii – if not, a renewal of the sacred rites and auspices? But, more than that, like men mindful of their old religious fervour, we have both brought in foreign deities to Rome and established new ones. Queen Juno was lately conveyed from Veii and enshrined on the Aventine, and how notable was that day, for the zeal of the matrons and the throng!

11–12

Aio Locutio templum propter caelestem uocem exauditam in Noua uia iussimus fieri; Capitolinos ludos sollemnibus aliis addidimus collegiumque ad id nouum auctore senatu condidimus; quid horum opus fuit suscipi, si una cum Gallis urbem Romanam relicturi fuimus, si non uoluntate mansimus in Capitolio per tot menses obsidionis, sed ab hostibus metu retenti sumus?	We ordered a temple to be built for Aius Locutius because of the voice from heaven clearly heard in the Nova Via. We have added to our annual festivals the Capitoline Games, and on the authority of the senate we have established a new college for this purpose. Was there any of these things we needed to have undertaken, if we meant to retire from the city of Rome along with the Gauls; if we remained not voluntarily in the Capitol, through so many months of siege, but constrained by fear of the enemy.

(continued)

13–14 De sacris loquimur et de templis; quid tandem de sacerdotibus? Nonne in mentem uenit quantum piaculi committatur? Vestalibus nempe una illa sedes est, ex qua eas nihil unquam praeterquam urbs capta mouit; flamini Diali noctem unam manere extra urbem nefas est. hos Veientes pro Romanis facturi estis sacerdotes, et Vestales tuae te deserent, Vesta, et flamen peregre habitando in singulas noctes tantum sibi reique publicae piaculi contrahet?	We talk of sacred rites and temples pray, what about the priests? Do you not realise what sacrilege will be committed? The Vestals surely have but that one abode, from which nothing ever caused them to remove but the capture of the city. The Flamen Dialis may not lie for a single night outside the city without sin. Will you make these priests Veientine instead of Roman? Shall thy Virgins forsake thee, Vesta, and the Flamen, as he dwells abroad, bring night after night, such guilt upon himself and the republic?
15–17 Quid alia quae auspicato agimus omnia fere intra pomerium, cui obliuioni aut neglegentiae damus? Comitia curiata, quae rem militarem continent, comitia centuriata, quibus consules tribunosque militares creatis, ubi auspicato, nisi ubi adsolent, fieri possunt? Veiosne haec transferemus? An comitiorum causa populus tanto incommodo in desertam hanc ab dis hominibusque urbem conueniet?	What about the other matters nearly all of which we transact, after taking auspices, within the pomerium? To what oblivion and neglect do we consign them? The curiate comitia which deals with the business of war, the centuriate comitia, where you elect the consuls and military tribunes– where can these be held, with due observance of the auspices, save in the customary places? Shall we transfer them to Veii? Or shall the people, for the sake of the comitia assemble with enormous inconvenience in their City, forsaken of god and man?
[53] Lines 1–2 At enim apparet quidem pollui omnia nec ullis piaculis expiari posse; sed res ipsa cogit uastam incendiis ruinisque relinquere urbem et ad integra omnia Veios migrare nec hic aedificando inopem plebem uexare. Hanc autem iactari magis causam quam ueram esse, ut ego non dicam, apparere uobis, Quirites, puto, qui meministis ante Gallorum aduentum, saluis tectis publicis priuatisque, stante incolumi urbe, hanc eandem rem actam esse ut Veios transmigraremus.	**LIII.** But you will say 'while it is obvious that everything will be polluted beyond all possibility of purification, yet the situation itself compels us to leave a City devastated by fire, and all in ruins, and migrate to Veii where everything is untouched, nor vex the poverty-stricken plebs with building here.' But that this is rather a pretext than a true reason is, I think apparent to you, Quirites, without my saying so; for you remember how, before the coming of the Gauls, when our roofed (buildings) public and private were unharmed and our City stood uninjured that this same proposal was urged of migrating to Veii.

(continued)

3–4	
Et uidete quantum inter meam sententiam uestramque intersit, tribuni. Vos, etiamsi tunc faciendum non fuerit, nunc utique faciendum putatis: ego contra—nec id mirati sitis, priusquam quale sit audieritis—etiamsi tum migrandum fuisset incolumi tota urbe, nunc has ruinas relinquendas non censerem. Quippe tum causa nobis in urbem captam migrandi uictoria esset, gloriosa nobis ac posteris nostris; nunc haec migratio nobis misera ac turpis, Gallis gloriosa est.	And consider, tribunes, how wide is the difference between my view and yours. You believe that even if the move was inadvisable then, it is inescapable now; I, on the contrary–and do not be surprised till you understand my meaning–am convinced that even if it was right to consider going while Rome still stood, to abandon her ruins now would be grievously wrong. For then our victory would have been a reason for migrating to a captured city– a reason glorious to ourselves and our posterity; but now such a removal is for us a wretched and humiliating course, and a glory to the Gauls.
5	
Non enim reliquisse uictores, sed amisisse uicti patriam uidebimur: hoc ad Alliam fuga, hoc capta urbs, hoc circumsessum Capitolium necessitas imposuisse ut desereremus penates nostros exsiliumque ac fugam nobis ex eo loco conscisceremus quem tueri non possemus. Et Galli euertere potuerunt Romam quam Romani restituere non uidebuntur potuisse?	For we shall be thought not to have left our country as victors, but to have lost it as men vanquished. It will be thought that the rout on the Allia, the capture of the City, the blockade of the Capitol, have compelled us to forsake our family gods, and sentence ourselves to banishment and exile from that place which we were powerless to defend. Have Gauls then been able to destroy Rome; and must Romans appear unable to restore her?
6–7	
Quid restat nisi ut, si iam nouis copiis ueniant—constat enim uix credibilem multitudinem esse—et habitare in capta ab se, deserta a uobis hac urbe uelint, sinatis? Quid? Si non Galli hoc sed ueteres hostes uestri, Aequi Volsciue, faciant ut commigrent Romam, uelitisne illos Romanos, uos Veientes esse? An malitis hanc solitudinem uestram quam urbem hostium esse? Non equidem uideo quid magis nefas sit. Haec scelera, quia piget aedificare, haec dedecora pati parati estis?	What more remains except for them to come again with fresh forces--we all know that their numbers surpass belief--and should wish to dwell in this City which they have captured and you have abandoned, but that you should suffer them? What if not the Gauls but your ancient foes the Volsci and Aequi should migrate to Rome? Would you wish them to be Romans and you Veientines? Or would you rather that this were a desert of your own than the city of your enemies? For my part, I do not see what could be more abominable. Are you ready to stomach these outrages, these infamies, because it irks you to build?

(continued)

8	
Si tota urbe nullum melius ampliusue tectum fieri possit quam casa illa conditoris est nostri, non in casis ritu pastorum agrestiumque habitare est satius inter sacra penatesque nostros quam exsulatum publice ire?	If in all the City no house could be put up better or bigger than is the famous hut of our Founder would it not be better to live in huts, as shepherds and rustics do, amongst our sacred monuments and our household gods, than go forth as one into exile?
9	
Maiores nostri, conuenae pastoresque, cum in his locis nihil praeter siluas paludesque esset, nouam urbem tam breui aedificarunt: nos Capitolio, arce incolumi, stantibus templis deorum, aedificare incensa piget? Et, quod singuli facturi fuimus si aedes nostrae deflagrassent, hoc in publico incendio uniuersi recusamus facere?	Our ancestors, refugees and herdsmen, at a time when there was nothing in this place/region but forests and marshes, built quickly a new City; and are we loath, though Capitol and Citadel are untouched and the temples of the gods are standing, to rebuild what has been destroyed by fire? And what each would have done for himself if his house had been burned, shall we refuse to do together after this common conflagration?
[54] Lines 1–2	**LIV.**
Quid tandem? Si fraude, si casu Veiis incendium ortum sit, uentoque ut fieri potest diffusa flamma magnam partem urbis absumat, Fidenas inde aut Gabios aliamue quam urbem quaesituri sumus quo transmigremus? Adeo nihil tenet solum patriae nec haec terra quam matrem appellamus, sed in superficie tignisque caritas nobis patriae pendet?	Well now, suppose that either through crime or chance a fire should break out in Veii, and that the wind should spread the flames, as may easily happen, until they consume a great part of the city; are we to quit it, and seek out Fidenae or Gabii, or any other city you please, and migrate there? Has the soil of our patria and this land which we call our mother so slight a hold on us? Is our love of patria confined to buildings and rafters?
3	
Equidem—fatebor uobis, etsi minus iniuriae uestrae [quam meae calamitatis] meminisse iuuat—cum abessem, quotienscumque patria in mentem ueniret, haec omnia occurrebant, colles campique et Tiberis et adsueta oculis regio et hoc caelum sub quo natus educatusque essem; quae uos, Quirites, nunc moueant potius caritate sua ut maneatis in sede uestra quam postea, cum reliqueritis eam, macerent desiderio.	And in truth I will confess to you–though I like not to recall the wrong you did me–that as often, during my absence, as I thought of my country, all these objects came into my mind: the hills and the fields and the Tiber and the region/landscape familiar to my eyes, and this sky beneath which I had been born and reared. And I wish these things may rather move you with affection, Quirites, to make you abide in your own seat, than afterwards, when you have left it, torment you with vain regrets.

(continued)

4	
Non sine causa di hominesque hunc urbi condendae locum elegerunt, saluberrimos colles, flumen opportunum, quo ex mediterraneis locis fruges deuehantur, quo maritimi commeatus accipiantur, mari uicinum ad commoditates nec expositum nimia propinquitate ad pericula classium externarum, regionum Italiae medium, ad incrementum urbis natum unice locum.	Not without cause did gods and men select this place for establishing our City– with its healthful hills; its convenient river, by which crops may be floated down from the midland regions and foreign commodities brought up; its sea, near enough for use, yet not exposing us, by too great propinquity, to peril from foreign fleets, a situation at the center of Italy– a spot/position, in short of a nature uniquely adapted for the expansion of a city.
5	
Argumento est ipsa magnitudo tam nouae urbis. Trecentensimus sexagensimus quintus annus urbis, Quirites, agitur; inter tot ueterrimos populos tam diu bella geritis, cum interea, ne singulas loquar urbes, non coniuncti cum Aequis Volsci, tot tam ualida oppida, non uniuersa Etruria, tantum terra marique pollens atque inter duo maria latitudinem obtinens Italiae, bello uobis par est.	This is proved by the very greatness of so new a city. It is now Quirites in its 365th year. Amongst all these ancient nations you have for so long a time been waging wars; and all this while– to say nothing of single cities–neither the Volsci joined with the Aequi, and all their powerful towns, nor all Etruria, with its enormous strength on land and water, and its hold/occupancy of the entire breadth of Italy from sea to sea, has been a match for you in war.
6–7	
Quod cum ita sit quae, malum, ratio est [haec] expertis alia experiri, cum iam ut uirtus uestra transire alio possit, fortuna certe loci huius transferri non possit? Hic Capitolium est, ubi quondam capite humano inuento responsum est eo loco caput rerum summamque imperii fore; hic cum augurato liberaretur Capitolium, Iuuentas Terminusque maximo gaudio patrum uestrorum moueri se non passi; hic Vestae ignes, hic ancilia caelo demissa, hic omnes propitii manentibus uobis di".	Since this is so, what a plague is the reason why you that have experienced these things should experiment with others? Granting that your valour may go elsewhere, yet surely the fortune of this place could not be taken along! Here is the Capitol, where men were told, when of old they discovered there a human head, that in that place should be the head/capital of the world and supreme power/imperium; here, when the Capitol was being cleared with augural rites, Juventas and Terminus, to the vast joy of your fathers, refused to be removed; here are Vesta's fires, here the shields that were sent down from heaven, here are all the gods propitious, if you remain".

(continued)

[55] Line 1

Mouisse eos Camillus cum alia oratione, tum ea quae ad religiones pertinebat maxime dicitur; sed rem dubiam decreuit uox opportune emissa, quod cum senatus post paulo de his rebus in curia Hostilia haberetur cohortesque ex praesidiis reuertentes forte agmine forum transirent, centurio in comitio exclamauit: “signifer, statue signum; hic manebimus optime”.

LV.

The speech of Camillus is said to have moved them, particularly where he touched upon religion; but the doubtful issue was resolved by a word that was let fall in the nick of time. It was while the senate, a little later, was deliberating about these matters in the Curia Hostilia; some cohorts returning from guard-duty were marching through the Forum, and as they came to the Comitium a centurion cried out, “Standard-bearers, fix your standard; it will be best for us to remain here”.

2–4

Qua uoce audita, et senatus accipere se omen ex curia egressus conclamauit et plebs circumfusa adprobauit. Antiquata deinde lege, promisce urbs aedificari coepta. Tegula publice praebita est; saxi materiaeque caedendae unde quisque uellet ius factum, praedibus acceptis eo anno aedificia perfecturos.

Hearing this sentence the senators came out from the Curia and shouted their acceptance of the omen, and the commons gathering round them signified approval. The bill was then rejected, and people began in random fashion to rebuild the City. The state supplied tiles, and granted everybody the right to quarry stone and to hew timber where he liked, after giving security for the completion of the structures within that year.

5

Festinatio curam exemit uicos dirigendi, dum omisso sui alienique discrimine in uacuo aedificant. Ea est causa ut ueteres cloacae, primo per publicum ductae, nunc priuata passim subeant tecta, formaque urbis sit occupatae magis quam diuisae similis.

In their haste, men were careless about making straight the streets, and paying no attention to their own and others’ rights, built on the vacant spaces. This is the reason that the ancient sewers, which were at first conducted through the public ways, at present frequently run under private dwellings, and the plan/appearance of the City resembles one that has been occupied (by settlers) rather than planned.

16 Urban recovery at the mall

Displacement and solace in Beirut's spaces of consumption

Judith Naeff

This chapter begins and ends in the reconstructed souks of Beirut's Central District (hereafter, BCD). When 15 years of civil wars came to an end in 1990, this area of Beirut was severely damaged. Given its symbolic value as the cosmopolitan heart of an otherwise divided country, reconstruction of BCD was prioritized over other parts of the country and the private real estate company Solidere, a French acronym for the Lebanese Society for the Development and Reconstruction of Beirut's Central District, was established to expropriate an area of 191 hectares for a decades-long mega-project, which has not yet been completed at the time of writing. From the very beginning, questions concerning expropriation, demolition, preservation, planning and design were subject to heated debates. Today, on the place where the traditional souks used to be, stands a well-designed fully pedestrian shopping mall, with shaded alleys and open plazas, punctuated by an abstract sculpture here and a fountain there (Figures 16.1 and 16.2). Excessively sanitized and highly securitized, the commercial floor space is largely occupied by international chains such as Gucci, Zara and Yves Saint Laurent as well as high-end cafés and restaurants. The residential floor space is largely vacant. When the apartments do have new tenants, they often do not live there year-round.

According to Saree Makdisi, we should see the souks as a synecdoche for the entire BCD project (1997, p. 686). Polished urbanism driven by profit resembles a shopping mall or entertainment park, more than a city, whether aimed at residential or commercial use, business or leisure. Following Makdisi, this chapter only loosely focuses on the shopping mall, zooming out to include other forms of sanitized consumerist urban spaces. The first part considers the role of displacement in the eerie nature of the sanitized mall in the heart of Beirut. The subsequent parts of the chapter explore forms of urban recovery that may develop within consumerist urban spaces like the Beirut Souks. Following a brief interlude on ethnographic studies of Middle Eastern malls, the chapter argues that in Rabee Jaber's *The Mehlis Report* and *Beyritus: Underground City* and in Hassan Daoud's *As She Once Was*, Beirut's spaces of consumption emerge as convivial sites offering solace in the face of loss, despite their implication in structures of displacement,

Figure 16.1 Beirut Souks, a pedestrian shopping mall

Figure 16.2 Beirut Souks, international chains

exclusion and exploitation. Building upon Walter Benjamin's ambivalent assessment of consumer capitalism, the chapter reads these works of fiction as attempts to recover the liberatory potential of human desire fostered in places wrought with inequality.

Beirut souks: A mall at the intersection of displacement and reconstruction

"Posh people occupy the restaurants and cafés; tourists stroll in the aseptic streets; the artisans of the old days have been dislodged; the regulars have lost their bearings; the memory places no longer exist" (Najjar, 2005, p. 458). This literary description of BCD ties together three interrelated forms of displacement: The expulsion of old tenants, the exclusive clientele and the erasure of memory places; in other words, actual displacement, socio-economic exclusion and cultural uprooting. Most residents and shop owners in BCD were displaced in the first two years of the civil war (1975–1976) and others in the subsequent years of violence, seeing their properties reduced to ruins. In later phases of the war, displaced families from other areas of the country moved into the damaged buildings. The first phase of Solidere's comprehensive reconstruction project involved the expropriation of all properties in the area and the eviction and relocation of approximately 30.000 squatters (e.g. Mango, 2003). Owners received shares in the company in return. This phase was a continuation, moreover, of previous rounds of demolition that took place illegally whenever there was a lull in the civil war violence (e.g. Verdeil, 2001). The experience of actual displacement was thus violently inscribed into the BCD reconstruction project, not only as its prehistory – the civil war destruction – but also in its history – the eviction, expropriation and demolition.

The second form of displacement at stake is socio-economic exclusion. Doris Summer explains that, in line with the neoliberalization of urban planning elsewhere, this exclusion concerns both the lack of citizen participation in the planning process and implementation, and the design of exclusive landscapes (2006, p. 8). Her fieldwork in Beirut and Amman shows that planners deliberately seek to exclude the lower strata of the population by designing the environment in such a way that both ambiance and pricing discourage anyone but the most wealthy residents and visitors to enter. In Sirine Fattouh's touching short video *Temps Morts*, a former bookstore owner returns to BCD complaining, "this is not a city center anymore. There are only cafés. In a city center there should be different shops, here a cobbler, there a… what's it called… anyway, a diversity of shops" (2005, min. 3:16–3:39). Likewise, in Yasmine Arif's ethnographic fieldwork, an old coffin maker responds to the architectural model of the new BCD that "one should be reasonable – they won't keep this kind of business here" (2002, p. 122). These shop owners have not only been displaced, but the reconstructed BCD also effectively casts them as outsiders along lines of age and lifestyle, as well as social class.

The third element of displacement is the loss of memory places and the concomitant sense of disorientation. I would like to argue for an understanding of displacement as a condition that emerges not only when people are forcibly removed from their place of belonging, but also when their places of belonging – whether residential, work or leisure – have transformed beyond recognition. While in some fields, such as in humanitarian aid provision, a more narrow definition of displacement might be desirable, I find it helpful to understand a shared sense of spatial disorientation and alienation caused by urban renewal in terms of displacement as well. The loss of memory places in BCD has of course to do with the large-scale demolition of buildings, many of which could have been restored (Fawaz and Ghandour, 2010). Many critics have understood this as part of what has been termed "collective amnesia," that is, as a deliberate attempt by the state to erase memories of a divisive past (e.g. El-Khoury, 1998; Tabet, 2001).

But it is not only the destruction and demolition of what was before, but also the type of urban design that came in its place, that is seen as profoundly unhomely. Described as "a strange impression to stroll in a décor" (Tabet, 2001, p. 56), "a 'mirage' of a new city" (Salam, 1998, p. 132), "a sense of Disney" (Cooke, 2002, p. 409) and "too artificial, like a theatre set" (Najjar, 2005, p. 458), the district turns city dwellers into consumers of a spectacle. Various critics argue that the district amounts to "nothing more than a postmodern pastiche" (Makdisi, 1997, p. 685), its cultural heritage reduced to "kitsch and sleazy consumerism" (Ragab, 2011, p. 111). Drawing upon thinkers such as Jean Baudrillard and Fredric Jameson, these scholars draw attention to the ways in which postmodern urbanism commoditizes all references to history or cultural identity (Jameson, 1997; Baudrillard, 1994). How does this cultural critique of the postmodern shopping mall relate to displacement?

At the heart of this cultural critique is the Marxist notion of alienation, a sense of estrangement resulting from commodity fetishism. In commodity fetishism, the production process is obscured in favor of a mysterious intrinsic or transcendent value of the object. Commodity fetishism leads to a feeling of estrangement from the world by both laborers and consumers (e.g. Musto, 2010; Billig, 1999). Billig argues that the commodity fetishism of late capitalism is grounded in the habitual forgetting or repression of the exploitative conditions that produce our consumerist dream worlds (1999). If place is understood as a location firmly embedded in a web of social relations and historical roots, the habitual repression of those relations implies a loss of place.

Marc Augé developed his term "non-place" to refer to very specific locations of late capitalism: The airport, the hotel chain and the shopping mall. Characterized by transitoriness, visitors are temporarily liberated from social and identarian relations such as class, tribe and family (Augé, 1995 [1992], pp. 101–104). The design is deliberately generic and the layout stimulates movement and consumption. As argued above, profit-driven and excessively polished urbanism could be seen as the expansion of the non-place specified by Augé to include parts of the city that offer residential, office and leisure facilities. Neoliberal urban planning thus inspires a sense of loss and disorientation that resembles displacement by repressing historical continuities and social relations, thus turning place into non-place. Processes such as privatization, commoditization and museumification, with their strategies of sanitization, polishing, atmospheric lighting, as well as securitization and high-end commerce, recast familiar surroundings as artificial. It invites temporary consumption, not durable dwelling. A sense of displacement is therefore fundamentally part of any shopping mall experience. Compared to suburban American malls, this process is imbued with heightened political, symbolic and emotional significance when entangled with other forms of often-violent displacement, such as in Beirut.

In late capitalist societies, this alienation is doubled by increased mediatization. Not the object itself, but the appearance of the object is fetishized. Walter Benjamin studied the nineteenth-century arcades as the moment where objects were no longer only fetishized as commodities-in-the-market, but also as commodities-on-display (Buck-Morss, 1989, p. 81). The arcades preceded the department store, which in turn formed the precursor to the shopping mall. If Benjamin described the spectacle of the nineteenth-century Paris as a phantasmagoria, a magic lantern show of optical illusions, Baudrillard describes the late-twentieth century as a world of images that no longer seem to refer to any reality outside of representation (e.g. Baudrillard, 1994). This, then, is the sense of Disney that the reconstructed BCD evokes (see Ragab, 2011, for a Baudrillardian analysis of BCD).

I understand urban recovery as the reemergence of urban liveliness after destruction; that is, the fostering of the peculiar sociality of cities, characterized by a multiplicity of sensual stimuli and fleeting encounters with strangers. How can a reconstruction project that expulses, excludes and alienates its residents offer the urban recovery so desperately needed by a city after the losses suffered from war violence? While urban liveliness is nurtured by commerce – the market, the café and the street vendor – it surely cannot be reduced to economic growth. Indeed, critics have argued that, instead of offering a site of recovery, the brutality of Solidere's reconstruction project forms a continuation of war violence (El-Khoury, 1998, pp. 183–184; Westmoreland 2008, pp. 241–242). In contrast, the following will argue that a space such as the Beirut souks, even if wrought by displacement, might accommodate and foster forms of urban life to which urban recovery aspires.

Shopping malls in the Middle East

The spread of shopping malls, or shopping-mall-like urban planning and design, is a quintessential manifestation of late-twentieth-century globalization. El-Sheshtawy argues that we should not look at globalization as solely externally imposed, but as produced by, and firmly embedded in, local conditions (2004, p. 18). We have seen that the souks in BCD may be generic in its unhomely planning and design, but also that this uprootedness is given unique meaning in its historical context of frontline devastation, its symbolical value as a potential site for reconciliation, and its political context of general amnesty and collective amnesia. However, this unique context may also have led to certain blind spots. In this section, I will review a number of studies on shopping malls in the wider Middle Eastern region, to present a more ambiguous approach to shopping malls than the amnesiac hyperreal non-places presented in the previous section.

Mona Abaza, in her research of Cairene shopping malls signals displacement at the level of socio-economic exclusion similar to that experienced in BCD: "With [modernization], a *baladi* way of life and culture is quickly erased to cater to the new Cairene yuppies" (2009, p. 100). But, she proceeds to identify a paradox. While security guards are expressly instructed to ward off lower-class visitors and to keep out the "riffraff" from the streets, visitors repeatedly describe the mall as an inclusive space of social mixing (2009, p. 213). If some malls have acquired a bad reputation as a place where young men pick up loose girls, more upscale malls are seen as spaces that allow modern and independent ways of being for youth and women in particular (Abaza, 2001). The mall is protected from the hot humidity, dusty traffic, and, to some extent, crime and harassment, but it is less restrictive than a private club for which visitors need to pay a membership fee. A "disguise" is desired from visitors of a popular background so as to partake in the mall's upper-middle class fantasy space and employees "masquerade," for example, by taking off their Islamic attire, before starting their work (2001, p. 118). Nevertheless, they express appreciation for the cleanliness and class of the shopping mall even though salaries are often lower than those in shops on the street (2009, p. 213).

Similarly, scholars have pointed out how in Iran and Saudi Arabia, where social conduct in public space is strictly policed, the shopping mall has an ambiguous function. On the one hand, the mall's hyper-surveillance has a triple function: Responding to crime, maximizing profit and protecting morality. In their study of camera surveillance in a mall in Riyadh, Alhadar and McCahill narrate how elderly women sitting on the floor were asked by security to take seat on chairs (2011, p. 322) fully in line with the erasure of *baladi* culture in Cairo described by Abaza. Most of those targeted by CCTV operators are young male singles who are suspected of flirting. In both Iran and Saudi Arabia, private security companies work together with morality police. Yet, despite this intense surveillance, young people use a variety of strategies to enter the mall and make contact with singles of the opposite sex, from eye contact to exchanging phone numbers. Why would they choose the mall for this kind of social mixing?

As seen in the Cairene malls observed by Abaza, the mall offers a protected atmosphere that provides a measure of both convenience and respectability that the hot and crowded streets of these Gulf cities do not. In addition, the particular layout of the mall is designed to stimulate constant movement as well as to entice shoppers to gaze at the commodities on display. Shahram Khosravi, in his study of the Gholestan mall in Tehran reminisces, "by being constantly on the move through corridors, floors, shops, stairs, we could not be stopped and separated from the opposite sex" (2008, p. 91). The fleeting placelessness of the shopping mall precisely offers the liberation from the restraints on social conduct that the Iranian youth seeks. Pedram Dibazar moreover observes in relation to Iranian shopping malls that the mall's "blurring of the distinction between the spaces for movement (corridors), purchase (shops) and display (window shops) [results in] the conflation between the subject positions of the stroller, the buyer and the commodity itself" (Dibazar, 2020, p. 126). Unlike the traditional shopping street with a more strict distinction between retail space and public space, the mall thus allows for a "scopic regime," i.e., the social configuration of modes of looking and being looked at, of which single young people in Iran can take advantage. While looking at objects on display, they can also look at each other and display themselves.

Despite various forms of exclusion and segregation, malls in the region are understood to have a democratic potential. Erkip suggests, for example, that malls in Turkey replicate social segregation according to the quality and pricing of the goods on offer, but nevertheless, provide "a more heterogeneous and democratic consumption site" by collecting these differences under one roof (Erkip, 2003, p. 1078). And similar to the way in which youth uses the mall to defy the rigid social norms of post-revolutionary Iran and Wahhabi Saudi Arabia, informal life unfolding in the cracks of the urban spectacle that is contemporary Dubai resists the highly commercial script prescribed by the urban lay-out (El-Sheshtawy, 2008).

What these studies from the region show us is that the critique of shopping malls as developed in Anglo-Saxon cultural theory and French philosophy fails to address the ambiguities of social life in the Middle Eastern mall. Far from the alienated consumer-zombie imagined by postmodern pessimists, those frequenting malls in different parts of the region praise it as a pleasurable site offering liberation and conviviality. Khosravi, therefore, dismisses the emphasis on false consciousness by critics following Jameson and Baudrillard. Instead, he follows Walter Benjamin's more ambivalent assessment of the Parisian arcades as simultaneously a capitalist dreamspace and a potential site for materializing collective energy (Khosravi, 2008, p. 92). While it has been suggested that Benjamin's notion of phantasmagoria is synonymous with commodity fetishism (Tiedemann, 1999, p. 938), it is also clear that Benjamin is not only interested in the more rigid Marxist endeavor of lifting the veil of false consciousness in order to show material objectivity. Inspired by the surrealists' faith in the liberatory potential of human desire, Benjamin is also interested in the progressive potentialities that the enchantment of "the new" might harbor despite its inevitable cooptation by capital (e.g. Tiedemann, 1999, pp. 939–940; Buck-Morss, 1989, e.g. pp. 115–120, 143–145).

Beirut is a very different place than Tehran, Cairo, Riyadh, Dubai or Ankara, but the ambivalence highlighted in the studies of these places is productive in the Lebanese context too. Such an ambivalence acknowledges that displacement does not mean erasure and that social life can thrive despite conditions of exclusion and exploitation. The following sections analyze literary representations of consumer spaces in post-civil war Beirut. They demonstrate an ambivalent approach similar to the one emerging from these ethnographic studies from the region, but tailored to the specific context of Beirut, as introduced in the first section of this chapter. Far from the hyperreal non-place lamented by critics such as Saree Makdisi and Tarek Ragab, the malls and other spaces of consumption in these novels are ordinary urban spaces, in which social life offers solace, despite being marked by displacement. The chapter is interested in the potential of these spaces' phantasmagoric enchantment to offer pleasure and foster sociality without disguising its means of production.

Rabee Jaber: The solace of spaces of consumption

With a pace of one novel each year, Rabee Jaber (b. 1972) is possibly the most prolific writer of Lebanon. Yet, apart from his prize-winning *Amrika*, his work remains largely unknown abroad. After growing up in Mount Lebanon, Jaber moved to Beirut in 1989 to pursue higher education, soon after which the war ended and the city suddenly transformed as if "in an enchanted world" (Sierra, 2017, citing al-Maqarrī, 2010). Various critics have drawn attention to the intersection of urban space and memory in his work (e.g. Aghacy, 2015; Seigneurie, 2011). For someone like Dalia Mostafa, Jaber's work is typical for post-civil war literature, whose "estranged protagonists [... suffering] from the distortion of identity and the dissociation of memory become part of the story of a city ravaged by war and contradictions" (2009, p. 234). Zeina Halabi, in contrast, draws attention to "the ways in which Jaber appropriates the narrative strategies and themes of his precursors in order to subvert them" (2018, p. 75). According to her, Mostafa misreads Jaber's novel *Ralf Rizqallah* as an allegory for the grand narrative of trauma and the nation. The novel rather points to the ways in which people may simply suffer from clinical depression amidst and against the grand strides of history. Such a gesture may subvert the towering figure of the Arab intellectual, as is Halabi's main argument, but it does not deny how such personal struggles may intersect with those of the city or country. If his precursors, such as Elias Khoury, lamented the impact of violence in grand impressions of catastrophic erasure, Jaber narrates how everyday life continues to unfold in a city scarred by war, and how grief and anxiety may punctuate an otherwise mundane existence in such a city.

The personal psychology of his characters resonates with, and responds to, an environment charged with past and imminent violence, without being fully defined by it, and without merely symbolizing it. This is also Zeina Tarraf's conclusion in her thoughtful reading of Jaber's *The Mehlis Report* when she argues that "the tense political moment weaves its way into the patterns of the ordinary, becoming absorbed by the perpetuation of the everyday" (2015, p. 205). It is on the level of the mundane that we begin to see the contours of urban recovery too, as I will attempt to show here.

The Mehlis Report follows protagonist Saman Yarid in the aftermath of the 2005 bombings. The title refers to a special UN investigation into the assassination of Rafiq Hariri. Restless, moody, tense and distracted, Saman seeks release in watching television and drinking at bars – while his disappeared sister Josephine roams a parallel realm of the dead situated within Beirut's geography. At the end of the novel, Saman dies of a heart attack before the Mehlis report has come out. As I argued elsewhere, the novel is therefore about life in a limbo: For Lebanon in the wake of the attacks; for Saman who lacks purpose and meaning in life; for his dead sister who has left the living but is unable to join the dead (Naeff, 2018, p. 118). "In a novel where those who leave are forgotten by a city endlessly changing and relentlessly transforming, the desire to stay attached to this place becomes valuable in and of itself," writes Ghenwa Hayek (2016, p. 597). Yet, Saman's relation with urban renewal is ambiguous. Saman is an architect and in the wake of the 2005 assassinations, he thinks that "these rising buildings are a good omen" (Jaber, 2013, p. 114). The narrator and the novel are thus marked by an ambiguity between the desire for attachment and the confidence in urban change.

This ambiguity is captured beautifully in the following description of a shopping mall in Achrafieh (Figures 16.3 and 16.4).

> He'd turned onto the street in front of the entrances to the ABC Mall. The city was buzzing that night. Tomorrow was Saturday: a holiday. The people had come out into the streets. A group of women had appeared. They were young, and were laughing and leaning on one another. They'd gone inside the mall and hovered around the fountain. On the other side of the street, men were lined up at the taxi stand. They were eating warm sandwiches from Sheikh Shawarma and from Falafel Fareeha. The two restaurants were next to each other, and the line of men stretched between them. Juice was dripping from the sandwiches onto the edge of the sidewalk, among the wheels of parked cars and bikes. The streetlights shone on their hands and mouths, and were reflected in their eyes. As they ate, the sandwich wrappers fell to the ground. They stared at the girls and women coming and going from ABC. The mall was buzzing too. The restaurants were packed, as were the cafés – even those underground. At first, people were angry about the mall being built here, here in the heart of Achrafieh. People were

Figure 16.3 ABC mall, Achrafieh, rooftop

angry because the developers had bought the sports club that used to be here and tore it down to build this mall. Ever since he was in college, since before he was in college, Saman had gone to the Sons of Neptune Club on a daily basis. When he heard ABC had bought the place and that it would be torn down, he'd felt a lump in his throat. But here the mall was: electric, teeming with people, its gorgeous white marble standing against the night. And that beautiful moon filling all the bodies with energy. The city was hopping, as if it hadn't been rocked by

Figure 16.4 ABC Mall, Achrafieh, interior

> explosions just a few weeks earlier. As if there weren't a threat of it being completely upended at any hour.
>
> (Jaber, 2013, pp. 34–35)[1]

The description shares with Alexandre Najjar's description at the opening of this chapter a sense of displacement: The people were angry and Saman mourned the loss of his club. Yet, the passage celebrates the liveliness of the consumerist sociality in the present, even if it is acutely aware of the inequalities of such a sociality. Significantly, the men at the shawarma restaurants are both separated from the ABC mall and implicated in its scene. The men are not only physically removed from the mall, but seem to belong to a lower class than the women who enter the mall. They are eating cheap food standing on the pavement at a taxi stand. At the same time, they are also implicated in the scene, as spectators looking at the girls and women walking to and fro, but also as part of the spectacle, because the author takes time to describe the men, the sandwiches, the dripping sauce and the positioning of the two restaurants. In Saman's experience of the city, and by extension the reader's, the men on the sidewalk are part and parcel of what makes the city vibrant and electric. Despite structural inequality running right through this spectacle, its extraordinary energy is expressly contrasted to feelings of loss and anxiety, to the "lump in the throat" when Saman's old club was demolished, and to the assassinations that charge the entire novel with a sense of numbed grief and imminent doom. The shopping mall in this section, therefore, offers solace in an imperfect world. The section encourages us to embrace life that flourishes within structures of inequality and violence.

Something similar can be seen in *Beyritus: Underground City*. The main narrator is a security guard, Butros, appointed by Solidere to safeguard a notorious civil war ruin in BCD. When chasing an intruder on a rainy night, he falls into a pit and wakes up in an underground city populated by pale-skinned people who once fled from the city aboveground. He spends a year with a host family in this parallel world, visited by memories of his life above, while he allows his body to recover sufficiently to manage the passage back to the other Beirut. Samira Aghacy reads the parallel city as an "oneiric place, an amalgamation of necropolis and pastoral, an antiquated and clandestine version of the upper city" (2015, p. 165). The function of this amalgamation is neither utopian nor dystopian. Like Beirut, Beyritus knows hospitality and distrust, care and neglect, grief for lost ones and fear for collapse. The underground city rather functions as Beirut's "shadowy double" (Aghacy 2015, p. 171). Elsewhere, I have read Butros's journey as the classical trope of the catabasis, with the underground space not so much a netherworld of the dead, as a combination of shelter and subconscious (Naeff, 2018, p. 209). Like his classical precursors, the hero descends to meet lost others, both in the memories that consistently haunt him, and in the climax when he sees his cousin Ibrahim, whom he thought dead, suffering among a group of blind beggars. But, as Ken Seigneurie astutely observes, unlike the classical catabasis, Butros does not speak to this lost loved one (2011, pp. 92–93). Instead, despite his sympathy, he remains silent and turns away, eager to return to his old world aboveground. His encounter with the dead is an encounter with the limits to his capacity to respond to the needs of others and an unambiguous embrace of his own life.

This chapter focuses on a specific theme of this journey: The ostentation, expenditure and noise of the world above in contradistinction to the silence, austerity and sustainability of the underground society. The novel opens with the frame narrative in which the first-person narrator, who, like the author, goes by the name Rabee, is having dinner on a roof top terrace above the Virgin store in BCD. The scene is described as plentiful. The city lights shimmer; the terrace is humming with languages from all over the world; odors of food, perfume, bodies and the salty sea mingle. Rabee and his friends eat grilled meat and pizza. They drink cold white wine and discuss how Beirut thrives from the many tourists from the Gulf that have flocked to the city for Summer (Jaber, 2006, pp. 9–10). It is in this setting that a strange-looking, thin man approaches and introduces himself as Butros. The guard wishes to share his story with the author, because he knows his novels, which revolve around real people and events (p. 15). With this opening, the author firmly positions himself within Solidere's district, enjoying the view, the food and the international audience. In contrast, Butros occupies a more ambiguous position. Employed by Solidere, he takes his duty to secure private property seriously. At the same time, he feels excluded when passing through the splendid nightlife of Rue Monot (Jaber, 2006, pp. 42–43) or when peering through the windows of an expensive restaurant (p. 72). Yet, as Seigneurie points out, during his time underground, Butros sheds his stupor and is reinterpellated as a desiring subject, which shifts his relation to urban spaces of consumption (2011, pp. 88–95). The following passage is exemplary for his transformation:

> That night I heard voices in Beirut, our city. Beirut above. I heard song and music (it was not foreign music. It was Arabic music. I am not sure which song, but it was a modern song. Not one of the old classics, not Umm Kalthoum, whom I love. Not Umm Kalthoum or Abdel Wahab or any of those. One of those common new songs which rise to stardom for two months, and the whole city pulsates with it, and after that we do not listen to it anymore. One of those songs I detested, with lyrics with no meaning – or with a meaning that was repeated endlessly until it no longer referred to anything but its own words, meaningless words, without spirit, senseless, with no value. It was a modern song, with music that did not resemble music, blending instruments and voices and I don't know what and I used to hate it, but then – while I was under the heavy, silent, dark earth – I thought it was the most beautiful song I had heard in my life.)
>
> (Jaber, 2006, pp. 68–69)

Like in the passage from *The Mehlis Report* on the ABC mall cited above, Jaber takes his time to describe the scene and uses this time to demonstrate an acute awareness of capitalist exploitation. If the mall is described in terms of class segregation, the song is described in terms of late capitalist acceleration. For Butros, as for cultural critics, such profit maximization of inflated stardom comes at the cost of meaning, value and durability. And yet, this passage celebrates the human desire for temporary pleasure.

When Butros emerges to the world above, leaving the people in Beyritus to their suffering, the first thing he sees is the terrace of the seaside restaurant Dbibu.

> This time it didn't bother me that they were drinking late in the night and bothering sleeping neighbors. This time I did not detest the noise. No. This time, I wanted to go up and sit together with them, drink arak and eat hummus and fattouche, bread and fried potatoes, with labneh and cheese and tomatoes.
>
> (Jaber, 2006, p. 238).

The closing paragraph thus effectively mirrors the opening of the novel. Both scenes describe the animated laughter, bright city lights and good food with gusto. They stand in sharp contrast to the silent and dark underground city whose inhabitants eat the same dreary fish every day. "It is when he experiences the alterity, the premodern and archaic, that the full significance of Beirut's brand of modernity becomes apparent and makes him risk his life to be part of it," observes Aghacy (2015, p. 172). The lure of Beirut's pleasure industry does therefore not stem from false consciousness, as Marxist critics would have it. On the contrary, it is only after Butros reflects upon his own experience with displacement, loss and social exclusion, and after his encounter with the suffering of others, that he can truly appreciate the phantasmagoria of Beirut's nightlife.

Hassan Daoud: De-fetishizing the urban spectacle

Unlike the younger Rabee Jaber, Hassan Daoud (b. 1950) belongs to the "war generation." This is a generation of authors that started writing during the Lebanese civil war. They often play a role as public intellectual, combining the production of fiction with non-fiction and journalism, and their work still dominates the literary scene (e.g. Lang, 2016). Daoud's Shiite working class background and leftist convictions can be detected in his keen awareness for class and sectarian dynamics. Inspired by the French nouveau roman, his novels are much more sec in style and uneventful in plot than Rabee Jaber's. The authors, however, do share their close attention to the rhythms of everyday life, and its sedimentation in urban space, as the site in which we can read the layered and ambiguous inscriptions of historical change.

The novella *As She Once Was* was published in 2013 in a special issue of the magazine *Portal 9*. The magazine presented itself as an "Arabic-English journal of stories and critical writing about urbanism and the city." It celebrated critical autonomy and creativity, but was also fully funded by Solidere, which assured editorial independence. The journal ran for several years with two thematic issues each year, but is out of print at the time of writing. The third issue of *Portal 9* was an experimental format: A carton case filled with separately bound contemporary short Russian and Arabic fiction, published in Arabic and English translation. Hassan Daoud's novella was especially commissioned for this edition of the magazine.

The context of the novella's publication is reflected in its narrative, which revolves around a first-person narrator, Qassem, who has recently taken on a job at a cutting-edge magazine, the office of which is situated in BCD. Nearing 60, the narrator feels out of place, both in the magazine's editorial office and in the neighborhood. In Qassem's view, the magazine consistently wastes time and money. The office is unnecessarily completely refurbished; designers take months to produce what his old colleague at the newspaper could do in a day; endless meetings circle around concepts without arriving at any concrete ideas or decisions. The chief editor, Talal, informs Qassem over Skype that the articles he has produced "are not what we are looking for" (Daoud, 2013, p. 90). In the district, too, he feels estranged from his surroundings.

> During my morning journey, passing through the scores of shops whose windows turn as I turn, I don't find myself looking at the clothes they display but at my own image, reflected in the glass. As if I am following my own shadow, step by step, observing how it walks, how my thighs move in that walking motion, wondering whether I appear like a transgressor to the hip pants designed for a generation no longer satisfied with the pants I always used to wear.
>
> (Daoud, 2013, p. 32)

The clean windows do not invite in, but, instead, reflect back Qassem's appearance, out of tune with the style and fashion of the surroundings, like the coffin maker from Yasmine Arif's ethnography.

The contrast between the narrator and his surroundings is presented as a clash between old and new. Qassem obsessively goes over the memories of his childhood love, Dalal. She emigrated before they even held hands and he has lost touch of her ever since. He has no physical reminder of her except for the abandoned ruin of her apartment building, which has miraculously survived the demolition and reconstruction of the surrounding blocks (Daoud, 2013, p. 152). He desperately calls back the minute details of their last encounters on a daily basis, which inhibits him from moving forward with his life and from committing to a new relationship. His clinging to an idealized past is mirrored by Talal's obsession in searching "the glow of the new, but without knowing what it was or what it looked like. Only the glow, or what he felt to be a glow" (Daoud, 2013, p. 108). Again, this reflects the district as a whole. "Everything is new here," observes the narrator about the souks (p. 18).

Despite the great gap the narrator experiences, both sides of the equation share utter stagnation. Qassem's memories of Dalal remain unchanged and he postpones any initiative to search for her. The magazine is stuck in a delusional pursuit of innovation, endlessly postponing deadlines. In the *Arcades Project*, Walter Benjamin writes that the new "is the quintessence of that false consciousness whose indefatigable agent is fashion. This semblance of the new is reflected, like one mirror in another, in the semblance of the ever recurrent" (Benjamin, 1999 [1927–40], p. 11). For Benjamin, the myth of progress – here represented by the editor and BCD – and the myth of the archaic – here represented by Qassem's desperate idealization of his childhood love – are the complementary antinomies from which "the dialectical conception of historical time needs to be developed" (Benjamin, 1982, vol. V, p. 178, D10a, 5; as cited in Buck-Morss, 1989, p. 108). If Qassem feels alienated, it is not only because of the hall of mirrors that is BCD's promise of the new, but also because of his own clinging to a mythical past of innocent love.

Yet, as much as Qassem may feel alienated, he is not tricked by BCD's dream image. Early in the novella, we read that "[BCD's] newness is a consistent effort, for the employees never stop wiping dust and polishing glass and oiling the metal of the escalators and hoisting up large billboards displaying provocative women dressed in lingerie or skintight pants" (Daoud, 2013, p. 18). Later, after repeating the theme of the glass façades as a hall of mirrors, he mentions that "workers are cleaning them as I walk past or have just finished cleaning them" (p. 40) (Figures 16.5 and 16.6). In truly Marxist fashion, the new is thus presented as a product of continual labor against inevitable decay. This de-fetishization of the phantasmagoria of the Beirut Souks precipitates the novella's central turning point, where the narrator finds his way out of stagnation (see also Naeff, 2018, pp. 156–157).

On his daily way to work, Qassem passes by three immigrant women sitting and chatting on a bench in the Beirut Souks. He assumes they are waiting for the shop where they are employed to open up. Qassem is fascinated by the confidence with which they compose themselves. "They even control their laxity, the way they free their feet from their slippers, swinging them from their toes, at once both shod and barefoot" (Daoud, 2013, p. 19). He admires their sense of style, which "differentiated them from the other girls who made the same journey," thus allowing them to work in this fancy district (p. 31). Most importantly, he loves "their indifference to that [the souks] which I still feel myself a mere addition to" (p. 32).

One day, Qassem decides to buy them croissants. The handing over of the box proceeds clumsily and he is so embarrassed that he tries to avoid them for the next few weeks. But when he finally does see them, one of them comes over and invites him to join them, taking his hand in hers and asking "where have you been?" (pp. 137–138). It is the start of an unlikely friendship. He discovers that the women are not working for a shop, but clean the district in the early mornings, and again after their long break on the bench. They are thus fundamentally part of the production of the district.

Figure 16.5 Sebastian Dahl, From the series Downtown Beirut

Figure 16.6 Sebastian Dahl. From the series Downtown Beirut

Qassem also realizes that their striking fashion is therefore not in the service of some shop owner to increase the sale of her products, but entirely of their own making (p. 164). This reflects Benjamin's ambivalent "metaphysics of fashion." If the editor's pursuit of the new is nothing but "the ritual by which fetish commodity wished to be worshipped" (Benjamin, 1982, vol. V, p. 51 [1935 exposé]; as cited by Buck-Morss, 1989, p. 98), then the fashion of these three women is conceived affirmatively as "irreverent toward tradition, celebratory of youth rather than social class, and thus emblematic of social change" (Buck-Morss, 1989, p. 98).

Confronted by the agency with which these women shape their own lives despite being part of the most vulnerable demographic of Lebanon, Qassem's disposition changes. He starts to like the district. He now takes initiative to visit one of the fancy cafés, to wander the alleyways, to enter the shops and look at the products on display without buying them (Daoud, 2013, pp. 160–163). His alienation has made place for a sense of positive anticipation, because he now knows "that it was possible to make the acquaintance of people I hadn't known before. And that there was the possibility of conversing further" (p. 163). Walter Benjamin predicted that dialectical time would emerge from the antinomy between the obsession with novelty and the eternal recurrent. For Daoud, such radical social change happens in the intimate details of everyday life. Qassem's transformation emerges from his confidence that it is still possible to forge meaningful social relations outside the workings of capital even if situated right in the middle of the most spectacular phantasmagoria of Beirut.

Conclusion

The three works of fiction discussed in this chapter show an acute awareness of the structures of displacement and social exclusion implied by consumerist urban spaces. Examples are the social segregation at the ABC mall, Butros's peering through the window of an expensive restaurant, and Qassem's feeling out of place. The novels also express the profound sense of loss that haunts these unhomely spaces in the context of post-civil war Beirut. Saman's city is in shock after the 2005 assassinations and the dead linger invisibly on the other side of the river. The underground city reminds Butros of the violence of his past and the people he has lost. Qassem keeps visiting the abandoned ruins of Dalal's old apartment. Yet, despite all of this, none of the novels rejects these spaces of consumption as the blank slates of forced amnesia lamented by many Lebanese intellectuals in the late 1990s and early 2000s. Rather, and more in line with Walter Benjamin's ambivalent approach to the phantasmagoria of consumer capitalism, they seek to recover the potential of human desire and social liveliness that are fostered in shopping malls and shopping mall like urbanism.

One way in which this ambivalence takes form is by denying the characters an outside position. The architect Saman, the security guard Butros and the immigrant cleaning ladies in *As She Once Was* are all directly implied in the production of consumerist urbanism in the post-civil war redevelopment of Beirut. Precisely because of this implication, they have access to these urban spaces that supposedly critical outsiders have not. On the one hand, these characters show the labor invested in the production of spaces of consumption and thus de-fetishize the urban spectacle in a Marxist sense. On the other hand, they refuse easy distinctions between good and bad, or inside and outside in such a complex configuration as urban renewal. Most importantly, they show how life may thrive despite the loss caused by displacement. From the dripping sandwiches in *The Mehlis Report* to the fashionable haircuts of the cleaning ladies in *As She Once Was,* the narrators in these stories derive both solace and hope from the liveliness of these spaces.

We have seen how visitors and employees of shopping malls in the wider Middle Eastern region forge new ways of being that they feel unable to develop on these cities' public streets. If Beirut's climate is more kind than Dubai's, and its social spaces more liberal than Iran's, these three novels show that its spaces of consumption also have something to offer. The sight of fellow citizens of different social strata embracing life offers comfort in a postwar context of shared – if unequal – loss. This then, may offer a starting point for imagining urban recovery. In the words of Saman Yarid:

> Empty cars flock there from the four corners of Beirut and leave filled to the brim with food, more evidence that the Civil War has ended, fifteen years of peace seem inconclusive, but the sight of Monoprix's kitchen and all the orders coming in during Ramadan, that's decisive proof.
>
> (Jaber, 2013 [2005], p. 92)

Note

1. THE MEHLIS REPORT, copyright ©2005 by Rabee Jaber. Translation copyright 2013 by Kareem James Abu-Zeid. Reprinted by permission of New Directions Publishing Corp.

References

Abaza, M. (2001). Shopping Malls, Consumer Culture and the Reshaping of Public Space in Egypt. *Theory, Culture & Society, 18*(5), 97–122.

Abaza, M. (2009). Egyptianizing the American Dream: Nasr City's Shopping Malls, Public Order, and the Privatized Military, in *Cairo Cosmopolitan: Politics, Culture, and Urban Space in the New Globalized Middle East*, edited by D. Singerman and P. Amar. Cairo: Cairo University Press, 193–220.

Aghacy, S. (2015). *Writing Beirut: Mappings of the City in the Modern Arabic Novel.* Edinburgh: Edinburgh University Press.

Alhadar, I. and McCahill, M. (2011). The Use of Surveillance Cameras in a Riyadh Shopping Mall: Protecting Profits or Protecting Morality? *Theoretical Criminology*, *15*(3), 315–330.

al-Maqarrī, Alī. (February 24, 2010) "Rabī' Jābir: Ikhtalaqtu qissa min ajl ta'shīra ila Amrīka [I Made up a Story for a US Visa] *Al-Ittihad*.

Arif, Y. (2002). Past Places/Future Spaces: Reconstructing Postwar Beirut. *Sarai Reader*, *2*, 117–125.

Augé, M. (1995[1992]). *Non-Places: Introduction to an Anthropology of Supermodernity*. Trans. J. Howe. London: Verso.

Baudrillard, J. (1994[1981]). *Simulacra and Simulation*. Trans. S.F. Glaser. Ann Arbor: University of Michigan Press.

Benjamin, W. (1982). *Gesammelte Schriften*, 6 vols., edited by Rold Tiedemann and Hermann Schweppenhäuser, with the Collaboration of Theodor Adorno and Gershom Scholem. Frankfurt am Main: Suhrkamp Verlag.

Benjamin, W. (1999[1927–40]). *The Arcades Project*. Trans. Howard Eiland and Kevin McLaughlin. Cambridge, MA: The Belknap Press of Harvard University Press.

Billig, M. (1999). Commodity Fetishism and Repression: Reflections on Marx, Freud and the Psychology of Consumer Capitalism. *Theory and Psychology*, *9*(3), 313–329.

Buck-Morss, S. (1989). *The Dialectics of Seeing: Walter Benjamin and the Arcades Project*. Cambridge, MA: MIT Press.

Cooke, M. (2002). Beirut Reborn: The Political Aesthetics of Auto-Destruction. *The Yale Journal of Criticism*, *15*(2), 393–424.

Daoud, H. (2013). As She Once Was. Trans. L. Mounzer. *Portal*, *9*, 3.

Dibazar, P. (2020). *Urban and Visual Culture in Contemporary Iran: Non-Visibility and the Politics of Everyday Presence*. London: Bloomsbury.

El-Khoury, R. (1998). The Postwar Planning of Beirut, in *Projecting Beirut: Episodes in the Construction and Reconstruction of a Modern City*, edited by P. Rowe and H. Sarkis. Munich: Prestel Verlag, 183–186.

El-Sheshtawy, Y. (2004). The Middle East City: Moving beyond the Narrative of Loss, in *Planning Middle Eastern Cities: An Urban Kaleidoscope*, edited by Y. Elsheshtawy. London: Routledge, 1–21.

El-Sheshtawy, Y. (2008). Navigating the Spectacle: Landscapes of Consumption in Dubai. *Architectural Theory Review*, *13*(2), 164–187.

Erkip, F. (2003). The Shopping Mall as an Emergent Public Space in Turkey. *Environment and Planning 35*(6), 1073–1093.

Fattouh, S. (2005). *Temps Mort*. Video 16 min. Beirut. https://vimeo.com/228545603 (accessed 10 July 2019).

Fawaz, M. and Ghandour, M. (2010). Spatial Erasure: Reconstruction Projects in Beirut. *ArteEast Quarterly*, spring. https://arteeast.org/quarterly/spatial-erasure-reconstruction-projects-in-beirut/?issues_season=spring&issues_year=2010 (accessed 10 July 2019).

Halabi, Z. (2018). *The Unmaking of the Arab Intellectual: Prophecy, Exile and the Nation*. Edinburgh: Edinburgh University Press.

Hayek, G. (2016). Beirut, in *The Palgrave Handbook of Literature and the City*, edited by J. Tambling. London: Palgrave Macmillan, 587–597.

Jaber, R. (2006). *bayrītūs: madīnah taḥt al-arḍ* (Beyritus: Underground City). Beirut: Dar al-Adab.

Jaber, R. (2013[2005]). *The Mehlis Report*. Trans. K.J. Abu Zeid. New York: New Directions.

Jameson, F. (1997[1991]). *Postmodernism and the Cultural Logic of Capital*. Durham: Duke University Press.

Khosravi, S. (2008). *Young and Defiant in Tehran*. Philadelphia: University of Pennsylvania Press.

Lang, F. (2016). *The Lebanese Post-Civil War Novel: Memory, Trauma, and Capital*. London: Palgrave Macmillan.

Makdisi, S. (1997). Laying Claim to Beirut: Urban Narrative and Spatial Identity in the Age of Solidere. *Critical Inquiry*, *23*(3), Front Lines/Border Posts: 660–705.

Mango, T. (2003). *Solidere: the Battle for Beirut's Central District*, MCP book, Urban Studies and Planning, Cambridge MA: Massachusetts Institute of Technology.

Mostafa, D. (2009). Literary Representations of Trauma, Memory, and Identity in the Novels of Elias Khoury and Rabīʿ Jābir. *Journal of Arabic Literature*, *40*(2), 208–236.

Musto, M. (2010). Revisiting Marx's Concept of Alienation. *Socialism and Democracy*, *24*(3), 79–101.

Naeff, J.A. (2018). *Precarious Imaginaries of Beirut: A City's Suspended Now*. London: Palgrave Macmillan.

Najjar, A. (2005). *Le Roman de Beyrouth*. Paris: La Table Ronde.

Ragab, T.S. (2011). The Crisis of Cultural Identity in Rehabilitating Historic Beirut-Downtown. *Cities*, *28*(1), 107–114.

Salam, A. (1998). The Role of Government in Shaping the Built Environment, in *Projecting Beirut: Episodes in the Construction and Reconstruction of a Modern City*, edited by P. Rowe and H. Sarkis. Munich: Prestel, 122–133.

Seigneurie, K. (2011). *Standing by the Ruins: Elegiac Humanism in Wartime and Postwar Lebanon*. New York: Fordham University Press.

Sierra, F.R. (2017). Fictional Boundaries in the "Journalistic Fiction" of Gabriel García Márquez and Rabee Jaber. *Alif: Journal of Comparative Poetics*, 37, Literature and Journalism, 199–228.

Summer, D. (2006). The Neoliberalization of Urban Space: Transnational Investment Networks and the Circulation of Urban Images: Beirut and Amman. *Villes et Territoires du Moyen-Orient*, *2*, 1–9. Electronic journal. https://halshs.archives-ouvertes.fr/halshs-00139204 (accessed 15 July 2019)

Tabet, J. (2001). La cité aux deux places, in *Beyrouth, la brûlure des rêves*, edited by J. Tabet. Collection Monde HS 127. Paris: Éditions Autrement, 42–57.

Tarraf, Z. (2015). *Affective Landscapes: Re-negotiating the Ordinary in Contemporary Lebanese Cultural Production*. PhD dissertation. Alberta: University of Alberta.

Tiedemann, R. (1999[1982]). Dialectics at a Standstill: Approaches to the Passagen-Werk, in *The Arcades Project*, edited by W. Benjamin, trans. Howard Eiland and Kevin McLaughlin. Cambridge, MA: The Belknap Press of Harvard University Press, pp. 929–945.

Verdeil, E. (2001). Reconstructions manquées à Beyrouth: la poursuite de la guerre par le projet urbain. *Annales de la recherche urbaine*, *9*, Villes et guerres, 65–73.

Westmoreland, M. (2008). Crisis of Representation: Experimental Documentary in Postwar Lebanon. PhD thesis, University of Austin, Texas.

17 Confiscated imaginaries

Notes on a work in progress

Omar Mismar

In the summer of 2018, I went to a Syrian refugee camp in the town of Qab Elias in the Beqaa Valley, where I grew up and went to school, and where my parents still reside, 40 minutes away from the Syrian border. I went devoid of any clear ideas of what I wanted to do but accompanied with a new good camera that was already indispensable, since the shortcut to getting a permission to work in the camp as a visual artist was stating that I was going to photograph and film in the camp. "Record what?" asked the officer suspiciously, "daily life" I answered softly. "Ok, as long as there is nothing political," he said with a vivid conviction.

Burdened by the urgency of a crucial historical moment that needs to be attended to, that is unavoidable in its proximity, but coupled with an ample dosage of skepticism bordering cynicism around the trendiness and sexiness of the Syrian war as a hot topic, I frequented the camp. Photographs and videos piled up, flooding memory cards and draining batteries. It helped that children, running off camera, now ran towards it and froze in front of the lens, their bodies internalizing the mechanism of the closing shutter. It was something of a habit for them to ask "sawwerni" (photograph me), which I did most of the time, not without an impatient sigh every once in a while. I was asked many times if I am part of the group who came last week to take pictures, or if I know the girl who was here a month ago, or if I am an affiliate of the UNHCR. The inhabitants of Camp 007 were subjects cognizant of the part they were playing or made to play. Urged, prodded, forced, or framed, they mostly obliged, performing their condition, their poverty and wretchedness to what amounts to an undistinguishable coterie of coveting onlookers.

Photographs and footage from my visits resembled most of what one encounters when reading on and researching about the Syrian refugee camps. They slipped effortlessly into a hackneyed discourse of visual representability, inundated with sensationalist rhetoric oscillating between victimization, witnessing, and humanitarian blackmail; indignation *tout court*. The camp as a site, a landscape, a makeshift habitat, is rich, and photogenic. But, such richness is deceiving; it appears swiftly fathomable, transparent, while it ramifies indefinitely and conceals opaquely, cunningly overwhelming its onlooker towards a leveling field of depiction, condemning by regurgitating. Soon after, I decided that I am not going to use or show any of the material I shot in the camp.

There

In *Here and Elsewhere*, the 1974-acclaimed film by Jean-Luc Godard and Anne-Marie Miéville, Godard's voiceover proclaims:

> We did what many others were doing. We made images and we turned the volume up too high. With any image: Vietnam. Always the same sound, always too loud, Prague, Montevideo, May '68 in France, Italy, Chinese Cultural Revolution, strikes in Poland, torture in Spain, Ireland, Portugal, Chile, Palestine, the sound so loud that it ended up drowning out the voice that it wanted to get out of the image.[1]

Dismissing the images I shot, lumping them with an already prevailing visual language around Syria—further turning the knob on a loud volume—was an easy way out. Instead of seeking voice out of the images, any images, I was not only drowning out the voice, but also hiding underwater. I resurface, re-scrolling through the images, without the illusion of discovering a self-congratulatory hidden gem but with a rehabilitated belief in making strange and slow looking. I was able to tune to one image, which became the trigger in developing the project further and forging the relationship between displacement and indexicality via a semiotic analysis. In fact, such semiotic lens arises more imposingly and vividly in the whirlwind of displacement, which stirs bodies in and out of altered time-space contexts, defamiliarizing their structures of meaning—of which stubborn residues endure—while mandating a perpetual readiness to read new settings; displacement thus as palimpsest of semiotic interpretations.

A man wearing a helmet raises two hands, protected by gloves, close to his face. The hands, sustained by sun-kissed forearms, hold a pair of binoculars in front of his eyes. He is upside down: the head rests on a tire, against a faintly blue sky anchored by few concrete blocks. A spray-painted number 6 in Arabic brackets the man left and right, while the binoculars peak into the distance, interrupted by the word "al watan" (home). Part of the "tah" letter of "al watan" is cut out to create a door.

The photograph is of tent no. 6 (Figure 17.1). Like many of the makeshift structures in the camps, the outer shell is made of flex material, the PVC commonly used for billboard printing for its ability to withstand harsh outdoor conditions such as heat and rain, conditions quite rampant in the camps. Refugees often repurpose discarded billboards and recycle printed flex to wrap the tent. On tent no. 6, the original billboard dates back to 2016, issued and stamped by the Lebanese Army to celebrate the 73rd Lebanese

Figure 17.1 Tent no. 6, Camp 007, Qab Elias, Bekaa Valley, Lebanon

Independence (Figure 17.2). An unwrapping of the billboard reveals a young boy, also with binoculars standing next to the man, a soldier, together glancing forward. The text also unravels to read: “wa baqiya al watan” (and home remained).

What remains two years after the issuing of this billboard, now displaced into tent skin, obliges a different reading. The soldier of the 2016 billboard, who’s on the lookout for anything that will threaten home, performing as an idol and armor for a generation to come, is now, on tent no. 6, a floating toppled head looking for home. In the near distance, he finds the word “al watan”: a signifier punctured by a door that is the passageway to an actual physical space; a signifier anchored by a home-like structure, *here* in the camp, around which it ironically clings, and a signifier that evokes a *there*,

Figure 17.2 Billboard issued by the Lebanese Army to celebrate the 73rd Lebanese Independence

Syria. An indexical relationship emerges troubled, with the here and there collapsing onto one another.

Theorizing the different types of signs—icon, index, and symbol—Charles Sanders Peirce (2012) posits the index as a sign affected by its referent via a causal or existential relationship (p. 103).[2] Within the category of the index Peirce grouped footprints, the weathercock indicating the direction of the wind, the pointing finger, and demonstrative pronouns like the word "this." In fact, demonstratives belong to the category of indexical terms called deixis, of which words like "this," "that," "here," and "there" are typical examples. Deixis is "reference by means of an expression whose interpretation is relative to the extralinguistic context of the utterance," such as who is speaking, when and where, the gestures involved, and the current location (Deixis, n.d.). As such, deictic expressions are referred to as shifters, since their meaning shifts; it is contingent upon the context in which they are uttered.

Linguist Roman Jakobson's describes the shifter as "'filled with signification' only because it is 'empty'" (Kraus, 1977, p. 69). Words such as "here" or "there" are hollow, waiting to be supplied by a referent each time they are summoned. Their emptiness is provisionally occupied, as what they signify mutates again in a new situation. On tent no. 6, "here" and "there" fossilize into referents that are hailed into a predominant meaning by each other: *here*, the camp, is now home, because *there*, Syria, is not home anymore.

Displacement holds onto the index in its deictic form, saturating the hollow shifters *here* and *there* towards a halt. No longer does the shifter await to be supplied with an existential presence of a referent. The inherently empty shifter bulks up: A *here* fixating on the place of exile, a *there* fixating on homeland. The index, Peirce reveals, "is in dynamical (including spatial) connection both with the individual object, on the one hand, and with the senses or memory of the person for whom it serves as a sign" (p. 107).

In discussing Ghassan Kanafani's novel *A'id 'ila Hayfa*, writer and anthropologist Susan Slyomovics (1998) focuses on the passage where protagonist Said S. returns to visit his home in Palestine. Said S. notices things with surprise: the smaller and damper entryway, the changed doorbell, the names carved on the palm trees, two of the five chairs that belonged to him, how the three new ones "looked crude and out of place" with the rest of the furniture, and the number of feathers in a vase. He remembers them to be seven feathers, but after walking up to the vase and counting them, they were only five. Slyomovics here, through Kanafani, reveals how "partial memory takes on a particular value because it transforms what remains into pure evocation and resonance" (p. 23). Fragmentary memory here fragments space and objects, turning them into "archaeological artifacts," into traces: a burdened *here, now* fossilized with the memory of the *there, then*. Home remained, indeed. With displacement, the indexical shifter is insistently occupied, squatted. It converts to a refuge harboring memory, hinting at a claim of dispossession.

A blueprint to replace them all

Back in the camp, last summer, the plot thickens; with it the index. Three months earlier, in April 2018, Bashar El Asad issues Law No. 10, which allows the government to designate areas inside Syria for redevelopment. Once an area is designated for reorganization, property owners within it need to file a claim of ownership. To claim property, damaged or destroyed, one must appear in person with his/her real estate documents, "if found," within 30 days to prove ownership and allegedly receive compensation or otherwise lose their rights. If such proof is unavailable, "owners can make a case within the 30-day period to the local authorities using other evidence. They must provide the property location, boundaries, and information about the shares they own, and the real estate category the property comes under as well as information about any lawsuits they have brought or outstanding suits against them" ("Q&A: Syria's New Property Law," 2018). The fact that the impossible 30-day time frame was extended to one year in November 2018 does not gloss over the sectarian ethnic cleansing that this law pushes forth, redrawing the map before the war has even ended.

While Decree 66 issued in 2012 designates specific zones of informal settlements or slums in southwestern Damascus for redevelopment, Law No. 10, in its vagueness, is omnipotent. People who left Syria because they opposed the regime certainly cannot return in person to claim their properties, nor can the thousands who fled military draft. But Law No. 10 allegedly has that covered. Owners can appoint designated agents or relatives up to the 4th degree to claim on their behalf. But this claim by proxy is hampered by the fact that the representatives need to prove that owners cannot make a claim personally[3] and that they require clearance by government security service, something unlikely to be granted and will put to risk someone whose even their 4th degree cousin is on a black list.

With such resolute impasse to the conception and presentation of proof—because of its loss or damage, because of the risk that an in-person return posits, because of the self-destructing claim-by-proxy alternative—Law No. 10 commands the index. Deeds, documents, clues, evidence, records, blueprints… all are species of the index. Law No. 10 promises their signifying vectors a legitimation, a landing point, while in effect it blocks these vectors. Unlike Ariadne's thread, which Theseus first unrolls as he enters the unnavigable labyrinth to conquer the minotaur and later uses it to find his way back, the thread here is cut, Theseus forever roaming the labyrinth.

Promulgated in the wake of the Syrian government's recapture of Eastern Ghouta, one of the last major rebel-held areas near Damascus, Law No. 10 targets areas with opposition stronghold, engineering a demographic change. The regime will be able to claim "unclaimed" or rather claim-controlled property and land, repopulating it with settlers of political and socioeconomic backgrounds of its choosing, attracting investor-allies under the rubric of redevelopment. Law No. 10 then emerges as blueprint for the reconstruction of Syria and a consolidation of the regime's post-war power. A blueprint that replaces, overtakes, all other blueprints by severing the indexical relationship and subjugating it to the proclivities of a god-like ruler, making the index an acheiropoiete.[4]

Literally meaning "not made by hands," *acheiropoiete* describes many of the celebrated and worshipped sacred icons. Faces of Christ and the Virgin, Veronica's veil, and the Shroud of Turin are instances of images that have descended from heaven without any intermediary. To claim that a human hand made them is "to weaken their force, to sully their origin, to desecrate them" (Latour and Weibel, 2002, p. 18). Law No. 10 hijacks claims of ownership—the indexes of homes dwelt, properties labored, lives thrived—into the transcendental jurisdiction of the almighty ruler. There, only the hands of this self-proclaimed divine manufacture truth without spoiling it, because they are not human hands. All other hands and fingers are to be intervened upon.

Curiously, in Arabic, the pointing finger or the index is referred to as *sabbaba*. The etymological root for *sabbaba* is *sabba*, which means to curse. *Sababba* then is the curser, the finger you use to point to that whom you want to cuss. That same cussing finger is also referred to as *mosabbiha*, from the origin *sabbaha*, evoked to praise god (*sobhan allah*) or in bearing witness to the oneness of God (*tashahhud*). In this usage, the index finger is raised up towards the sky. In one instance, the index must be done without, amputated; it is an accusatory angry cursing index. In another instance, the index is sanctified; it must be preserved as an indicator and reminder of the one. Law No. 10 bends the cursing index into a praising index; the slightest defiance breaks it off.

In displacement, the hollow indexical shifters are filled, *here* and *there* fixating on exile and home, instancing in their utterance a claim of dispossession. Law No. 10 augments such fixation and closes the lid on the shifters by demanding proof while manipulating the indexical constituent of that proof, imbuing it with godly authority that eradicates humans, hands, and fingers. In closing the lid, the utterance of dispossession is legible only as cursing, forcibly transmuted as shouting in an echo chamber.

Vectors, prototypes

A clicking sound echoes from an apparatus (Figure 17.3). A magazine holds vertical transparent plexiglass sheets that sandwich blue translucent documents. At set time intervals, the magazine glides nimbly and randomly stops right before a light bulb to its left, to drop one of the slides from a vertical to a horizontal orientation (Figure 17.4), thus revealing the content of the blue document now legible against the light (Figure 17.5). Floor plans and maps feature houses, quarters, and rooms signed off with initials, followed by the name of the town: Reef Halab, Khanasir, Deir el Zour, Tal Abyad, Al Qusayr, Raqqa... (Figures 17.6–17.10). Like an index of indexes,

Figure 17.3 Dubious Prototype (Confiscated Imaginaries) Photo by Tony Elieh.

Figure 17.4 Dubious Prototype (Confiscated Imaginaries) Photo by Tony Elieh.

Figure 17.5 Dubious Prototype (Confiscated Imaginaries) Photo by Tony Elieh.

Figure 17.6 Tal Abyad, blueprint from Dubious Prototype (Confiscated Imaginaries)

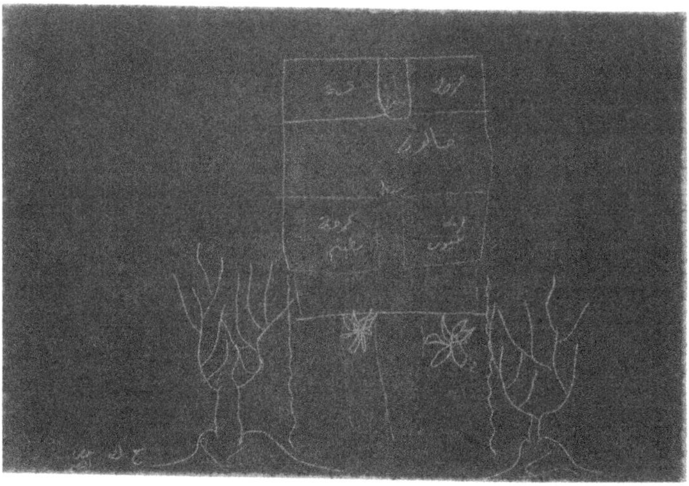

Figure 17.7 Al Qusay, blueprint from Dubious Prototype (Confiscated Imaginaries)

blueprints drawn by hand here, imagine homes there. Like a carousel projector, the apparatus keeps slides, but it confiscates the projection, defers the slideshow.

Against the acheiropoietia of Law No. 10, this gesture marks the return of the hand, in fact registers the human hand at work, making, thus instancing a poetics rooted in the Greek *poiein: to make*. But poetics makes us unhappy, Maurice Blanchot and Ann Smock (1995) ruminate in *The Writing of the Disaster*. "Criticism becomes warlike," he says, "because political

Figure 17.8 Raqqa, blueprint from Dubious Prototype (Confiscated Imaginaries)

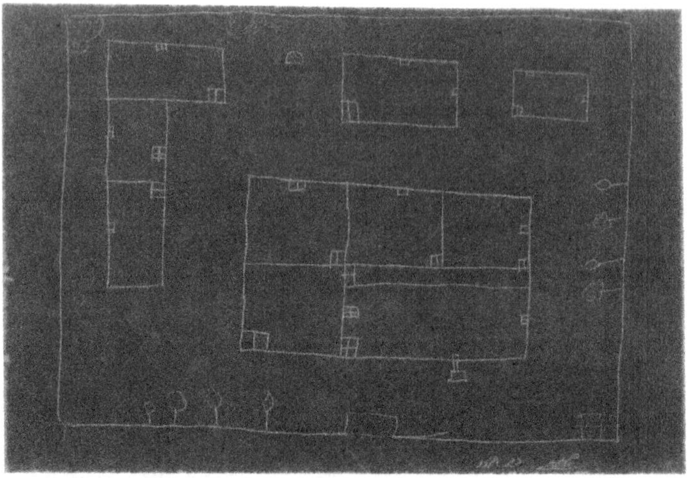

Figure 17.9 Deir el Zour, blueprint from Dubious Prototype (Confiscated Imaginaries)

impatience has won out over the patience proper to the 'poetic'" (p. 78). He argues that writing—and art making we must add—persists in a relation of irregularity with itself and the other, that it does not know what will become of it politically. Its necessarily indirect relationship to the political—its infinite detour which we try to fathom as its being, "out of phase or belated, as incertitude or chance (and also invention)"—displeases us (p. 78). Paralyzed or propelled (or both) by the urgency of human-inflicted

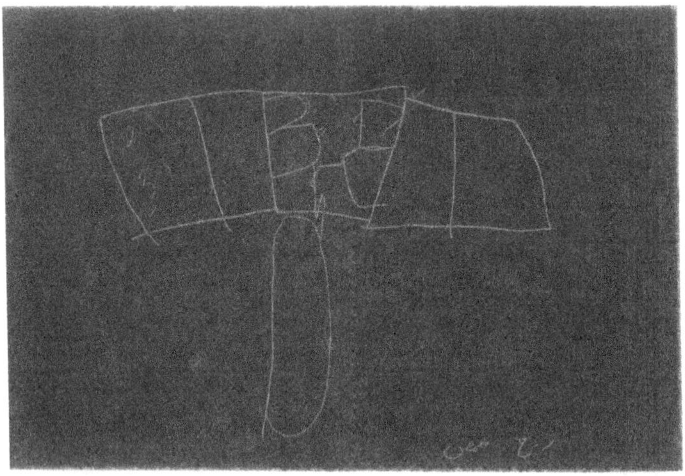

Figure 17.10 Homs, blueprint sample from Dubious Prototype (Confiscated Imaginaries)

disasters, we strive to proceed in a direct fashion towards the goal of social transformation and recovery.

Blanchot maintains that we must not turn away from the necessity to act and of language becoming action, but neither must we stop at it. His movement is towards "another sort of language entirely—the language of writing, the language of the other always other whose imperative does not develop at all" (p. 79). It is, as I read it, an attempt at a poetics of the insatiable, where the other always lets itself be excluded, never annulled nor reconciled; an "indelible effacement" that orients our direction. I glimpse such poetics in Adorno's famous essay on commitment (1978), where he asserts that "as eminently constructed and produced objects, works of art, even literary ones, *point to* [my emphasis] a practice from which they abstain: the creation of a just life" (p. 317). Here, the indirect, non-didactic, and seemingly politically insufficient bent of works of art—which he calls autonomous, opposes to the committed, and finds in Kafka and Beckett—is championed precisely for its disavowal of a "message" and its distance from subordination of "meaning," thus avowing for a jolt in signification which in turn allows for revaluating the discrete and the habitual. In such conception of art in relation to disaster and conflict, a poetics of anti-closure thrives, of rising to the occasion and never quite reposing, not as an indulgent structuring self-loathing aphorism, but as a compendium of gestures attempted towards the occasion; as an "it should be otherwise" hidden in the work, as Adorno has it, but also directed at the occasion itself.

"Thus the poet is a tragic figure. The poem is always a record of failure," declares poet and essayist Ben Lerner (2016) in *The Hatred of Poetry* (p. 8). Lerner, reading poet Allen Grossman, emphasizes the disconnect between what one is called upon to create, "that transcendent impulse," and between what one creates from the finitude of terms and materials. The move from one realm to the other is conflictual and never seamless; its opponents are the desire to sing another world on the one hand, and the "resistance to alternative making inherent in the materials of which any world must be composed," on the other hand (p. 8). The tragedy and the failure, again, are not to be indulged in as the stagnant state of affairs, as defenses and ready descriptors to poor work or no work. Rather, they bestow humility onto the work—one absent from those that simplistically feign politicalness and commitment—and they propel it forward, in iterations of itself that flicker between what one is stirred to do and what one can do, as rehearsals for a play that keeps getting postponed until the rehearsals become the play.

Despite the displeasure, the it-should-be-otherwise, the tragedy, and failure, poetics endures in the space between the self and the other, as a distancing act of play. In her seminal book *The Poethical Wager*, critic and poet Joan Retallack (2003) states that "to wager on a poetics of the conceptual swerve is to believe in the constancy of the unexpected—source of terror, humor, hope" (p. 4). This triad emits energy that keeps the mind in motion amidst chaos. Such motion is key: it requires "tolerance for ambiguity;

willingness to move forward with uncertainty; willing suspension of both belief and disbelief; willingness to wade purposefully, playfully into the mess" (p. 29). It turns predictable trajectories into vectors and delimiting systems into "dubious prototypes."

I stand in front of this apparatus, this commissioned body of evidence, with uncertainty on how to proceed and ambiguity on what it actually is. It has become a burden; but I do want to close on a hopeful, patient, note. I doubt that in the trace things speak themselves and are not spoken (Doane, 2007, p. 3) especially under the canopy of Law No. 10, where the index is ventriloquized. Against the monopolizing voice of ventriloquism, I consider the trope of prosopopoeia, the figure in which a speaker artificially endows inanimate objects with a voice.

In *Mengele's Skull: The Advent of a Forensic Aesthetics*, Eyal Weizman and Thomas Keenan elucidate that forensics is derived from the Latin *forensis*, with roots in the word forum, i.e. the skillful practice of making an argument before a gathering, be it professional, political, or legal. The forum is not only about findings of science and evidence lying there; it is also about the presentation of the latter and its use for persuasion. The forum then is not a given unchanging space, even though it can become institutionalized within fixed structures. Rather, it is a space produced via "a series of entangled performances" (Weizman and Keenan, 2012, p. 29) that can proceed, and not only precede, the presentation of evidence and the protocols of appearance, evaluation, and argumentation within it. The authors argue that Josef Mengele's skull marked the advent of a new type of evidence, neither the document nor the witness, but the thing, often bones and human remains. Now because objects do not speak for themselves, a translation, mediation, or interpretation between the "language of things" and that of people becomes pressing. Enter prosopopoeia. Objects take on agency via their interpretation, and speak by virtue of their "friends," the people who gather around them and read them. Weizman and Keenan follow Miguel Tamen quoting that "this gathering, a 'society of friends' or of advocates, in a legal sense, constitutes a sort of forum for interpretation and debate" (Ibid., pp. 26–28).

For Tamen (2004), objects become interpretable in an intentional way only in the context of a society of friends; interpretable because of the groups of people who count them as such and deal with them in recognizable ways.[5] He describes interpretation as "a set of complex actions performed in connection to a motley collection of things rather than as an attempt to capture the intrinsic properties of things" (Tamen, 2004, p. 131). This is crucial since capturing the intrinsic properties of things exiles them to the realm of representation, while performing actions in connection to them opens up the realm of participation. The drawings in the apparatus then move beyond representing what they denote towards participating in what they denote. Rather than filing the drawings into the gridlock of memory as witness, into a legitimation and reenactment of such memory, I—following Retallack

(2003)—pursue the poetics of memory, "what is made of it now... the language by which it is evoked" (p. 11), as a present form of living and claiming that can rearrange the actively present constituents of such a historical disaster.

From here, I conceive of recovery as a process of poeticizing memory into a present form of making meaning, one sparked by the geographical here/there and temporal pre/post cusps, but that is in no way limited to them. In the incessant struggle over meaning, recovery defies state-sponsored common sense, unsettling certain ideological chains, disarticulating them from the signifying structure to then rearticulate their associations into new arrangements. It is a practice of reclaiming and reading signs anew: under different lights, across accumulated layers, through alternative frames, and with others. And, this is not an esoteric semiotic exercise, but one rooted in the social. In his re-reading of Althusser, Stuart Hall (1985/1996) reminds us of the former's claim that "ideas don't just float around in empty space. We know they are there because they are materialized in, they inform, social practices. In that sense, the social is never outside of the semiotic. Every social practice is constituted within the interplay of meaning and representation and can itself be represented" (p. 23). To ignite the process of recovery is to retrieve oneself as a constituting and participating subject in this interplay, striving to maintain its relevance to how one makes the world intelligible.

What I am proposing at the end of this chapter is not only representation, but also participation; it is the prospect of these documents as evidence, as the *thing* in the forum around which I call on lawyers, policy makers, artists, designers, and writers—to name a few—to gather, perform, and interpret, to disarticulate the claim to ownership ventriloquized by Law No. 10 and rearticulate it through prosopopoeia in order to give voice to these indexically suspended documents of absence, and more importantly, of presence.

Notes

1. Writer Irmgard Emmelhainz deftly takes this excerpt to distill that we should comprehend sound as referring to militant ideology and image as referring to art, "that the ideology that informed the discourse of political art-making overpowered the image." See Emmelhainz (2012). Between Objective Engagement and Engaged Cinema: Jean-Luc Godard's 'Militant Filmmaking' (1967–1974), Part II. *e-flux, 35.*
2. Peirce posits the index as "a sign which refers to the Object that it denotes by virtue of being really affected by that Object."
3. Which will beg an interrogation of the reasons they can't, ensued by a sequence of assumptions and measures.
4. While this term is generally evoked to describe icons, it certainly applies to indexes, because Peirce was aware that any single sign may display some combination of iconic, indexical, and symbolic characteristics, where he asserts how difficult it would be, "if not impossible, to instance an absolutely pure index, or to find any sign absolutely devoid of the indexical quality." See Peirce (2012, p. 108).

5. In these groups, these societies of friends, "friends" does not suppose a special doctrine on friendship, but only some necessary measure of agreement, like that which holds cities together as Aristotle once remarked. See Miguel, T. (2004). *Friends of Interpretable Objects*. Cambridge, MA: Harvard University Press. 3.

References

Adorno, T. (1978). Commitment. In *The Essential Frankfurt School Reader*, edited by A. Arato and E. Gebhardt. New York, NY: Urizen Books, 300–318.

Blanchot, M. and Smock, A. (1995). *The Writing of the Disaster*. Lincoln, NE: University of Nebraska Press. (Original work published in 1980).

Deixis. (n.d.). In *SIL Glossary of Linguistic Terms*. Retrieved January 17, 2019, https://glossary.sil.org/term/deixis.

Doane, M. (2007). Indexicality: Trace and Sign: Introduction. *Differences: A Journal of Feminist Cultural Studies*, *18*(1), 1–6.

Emmelhainz, I. (2012). Between Objective Engagement and Engaged Cinema: Jean-Luc Godard's 'Militant Filmmaking' (1967–1974), Part II. *e-flux*, *35*.

Hall, S. (1996). Signification, Representation, Ideology: Althusser and the Post-Structuralist Debates, in *Cultural Studies and Communications*, edited by J. Curran, D. Morley, and V. Walkerdine. London: Bloomsbury Academic, 11–33. (Original work published in 1985).

Kraus, R. (1977). Notes on the Index: Seventies Art in America. October, *3*(Spring), 68–81.

Latour, B. and Weibel, P. (2002). *Iconoclash*. Karlsruhe: ZKM.

Lerner, B. (2016). *The Hatred of Poetry*. New York, NY: Farrar, Straus and Giroux.

Peirce, C. S. (2012). Logic as Semiotic: The Theory of Signs. In *The Philosophical Writings of Peirce*, edited by J. Buchler. New York, NY: Dover Publications, 98–119. (Original work published in 1955).

"Q&A: Syria's New Property Law." (May 2018) Human Rights Watch, www.hrw.org/news/2018/05/29/qa-syrias-new-property-law.

Retallack, J. (2003). *The Poethical Wager*. Berkeley, CA: University of California Press.

Slyomovics, S. (1998). *The Object of Memory*. Philadelphia, PA: University of Pennsylvania Press.

Tamen, M. (2004). *Friends of Interpretable Objects*. Cambridge, MA: Harvard University Press.

Weizman, E. and Keenan, T. (2012). *Mengele's Skull*. Berlin: Sternberg Press.

Index

Note: Page numbers in *italics* indicate figures and **bold** indicates tables in the text.

For Product Safety Concerns and Information please contact our EU representative GPSR@taylorandfrancis.com
Taylor & Francis Verlag GmbH, Kaufingerstraße 24, 80331 München, Germany

www.ingramcontent.com/pod-product-compliance
Lightning Source LLC
LaVergne TN
LVHW021115300926
849160LV00010B/747
9780367550431